P9-BJL-486

Plant hormones and plant development

To all my great teachers, but especially to
Ralph Hartley Wetmore,
Frits Warmolt Went, and
Jane Shaw Jacobs

Plant hormones and plant development

WILLIAM P. JACOBS
Professor of Biology, Princeton University

CAMBRIDGE UNIVERSITY PRESS
Cambridge
London New York Melbourne

Published by the Syndics of the Cambridge University Press
The Pitt Building, Trumpington Street, Cambridge CB2 1RP
Bentley House, 200 Euston Road, London NW1 2DB
32 East 57th Street, New York, NY 10022, USA
296 Beaconsfield Parade, Middle Park, Melbourne 3206, Australia

First published 1979

Printed in the United States of America

Typeset by Bi-Comp, Inc., York, Pa.
Printed and bound by Vail-Ballou Press, Inc., Binghamton, N.Y.

Library of Congress Cataloging in Publication Data

Jacobs, William Paul, 1919–

Plant hormones and plant development.

Includes bibliographical references and index.

1. Plant hormones. 2. Plants – Development.
I. Title.
QK731.J32 581.3 78-54580
ISBN 0 521 22062 9

Contents

Preface

My intention in writing this book is to describe the hormones of vascular plants and to discuss the ways in which hormones control and coordinate vascular plant development. The emphasis is on physiology and anatomy, the areas of my training and experience. I wrote the book for advanced undergraduates and beginning graduate students of biology.

This book exemplifies the approach and expands the coverage of an upperclass course that I have taught for several years at Princeton University. The approach is critical, developmental, and organismic. The success of this course with our college students led me to think that the approach might be useful to others also, particularly since no such textbook was available in the field of plant hormones. I was happy, therefore, to accept the invitation of Cambridge University Press to write such a book.

My experience in teaching this material has shown me that it is just as important to describe the development of research ideas as to describe the development of plants. The ongoing generation of theories constitutes a major appeal of experimental science, and it is therefore important to present material to students in a way that conveys the excitement and contradictions of successive discoveries. The presentation only of the current opinion as to the "finished" theories distorts the students' understanding of the field as a whole because it presents a fixed and frozen body of knowledge, whereas the knowledge of science is in fact fluid, challengeable, and incremental. The historic approach also facilitates learning: the thousands of current observations can be remembered and assimilated more easily when they are seen as points on developing lines of thought.

Even more important as an appeal of experimental science is the satisfaction it can give to those who hunger and thirst after perfection. Aside from art, there is probably no human activity where one can approach as close to perfection as in a beautifully planned, executed, and interpreted series of scientific experiments. The expression of this ideal, as well as the breadth of my topic, demand that I restrict my discussion to pertinent papers of high quality. The criteria used in selecting papers were that the research be reproducible, technically and logically sound, as quantitative as possible, supported by statistics, and published with data in a journal that uses outside reviewers for each paper. The fewer of these criteria that a paper meets the less reliable I consider it to be. Papers and abstracts lacking data are not reliable enough to cite. Doctoral theses present a special problem. Traditionally, they are cited, but they are of questionable value: the "readers" rarely review them as critically as a paper would be reviewed in a first-class journal. Their relative or complete inaccessibility to outsiders further limits their value. If a thesis contains information worth citing, I feel it should be published separately in a journal that uses outside reviewers. Direct evidence I routinely consider more convincing than indirect (e.g., as evidence that a given hormone has moved to a given location, direct extraction of radioactively labeled hormone from that location is more convincing than a changed growth rate in that location).

I have dealt with the touchy problem of priority as follows: the first person known to have published a paper of high quality on a topic is cited first; later confirmations are listed afterward. (The date of submission of the manuscript is used rather than the date of publication.) If later authors do not cite the pertinent paper, their dereliction in scholarship is described euphemistically by the phrase "they discovered, apparently independently . . ." (I apologize for the inevitable derelictions in my own scholarship, and I hope they will be pointed out to me by my readers.)

The combination that I have attempted – of critical discussion of experiments and historical development of hypotheses – inevitably resulted in too little space remaining to treat in depth some important areas of hormonal physiology. Geotropism is one of the currently most actively investigated such areas. Apical dominance is another. The role of hormones in fruit and seed development we have touched on only briefly. The gas ethylene does not seem to be a hormone in the strict sense, because of lack of evidence that it moves within the plant from site of production to site of action. However, it acts at hormonal levels on various developmental processes, such as abscission, often with striking interactions with the true hormones. The burst of research on ethylene that followed Burg's application of gas chromatography to its measurement has been recently summarized in Abeles's book. Entrees to the literature on these topics are given following the References so that the interested reader can pursue them as he desires.

The basic plan of each chapter is to describe and discuss a major developmental problem and the progress that hormonal physiologists have made in understanding it, followed by a discussion of pertinent techniques. Critical evaluation of techniques is inextricably bound up with progress in the field. Knowledge of the strengths, weaknesses, and limitations of a given technique are essential both for further progress in the field of hormonal physiology and for understanding the difficulties earlier investigators encountered.

As for thanks, what scientist in his 50s can ever mention all the people to whom he owes a debt of gratitude? In addition to my three major educators mentioned in the dedication – Professors Wetmore and Went and my wife, Jane – many others have helped greatly at various stages. Professor Kenneth V. Thimann was most generous with his time and broad knowledge when I was a young post-doc and Junior Fellow at Harvard. Professor John Tyler Bonner has been a stimulating and sympathetic colleague for many years. Princeton University has also been generous in its support. Of course, it has been funds from the major U.S. granting agencies that have made most of my own research possible. The National Science Foundation, the Office of Naval Research, the National Aeronautics and Space Agency, and the Department of the Army should be mentioned especially, although help from private agencies like the American Cancer Society and the Guggenheim Foundation has been valuable, too. I am most grateful for their help and encouragement.

For more specific help with this manuscript, my warmest thanks to my daughter, Anne Jacobs, for critically reviewing the manuscript from the point of view of a student of humanities and to my son, Professor Mark Jacobs, for reviewing it as a fellow plant physiologist.

14 March 1979 W. P. J.

Abbreviations

ABA = abscisic acid
DNP = day neutral plant
GA-13 = (or other numbers) = gibberellin-13 (or other numbers)
GC-MS = gas chromatography and mass spectrometry
GLC = gas–liquid chromatography
IAA = indole-3-acetic acid
IA-Asp = indoleacetyl-aspartic acid
IBA = indole-3-butyric acid
LDP = long-day plant
LSDP = long-short-day plant
MCS = meter-candle-seconds
MS = mass spectrometry
NAA = 1-naphthaleneacetic acid
NPA = 1-N-naphthyl phthalamic acid
Rf = *R*atio of the distance from the origin of a particular spot to the distance to the *f*ront of a chromatogram
^{Rf}IAA^{Rf} = at the Rf typical of IAA, some IAA-like property was manifested (whether bioassay activity, ^{14}C counting, indole color test, or fluorescence)
SDP = short-day plant
SLDP = short-long-day plant
TIBA = tri-iodo-benzoic acid
2,4-D = 2,4-dichlorophenoxyacetic acid
2,4,5-T = 2,4,5-trichlorophenoxyacetic acid
[9:*15*] = a 24-hour light cycle with 9 hours of light, 15 hours of darkness.

1

Polarity, phototropism, and the discovery of auxin

As applied to living organisms, polarity refers to a difference along some axis of the organism. In fish, for example, there is a longitudinal polarity manifested in the head versus tail differences. In higher plants longitudinal polarity is manifested in the growth of the shoot at one end of the longitudinal axis and the growth of the root at the opposite end (Figure 1-1). Although biologists have tended to concentrate on such longitudinal polarities, radial polarities also occur (e.g., the differences in cell differentiation along a radial axis from the center of a root to the exterior).

Polarity is a very general property of multicellular organisms, whether plant or animal. Even sidelines of evolution like the *Volvox* group, members of the grass-green algae that have evolved longer and larger hollow spheres of cells (Bonner 1952), have developed a front-to-rear polarity in the spinning balls of cells that constitute the larger species of *Volvox*. Because the polarity of living organisms is so ubiquitous and so striking, biologists have been interested for many years in trying to discover what factors initiate and maintain polarity. The initiation of polarity is important to understand because in so many species longitudinally polar adults develop from fertilized eggs that show no apparent longitudinal polarity. Understanding the maintenance of polarity is important because the ability of higher plants to act as organisms seems closely correlated with their polarity.

Examples of developmental polarity

By the 1870s various developmental phenomena that could be considered to be manifestations of longitudinal polarity had been observed by both experimentalists and practical horticulturists. Some of these phenomena, the main hypothesis to which they gave rise in

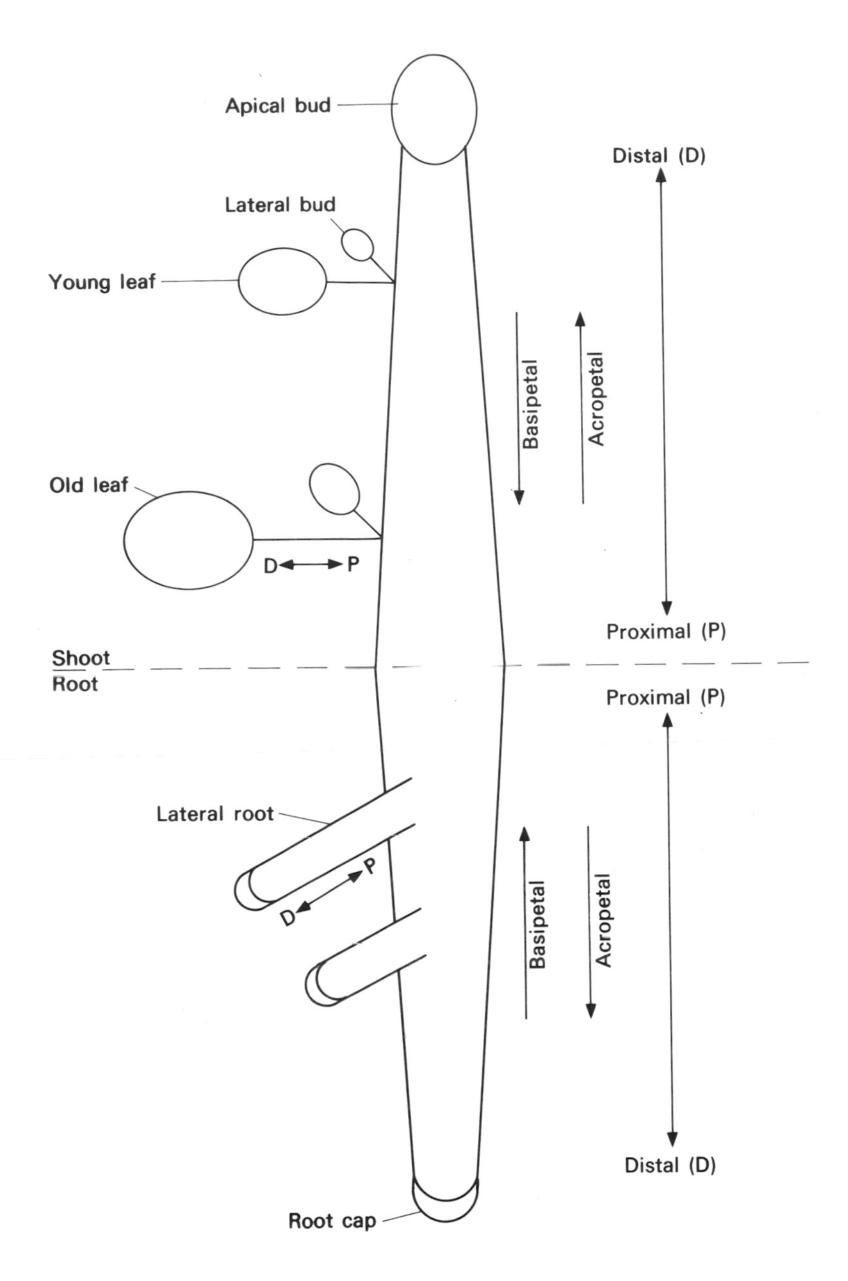

Figure 1-1. Topography of the shoot and root of a plant, illustrating the use of the terms "distal" (away from the main body of the plant) and "proximal" (toward the main body), as well as basipetal and acropetal.

1880, and, in more detail, the experimental investigations that followed are all described.

The polar regeneration of roots on stem cuttings was particularly well known. When a stem midsection was cut from the plant and kept in a moist environment, if regeneration occurred, roots typically formed at the base of the cutting (at the original root end). Roots hardly ever formed at the original shoot end, even if the plant was held upside down, indicating that gravity was not the controlling factor. After roots regenerated in this polar fashion, close to the basal cut, buds regenerated (if at all) at the other end of the cutting, usually not so limited to the immediate neighborhood of the cut. Roots formed in greater numbers if buds remained in the stem cutting. This fact led to the surmise that buds supply something needed for root regeneration.

Root cuttings also regenerated in a polar fashion. New roots formed at the distal cut (nearer the original root tip) and hardly ever formed at the proximal cut (nearer the original root base).[1] If buds regenerated, they formed later and proximally.

A prerequisite of successful root and stem regeneration was that the cutting be long enough to provide the regenerating organ with the material for differentiation and growth. An interesting phenomenon was reported if a very short cutting was used: the cutting typically formed buds at both ends and roots at neither. However, with long enough cuttings, an intact plant could be regenerated from a cutting of either a stem or a root – with roots regenerating at what had been the root-tip end of the cutting and buds regenerating from what had been the shoot-tip end. Such cuttings were said to show polar regeneration.

Apical dominance presented a different manifestation of polarity. The presence of the apical bud on a shoot was known to inhibit the growth of lateral buds (see Figure 1-1); the apical bud exercised apical dominance over the more basally located lateral buds. If the apical bud was excised, the lateral buds elongated faster, having been released from apical dominance. Releasing lateral buds from apical dominance is the physiological reason that pruning a hedge makes it grow thicker. The apical buds are excised so that the lateral buds will elongate and increase the impenetrability of the hedge.

[1] "Distal" and "proximal" (meaning "away from the main body" and "toward the main body," respectively) are often used as brief directional indications in plant biology (Figure 1-1). The junction of root and shoot is conveniently taken as the reference point. "Basipetal" and "acropetal" are related words, indicating "toward the base (or proximal end)" and "toward the apex (or distal end)," respectively.

(Other polar phenomena, such as polar cell regeneration, polar movement of the phototropic curve, and polar reactivation of cambium, we discuss in detail later on.)

Sachs's hypothesis

Although most of the preceding observations were made 100 or more years ago, finding the controlling factors was difficult and delayed. In 1880 Sachs presented the first valuable hypothesis concerning the internal factors that might control polar phenomena in plants. He assumed that organ-forming substances existed and that these substances traveled in a polar way. Thus, in the case of shoot cuttings, he assumed that root-forming substances existed that moved in a basipetally polar manner, collecting at the original root end of the cutting. He also hypothesized that shoot-forming substances existed that moved in the opposite direction in a piece of stem. Thus, he explained the polar regeneration of roots and shoots. Unfortunately, no one could find any of these substances, although the search continued for many decades.

(The student of biology should note that, although biologists usually look for the stimulators of a process, they could just as logically look for inhibitors. Thus, in the case of root regeneration, the observed phenomena could be explained as logically by inhibitors of root regeneration moving acropetally as by stimulators moving basipetally.)

Phototropism and polar movement of the response to light

The first break in the impasse of seeking the polarly moving substances in plants came from the study of what seems to be a quite different phenomenon, positive phototropism of shoots.

Tropisms are growth responses of plants in which the direction of growth is determined by the direction of external stimulus. The two most investigated forms of tropism are *phototropism*, a phenomenon in which the plant grows toward or away from light, and *geotropism*, a phenomenon in which it grows toward or away from the pull of gravity. When growth starts in a typical seedling, the shoot will show positive phototropism, it will grow toward the light; and the root will show positive geotropism, growing toward the pull of gravity. These tropisms are of obvious adaptive value in pushing the shoot system out of the ground, into the light and in plunging the root into the soil where the root will anchor the plant and absorb the minerals and water needed for further growth of the young plant. Tropisms are not always positive. Shoot systems show negative geotropism: they grow away from the pull of gravity. Obviously, positive phototropism and negative geotropism reinforce each other in the normal shoot.

Darwin and other early investigators of phototropism

The extraordinary Charles Darwin made the first big advance in the study of phototropism in 1880. A specialized and ephemeral organ in the seedlings of grasses, the coleoptile, is sensitive to light. The coleoptile is apparently an evolutionarily modified leaf. This hollow cylinder of tissue, closed at the tip, orients the shoot system of the young grass seedling toward the light (Figure 1-2). Its sensitivity to light is about 10 times greater than that of the cones that receive light for color vision in the human eye. Once the coleoptile has oriented toward the light by positive phototropism, its growth ceases and the first true leaf of the seedling grows through the coleoptile into the light. Darwin concluded that the tip of the coleoptile is the region that perceives light. His experiments supporting this conclusion showed that if the tip were excised, no phototropic curve resulted from one-sided illumination; and if only the tip were illuminated, by masking the rest of the coleoptile, a full phototropic curve nevertheless occurred. Darwin also presented evidence indicating that the light response was transmitted basipetally from the illuminated tip of the coleoptile.

Darwin attached men's wandering curiosity about phototropism to the bending tip of grass coleoptiles, where it has remained for almost 100 years. The use of coleoptiles in phototropic studies provides a striking illustration of how an organism is selected for research: one tends to use the organism one's predecessors used. As seedling organs, coleoptiles do have the distinct advantages of being cheap: three days' growth in the dark provides seedlings with coleoptiles of usable length. No greenhouse or long-term expensive care is required. On the other hand, coleoptiles are not unusually sensitive to light (Steyer 1967). In addition, one would be hard put to explain the logic of trying to understand the mysteries of phototropism in the 170,000 species of angiosperms by studying only the phototropism of an organ in the seedling of the grasses, one of the most highly specialized and highly evolved of angiosperm families – particularly when that organ itself is so specialized that it does not even exist in other angiosperms! Nevertheless, the over-

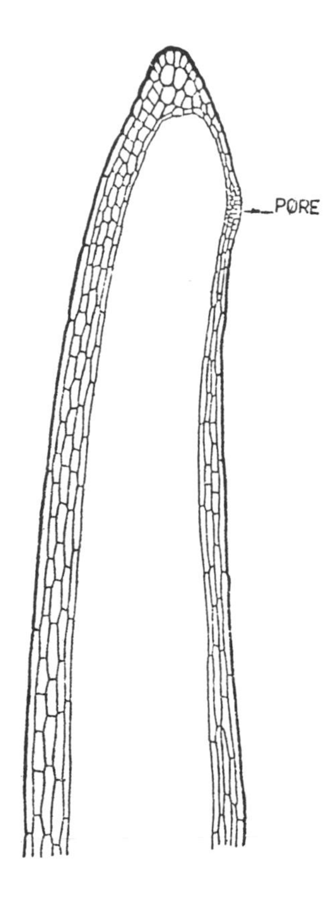

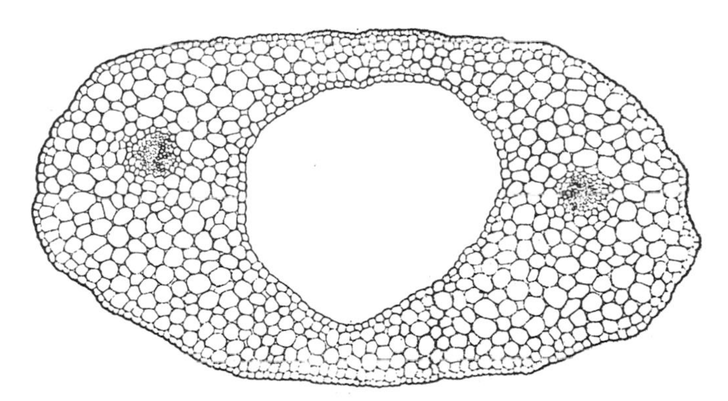

Figure 1-2. The coleoptile of a grass seedling. Histological sections through the *Avena* coleoptile (top: longitudinal section through the tip of a 46-mm long coleoptile; bottom: transverse section through the base of the coleoptile). Total length of top section was 2.5 mm. The first true leaf grows up into the space in the middle of the coleoptile. The sections are from Avery and Burkholder (1936).

whelming majority of papers on phototropism have used coleoptiles as research material.

By what mechanism did the coleoptile tip mediate phototropism? Boysen-Jensen addressed this question (1910, 1911, 1913). He confirmed Darwin's work showing that excising the tip prevents the phototropic response. In addition, he found that the excised tip, replaced on the cut surface, could restore a large part of the phototropic curve. In fact, he could interpose, between the tip and the base of the coleoptile, a layer of gelatin or a piece of living stem from another genus and the phototropic stimulus from the tip would still pass through the intervening material to cause a phototropic curve in the lower region of the coleoptile. This response could not pass through a thin layer of mica. His conclusion was that transmission is the result of a diffusible substance.

The phototropic mediator was a substance that originated in the coleoptile tip and moved basipetally to cause asymmetric growth in the underlying tissue. Did light stimulate production of this substance? One of the most aesthetically satisfying experiments in the field of plant hormones was conducted by Paál (1919) to answer this question. His experiment provides a marvelous example of how much information can be obtained from a simple experiment when the creative, intellectual content is high. Paál excised the tips of coleoptiles and replaced them, but in contrast to Boysen-Jensen, he replaced the tips offset from the center and he kept the plants in the dark. He found that the offset tip caused the coleoptile below it to grow, giving a curve that resembled a phototropic curve even though these plants had grown entirely in the dark. His conclusion was that the coleoptile tip forms growth substances with or without light and that light merely determines the distribution of these growth substances to either of the two sides of the coleoptile. This new concept differed drastically from the earlier idea that light exposure causes production of a new growth substance that then descends from the tip and causes differential growth. This finding of Paál was confirmed by many workers (Nielsen 1924; Snow 1924; Beyer 1925). In the early 1920s many workers sought to isolate the active material (Stark 1921; Nielsen 1924; Seubert 1925), but all failed.

Stark (1921) modified the experiment of Paál by offsetting not the coleoptile tip but a piece of agar or gelatin to which he added crushed coleoptile tips. Stark's hope was that the growth substance, hypothetically formed in the tips, would exude into the agar and then diffuse from the agar to the coleoptile portion directly beneath, causing growth there as had the offset coleoptile tip of Paál. Stark never obtained the expected growth.

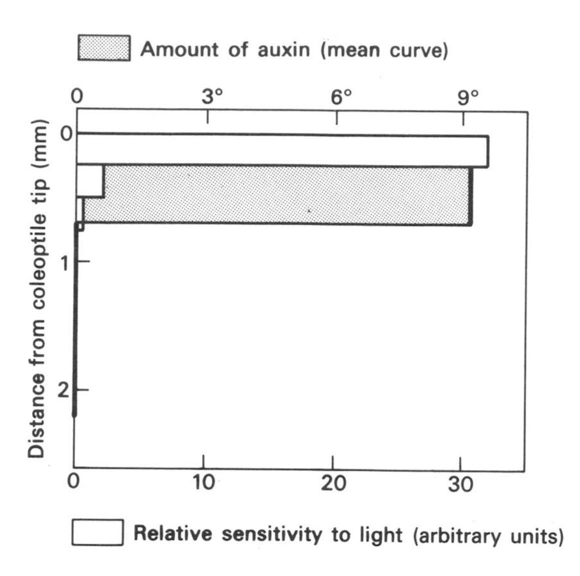

Figure 1-3. The distribution in the coleoptile tip of sensitivity to side lighting (open bars) and of diffusible auxin (solid bars) (Auxin data from Went 1928, phototropic sensitivity data from Lange 1927).

Went's discovery of the first plant hormone

Went (1928) modified Stark's technique in the following way: instead of crushing the tips, he merely placed the excised tip directly on top of the gelatin block so that the material from it could diffuse across the single cut surface. Went found that the more tips he placed in this way on an agar block, the greater the curvature he obtained when this block was offset on the tip of a decapitated coleoptile: the curvature was proportional to the number of tips that had been placed on the block and presumably to the amount of growth substance diffusing from the tips. Thus, Went could use the average curvature as a measure of the amount of substance in the agar blocks. This is the basis of the *Avena* curvature bioassay for growth substance. (See the following discussion of bioassays for details of the *Avena* curvature assay.) (Went called the growth substance *Wuchsstoff*. It is now called *auxin,* the term we use hereafter.) Having developed a bioassay to measure auxin, Went could quantitatively investigate the relation of auxin to phototropism. By bioassaying different portions of the coleoptile, he found that the tip is the only source of diffusible auxin; if auxin mediates phototropism, this could explain why the tip is so much more sensitive than other

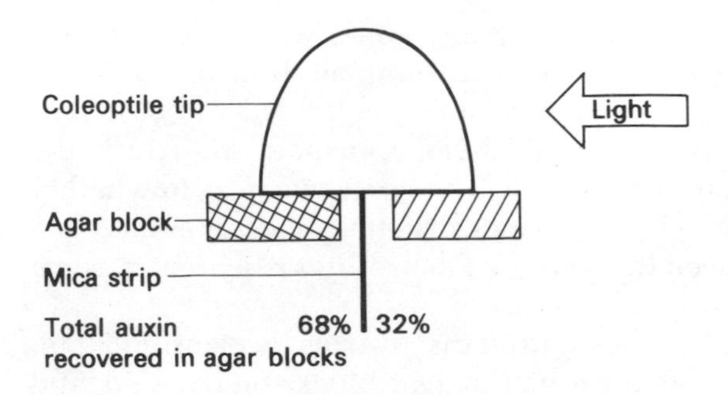

Figure 1-4. Went's (1928) experiment demonstrating that phototropically active light, impinging on an excised coleoptile tip, causes more auxin to move basipetally on the side away from the light.

regions to unilateral light, as earlier workers had reported (Figure 1-3). Went placed agar blocks containing auxin diffused from coleoptile tips symmetrically on decapitated coleoptiles and found that auxin stimulated straight growth also. (Auxin-agar resulted in 6.1% growth, compared to zero growth in the decapitated coleoptiles that had no auxin added.) This led him to conclude, as Paál had previously, that auxin stimulated both regular growth and that associated with the phototropic curve.

Another crucial part of Went's 1928 paper was the evidence he adduced that there was more auxin on the shaded side of unilaterally illuminated coleoptiles. Went excised the coleoptile tips and placed each tip astride a mica strip which served as a diffusion barrier between the halves. He then rested each half of each base on a separate agar block and illuminated the tips from one side with an intensity of light and a duration of exposure sufficient to cause a phototropic curve in the intact plant (Figure 1-4). The blocks of agar collecting auxin that had diffused from the illuminated sides and from the shaded sides were then removed and bioassayed separately in the *Avena* curvature test. As usual, the average curvature was taken as a measure of the amount of auxin present. Went found that there was more auxin in the blocks collecting from the shaded half of the coleoptile tip. The actual values were 68% of the total auxin coming from the shaded side, 32% from the illuminated side.

Went's conclusion was that unilateral light causes a lateral transport of auxin. This conclusion is not the only possible one, given the data: light could also be causing a gradient of auxin destruction. An obvious way to obtain evidence on this is to see if the total amount of auxin collected from tips that have been unilaterally illuminated equals the amount collected from tips that are kept in the dark.

Went's own data indicated that 16% less auxin was collected from the illuminated tips than from the unilluminated ones. But Went minimized this difference in the total auxin collected and based his interpretation on the premise that lateral transport of auxin is the main cause of the phototropic curve. The controversy as to whether phototropically active light causes a destruction of auxin or merely a lateral transport has been the source of many later researches, as we shall see.

Went made one of his most exciting discoveries by excising 2-mm sections of coleoptile, placing auxin in agar blocks on one end, and collecting the auxin that passed through the sections into receiver blocks at the other end. By assaying the receiver blocks in the *Avena* test, he discovered that auxin only moved from the original apex to the original base of the coleoptile sections; that is, there was strictly polar, basipetal movement of this plant hormone. The polar movement was not due to gravity acting in one direction during transport, because he found no acropetal movement of auxin even when he inverted the sections. The actual values from his experiments were an average bioassay curvature of 5.3° in the basipetal direction and an average curvature of 0.2° in the acropetal direction. In both cases gravity acted in the direction of potential movement. (An average curvature of 0.2° in the *Avena* curvature bioassay indicates a very small amount of auxin or none at all, since plain agar blocks will often give this tiny average curvature.)

Went tried to define the characteristics of the auxin obtained by diffusion from coleoptile tips and found that it was stable to heat and to light. He attempted to estimate the molecular weight of auxin by substituting in the formula $\sqrt{M} = K/D$ where M is the molecular weight, D is the diffusion coefficient, and K is a constant. Went calculated the diffusion coefficient by using the *Avena* curvature bioassay to measure how quickly auxin diffused through several slabs of agar. From earlier literature on physical chemistry, K was to equal 7.0 if there was no dissociation of the substance being tested and if there was only one substance in the agar. Using this value and these assumptions, Went estimated that the molecular weight of auxin was approximately 376. (As we see in Chapter 3, this estimate of the molecular weight misled people about the true chemical nature of the endogenous auxin.)

Went's 1928 thesis was astounding. It reported the first isolation of a plant hormone; it described the creation of a quantitative bioassay for auxin, an assay that is so reliable that it is still used in the 1970s; it demonstrated for the first time in either animals or plants the strictly polar movement of a hormone through tissue; it provided detailed evidence that the phototropic curve results from more auxin being on the shaded side of the unilaterally illuminated coleoptile.

From the earlier qualitative work of Cholodny (1924, 1926), mainly on root geotropism (see Chapter 10), and Went's work on phototropism in the coleoptile, the Cholodny–Went theory of tropisms was developed. This theory states that tropisms, whether of shoots or roots, are due to unequal distribution of the growth hormone auxin under the influence of gravity or light, with the redistribution of auxin being followed by unequal growth of the two sides.

The rate of movement of auxin and the phototropic curve

Plant physiologists instantly realized the importance of Went's results. Many other workers at Utrecht University began work on various aspects of auxin physiology. Dolk (1930) studied the time course of movement of the phototropic curve down the *Avena* coleoptile. Using the clearest data from his Figure 21, one can estimate that the phototropic curve moves basipetally with a velocity of approximately 6 mm/hour. Van der Weij (1932) provided evidence that auxin moves basipetally through coleoptile sections at roughly the same rate, thus confirming in yet another way that auxin is the mediator of the phototropic curve. The general principle of van der Weij's research is illustrated in Figure 1-5. If he used a 5-mm-long section of coleoptile, he found that auxin would collect in the receiver block after about 30 minutes and that the increase over a period of time would be apparently linear. By extrapolating the presumed straight line back to the time axis, van der Weij could estimate when auxin had first come through the section and appeared in the receiver. If van der Weij, testing a 5-mm section, found that the time intercept was at 30 minutes, this would indicate that the main front of auxin moved through the section at the rate of 5 mm/30 minutes or 10 mm/hour. Although many of his other deductions about auxin movement in the coleoptile have been shown by later work to be incorrect, van der Weij's estimate of the velocity of auxin movement through coleoptile sections has been confirmed by later workers (e.g., Went & White 1939; Hertel & Leopold 1963). It is quite surprising that these excised sections, separated from their normal sources of nutrient and water, would show such meaningful parallels with phenomena in the intact plant. Laibach and Kornmann (1933) added auxin to the uncut epidermis of intact coleoptiles and reported that elongation was greater than in control

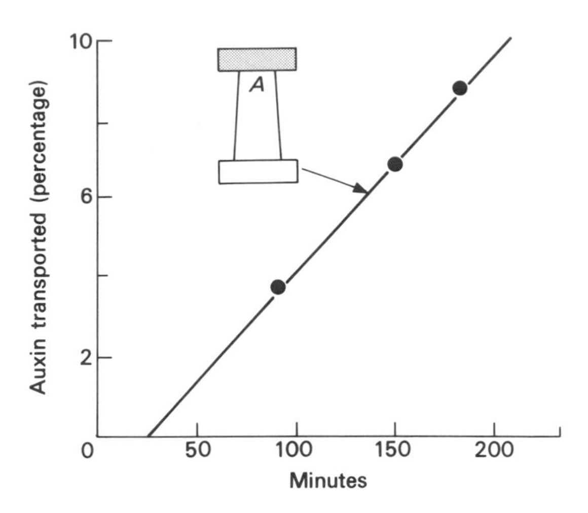

Figure 1-5. Graph illustrating the principle of van der Weij's (1932) auxin-transport experiments. Auxin added in an agar block at the apical (*A*) end of an excised coleoptile section moved basipetally through the section and accumulated in the basal receiver block. Van der Weij felt that the accumulation in the receivers was linear and interpreted his results accordingly after drawing lines by eye. (The data points are from van der Weij's experiments on sections held at 6°C, but the linear regression was calculated by Mer for Fig. 5 of Gregory & Hancock 1955.)

coleoptiles, but only proximal to the site of auxin addition. The latter point supported the view that auxin moves only basipetally in the intact coleoptile (as Went and van der Weij had reported for coleoptilar sections). The former point indicated that endogenous auxin is not at an optimal level for elongation (so that lateral transport of auxin under the influence of light would cause greater elongation on the shaded side). Went (1935) repeated this experiment and confirmed the earlier findings, but added that the region of extra growth induced by the externally applied auxin moved basipetally at 10 mm/hour – essentially the same rate at which auxin moved through the excised sections.

Substitution of auxin for the coleoptile tip

If the sole function of the tip of the coleoptile is to provide a source of auxin for the more basal regions, then one should be able to replace the actual tip of the coleoptile with a block of agar that contains auxin and thereby restore the phototropic curve to a unilaterally illuminated coleoptile. This has been tried by several people and to some degree the phototropism can be restored by substituting auxin-agar for the intact tip (Boysen-Jensen 1933a; Reinders 1934; Bara 1957; von Guttenberg 1959). Bara compared the phototropic curvatures of decapitated *Avena* coleoptiles, intact coleoptiles, and decapitated coleoptiles topped by auxin-agar. His data showed that 3 hours after treatment the decapitated coleoptiles had reached a degree of curvature that was 13% of that in the intact controls. In comparison, the auxin-topped coleoptiles had curved 45% as much as the intact controls. Six hours after treatment the decapitated coleoptiles showed a curvature that was 54% of that formed in the intact control, whereas the auxin-treated coleoptiles achieved 80% of the control's curvature. Why does not substitution of auxin-agar for the tip restore more of the intact coleoptile's response? It is probably because the function of the tip is not only to supply auxin but also to serve as the locus in which a lot of the lateral transport of auxin occurs – if lateral transport is, in fact, a major factor in the phototropic response. The smaller curve may also result from an insufficiency of substances other than auxin that the tip may normally provide.

Summary

Went's development of the *Avena* curvature test as a quantitative bioassay for auxin made possible the detailed evidence that auxin is the hormone that mediates phototropism in the coleoptile. By measuring amounts of auxin in the bioassay, Went showed that (1) the tip was the only source of auxin in the coleoptile (thus explaining why the tip was required for phototropism, as Darwin and others had reported); (2) auxin moved with strict basipetal polarity through coleoptile sections (as did the phototropic curve in the intact coleoptile, according to van der Wolk [1912]); (3) auxin caused growth of the coleoptile, both straight growth (when auxin was applied symmetrically) and curvature (when auxin was applied asymmetrically as in the bioassay), and the amount of growth was proportional to the amount of auxin added; (4) unilateral illumination of the coleoptile with light sufficient to induce phototropism resulted also in relatively more auxin on the shaded side. Went's conclusion was that this asymmetry in auxin levels on the two sides of the coleoptile caused the asymmetric growth that resulted in the phototropic curve toward the light.

If auxin mediates phototropism, what mediates the light-induced redistribution of auxin? And is the differential in auxin levels on the shaded and illuminated sides the result of lateral transport with no destruction (as Went concluded) or of auxin destruction on the illuminated side? These questions we consider in the next chapter.

Some form of quantitative bioassay for auxin was an obvious prerequisite for Went's discoveries. Bioassays are valuable in all research on hormones. In the section on techniques that follows, bioassays are discussed with emphasis on the *Avena* curvature test.

Techniques of auxin bioassay

A bioassay is a method of determining the presence or (more often) the amount of a substance by measuring the substance's effect on a designated biological process. Hormones and vitamins are particularly likely to be measured by bioassays, because they occur at concentrations too low to be detectable by the usual physical or chemical methods. In the early stages of investigation of a new hormone, before its chemical structure is known, bioassays are indispensable; with their aid the biological activity can be traced through the various steps in attempted purification. However, quantitative bioassays are onerous enough so that, as soon as the chemical has been identified, physical or chemical methods of measuring are substituted for the bioassays.

Many bioassays for auxin have been developed since 1928, but only a few are widely used. The original one – Went's *Avena* curvature assay – is described in detail, with an emphasis on general principles of bioassay design; the frequently used *Avena* coleoptile section bioassay of Bonner (1933) is then more briefly described.

The *Avena* curvature bioassay

The *Avena* curvature bioassay for auxin, devised by Went in his thesis and refined in the 1930s, is of such good design that it is still used in the 1970s. In broad terms it consists of starving the coleoptile of auxin by excising the auxin-producing tip, then placing an agar block containing auxin on just one side of the cut. The added auxin moves predominantly down that one side of the coleoptile, causing extra elongation of that side and resulting in a curvature (a sham phototropic curve). Within limits the resulting curve varies with the concentration of auxin in the original agar block.

The detailed schedule of the classical *Avena* coleoptile curvature bioassay, as worked out by Went and others in the early years of auxin research and as used subsequently by most of their students and co-workers, follows.

I. Remove the husks (by hand) from approximately 500 seeds of *Avena sativa* L. of the Victory strain. (This will provide an eventual 10–12 rows of fairly closely matched seedlings, with 12 plants in each row.)

II. *First day* (e.g., Monday):

1. 9 A.M. Soak the dehusked *Avena* seeds in distilled water.
2. 11 A.M. Lay the seeds in rows on wet filter paper in large Petri dishes so that the groove side is down and the embryo ends are all pointing in the same direction. Place seeds about 3 mm apart. Replace glass cover on Petri dish.
3. Place the filled Petri dishes in the *Avena* room and give red light for 1.5 hours. (Red light inhibits the elongation of the first internode, which is beneath the coleoptile; if this internode is not inhibited, it is difficult to obtain straight enough plants for the assay [see Fig. 36 of Went & Thimann 1937].) The *Avena* room is a darkroom held at 25°C, with controlled relative humidity of 85–89%., After the red illumination, leave the Petri dishes in the dark *Avena* room. All subsequent manipulations are carried out with the red safelight only.

III. *Second day* (Tuesday):

Add a few drops of distilled water to the *Avena* seeds, if the filter paper seems too dry.

IV. *Third day* (Wednesday): A.M.

Select the largest seedlings from the dishes and place them in the glass holders that Went designed (Fig. 2 of Larsen 1955). Be sure their roots reach the distilled water in the water trough.

V. *Fourth day* (Thursday):

1. 9 A.M. Select plants with straight coleoptiles of uniform height and rearrange them in rows.
2. Prepare the agar blocks of 1.5% agar. (See the following.)
3. 11 A.M. Cut off the top 1 mm of each coleoptile with a scalpel, razor blade, or fine scissors (the "first decapitation").
4. 2 P.M. Make final selection of plants for uniformity and cut off the topmost 4 mm of the remaining coleoptile. (This "second decapitation" is done with special scissors that can be adjusted to cut through the coleoptile cylinder but not to cut the primary leaf inside the coleoptile. The primary leaf is then pulled upward, gently, with forceps, until it snaps loose at its base. The detached leaf is used to help support the agar blocklet, as Stark and Drechsel proposed in 1922.)
5. Within 30 minutes (at most) after the second decapitation apply an agar blocklet to one side of the upper cut surface. (Each blocklet is typically taken up on the blade of a small spatula, then

applied to the coleoptile by drawing the spatula down at a 45° angle across the coleoptile cut. This movement pushes the blocklet against both the cut and the supporting primary leaf.)

6. Check that each coleoptile is straight and provided with a blocklet, then leave the racks with their water troughs in the darkened *Avena* room for 1.5 hours.
7. At exactly 1.5 hours from the time the last blocklet was applied to a given rack, remove the rack from the trough and make a photographic record of the curvature of the seedlings with their agar blocklets still in place. To record the curvatures place high-contrast enlarging paper (such as Kodabromide) behind a glass plate, position the rack holding the 12 *Avena* seedlings against the glass, then give an exposure of several seconds with a white lamp bulb placed a meter or so away. Protect the remaining seedlings from exposure to the white light. There is usually time to pencil a notation of the treatment on the back of the paper, to develop it, and to place it in fixative before the next rack needs to be photographed. The "photograph" is actually a silhouette of the seedlings, which after washing and ferrotyping can be used as a permanent record of the bioassay results.

VI. At your convenience measure the individual curves from the dried shadowgraphs (diagrammed in Figure 1-6) and calculate the average curvature and standard error for each row.

Explanations of procedures. The rationale for portions of this bioassay is as follows. The assay is run in a darkroom because the coleoptile is less sensitive to auxin if the seedlings have been grown in the light. The high relative humidity is required to prevent the 1.5% agar blocklets from drying out during the 1.5 hours in which the curve develops. If the blocklets do dry out, less curve is obtained from them.

The first decapitation removes the major source of endogenous auxin in the coleoptile so the remainder will be auxin starved. However, after 2.5–3 hours the top portion of the remaining coleoptile "regenerates a physiological tip"; that is, auxin is produced there and accelerates the elongation of the remaining coleoptile (Söding 1925; Dolk 1930; Went 1942; Anker 1967). Hence, the second decapitation in the bioassay serves to remove this "regenerated" auxin and thereby to maintain the auxin-starved condition of the coleoptile stump. The use of the 1.5-hour interval between adding the blocklets and photographing the seedlings is based on observations that the maximal curve obtained from a given dose of auxin occurs at that time (see Fig. 20 of Went & Thimann 1937).

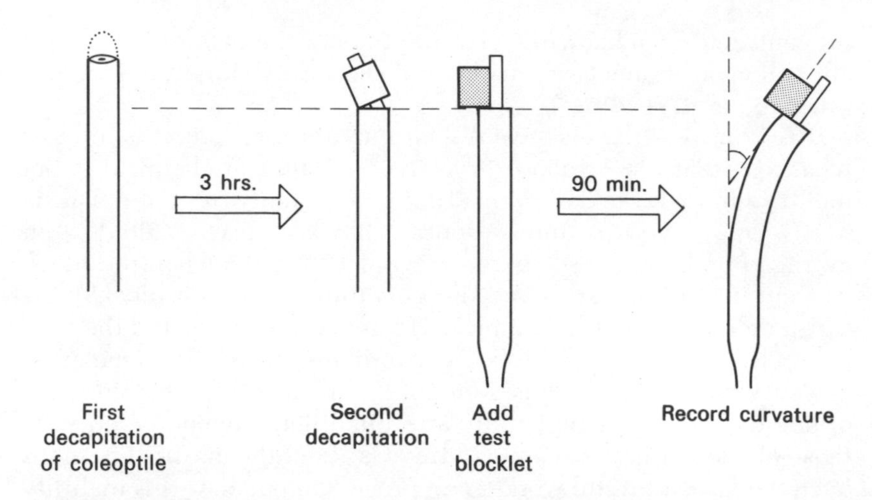

Figure 1-6. The procedures and timing of the *Avena* curvature bioassay. The shadowgraphs are made 90 minutes after the test blocklet was applied (after Went & Thimann 1937).

The block of 1.5% agar, which is typically used for each treatment, is 8.0 × 11.0 × 1.5 mm in size and is formed by pipetting somewhat more than 132 mm^3 of warm liquid agar into a metal mold that has been chilled on a glass microscope slide set on an upside-down tray of ice cubes. After the agar has gelled, the excess agar is sliced off the top with a double-edged razor blade. Each such block is cut into 12 blocklets of equal size (each 11.0 mm^3), using a metal cutting frame like the one in Figure 11 of Went and Thimann (1937). (Agar concentration and blocklet size affect the curve obtained, as discussed in detail by Larsen [1955].)

Red light is used routinely in the classic curve bioassay both because it inhibits the otherwise bothersome elongation of the first internode and because red light (unlike blue) results in no sporadic and obfuscating phototropic curvatures.

Improvements on the classical *Avena* curvature bioassay. Before measuring the angles from the shadowgraphs of the *Avena* seedlings, look over the shadowgraphs with a hand lens of magnification 10× or higher. Any plant whose agar blocklet appears solid white should be discarded. (The block has dried out too much and will give an aberrantly low reading.) Discard also any plant in which the blocklet is not resting solidly on the upper cut of the coleoptile. To eliminate the possibility of subconscious bias, all the shadowgraphs should be coded before measuring the angles. This prevents one from knowing what treatment is being measured until after all measurements are made (and should, of course, be standard practice for all readings).

It is typical of the classical *Avena* curvature test, as it is of most bioassays, that the response to a given amount of chemical is not identical from day to day. Went (1928), for instance, reported that six *Avena* coleoptile tips diffused onto a block of agar resulted in an average curvature (± standard error) of 11.2 ± 0.5 when the block was cut into 12 blocklets and the curvature from each blocklet assayed on one coleoptile of a rack. However, six tips tested the next day in the same way would be very unlikely to give the same average curvature. This obviously meant that results from the bioassays of one day could not be validly and quantitatively compared with those of another day. Because 25 trays is about the maximal number even the fastest careful worker can process in one day, this inability to calibrate was a serious limitation, even for rough comparisons. (Of course, the lack of calibration also meant that absolute values of auxin could not be determined.) After indole-3-acetic acid (IAA) became known, most auxin researchers calibrated their assays every day with one (see Leopold 1955, p. 26) or two known levels of synthetic IAA, then expressed their experimental curvatures as if they resulted solely from IAA. That is, if their unknowns gave a 10° curve and 25 μg IAA gave 10°, the unknown would be said to be equivalent to 25 μg IAA.

In actual practice, in the United States at least, the *Avena* curvature bioassay has not been used to optimal effect because of the historical accident that the assay was devised just a few years before the independent development of bioassay statistics. The founders of auxin research did not incorporate the new statistics into the curvature bioassay, nor did most of their students.

The minimum requirements for adequate quantitative bioassays are described in detail in Finney's (1964) book, *Statistical Method in Biological Assay*. In order that the *Avena* curvature assay be statistically valid at least two calibration concentrations (known concentrations of IAA) and at least two concentrations of the unknown must be used. Three of each is preferable.) The fact that such "four-point" or "six-point" designs have so seldom been used with the *Avena* test is probably the main reason why so relatively few quantitatively useful data have resulted from the assay. The neglect of modern bioassay statistics has also, of course, meant that biologists derived much less information from the results of their labors than they might have.

Diluting the unknown not only serves to balance the statistical design but serves the important additional purpose of providing at least an indication as to whether the curvature obtained from the concentration is really due solely to IAA or is due to a mixture of a larger concentration of IAA and some inhibitor(s) or to some quite different auxin. The activity from a mix of auxins and inhibitors is not likely to give the same straight line on double dilution as does IAA by itself.

There is an additional necessity for diluting the unknown at least once, a necessity that results from the characteristics of the response of the *Avena* coleoptile itself. Although there is a steady rise in average curvature with increasing concentrations of IAA, the response eventually plateaus and then often declines at still higher concentrations (see Fig. 19 of Went & Thimann 1937). Obviously, a single determination of an unknown gives no clue as to whether you are dealing with a low concentration or one so high that its effect is decreased. Dilutions are essential to answer this question. (This is not a theoretical problem only. Von Guttenberg and his co-workers published a series of papers in the 1940s and 1950s in which they concluded that the native auxin was not IAA, only to discover later that, among other problems, they had not diluted their samples sufficiently [Wiedow-Patzold & von Guttenberg 1957] or were comparing high concentrations of IAA with low concentrations of their unknowns [as Terpstra et al., 1962, pointed out].)

The red safelight of the classical *Avena* room has been replaced in several laboratories by a green one. (Although red light does not give a phototropic curve in coleoptiles – that being the original reason for using it [Went 1928] – it was later found to decrease sizeably the phototropic sensitivity of coleoptiles [see Briggs 1963a and Chapter 2].)

The Victory variety of *Avena sativa* (also called *Siegeshafer*) has long been the standard oat variety, agreeing with the general principle in bioassay design of using an inbred line or a clonal stock (to reduce genetic variability in the assay). A few investigators have substituted a hull-less *Avena* variety such as Brighton (Nitsch 1956); others have found the dehusking chore can be omitted even with the Victory variety (Ferguson 1971).

Specificity is an important asset for a bioassay. The fewer chemicals the bioassay responds to, the less purification is needed before applying the unknown. (No statistical tests can distinguish the bioassay response to IAA from that to any other substance that has the same effect on the bioassay.) The *Avena* curve assay is unusually valuable in this respect. It is very specific for auxins, presumably because an active substance must not only cause elongation of the auxin-depleted coleoptile stump but also be able to move down into the stump from the agar blocklet at a velocity fast enough to manifest its presence within the brief 90 minutes of the assay. It must also show little lateral spread while moving basipetally, or the differential growth that produces the measured curve will not occur. This double requirement (for growth and transport) apparently provides the specificity.

The early workers were troubled by the variation in average curves obtained from a given dose of IAA. Although it is a general feature of bioassays that day-to-day variation occurs (requiring a new calibration of the organisms' response with each run of the bioassay), much of this variation at California Institute of Technology could be removed by filtering the smog-polluted air of the Los Angeles basin (Hull et al. 1954). As one who spent too much of my younger years in the dark, moist holes that were optimistically called "constant environment *Avena* bioassay rooms," I can say that one does not need to invoke the lack of bioassay statistics as the sole explanation for the paucity of usable quantitative data from such laboratories. The conditions were often deplorable. It was like trying to do fine manipulations in a tropical rainforest at night, with only the dim red safelight to help you see the tip of the coleoptile onto one side of which you were trying to place the tiny blocklet of clear agar. Only a masochist would continue doing such tests after escaping from the semipeonage of graduate student status; and the literature, correspondingly, shows few investigators who have continued doing *Avena* curve bioassays long past their postdoctoral assignments. However, the bioassay does not need to be done under such repellent conditions. The requirement for high humidity is for the agar blocklets and the coleoptiles, and the humans can stay conveniently outside the enclosed merry-go-round chamber devised by Avery et al. (1939). The main disadvantage of this chamber is that it holds only nine trays. For larger or more frequent experiments a room with controlled relative humidity is needed. The general American experience has been that air conditioning firms will readily sign contracts for the construction of such rooms – blithely stating, "Of course we can provide stable 90% R.H.; we deal with the control of humidity all the time!" However, they then spend a year or so trying to get the high-humidity controls to work as well as the low-humidity ones they are familiar with. I have circumvented this problem by having my *Avena* rooms built as prefabricated units, controlled by Aminco portable humidifying units, and with demonstration required at the factory that specifications were met on the assembled rooms before the rooms were accepted. The higher cost was more than compensated for by the fact that the rooms could be put into immediate use on delivery, without the usual loss of a year of the investigator's time. The efficiency of control is such that much of the bioassay variability, which others reported and which Scott and I had experienced at Harvard, Stanford, and the California Institute of Technology, was removed in these units. Scott and I (unpublished) found, for example, that the mean curve (± standard error) from the averages of six experiments run on different days using 25 μg IAA/ liter was $10.8° + 0.42$. The very small standard error is an indication of how uniform the *Avena* room is if adequately controlled, if the air is relatively unpolluted and if the assayist is an expert.

Noting that the literature on the statistics of bioassays reports that most bioassays give a straight-line response when plotted against the *logarithm* of the dose, I was struck with the apparent anomaly of the *Avena* curve test, in which a straight-line response has been reported even without transforming the dose into logarithms (Fig. 19, Went & Thimann 1937). Accordingly, Scott and I reinvestigated the dose response curve in the improved *Avena* rooms described earlier. The mean curves showed a good fit to the logarithm of the dose from 5 to 100 μg IAA/liter. The dose response of the untransformed data was clearly not linear over this same range (Figure 1-7).

After we had found that plotting the logarithm of the dose against mean curve gave better results and was also thereby in agreement with more bioassays for other substances (see Finney 1964), we found that Bottelier (1959) had fitted various curves to data in the literature for six different assays (including the *Avena* curve test) and had concluded that the logarithm of response against logarithm of dose gave the best fit to all the assays. The next best fit to the data that he used came from plotting response against logarithm of dose (as in Figure 1-7). Bottelier emphasized that, so long as a limited range of doses was being considered, the experimental error in the bioassays was large enough that any one of the four equations he used would give a fit within those limits of error.

Summary. The *Avena* curve bioassay should be run as a six-point (or, minimally, as a four-point) assay using the terms of bioassay statistics, with the mean curvature plotted against logarithm of the dose, thus agreeing with the dose-response curves of most other bioassays.

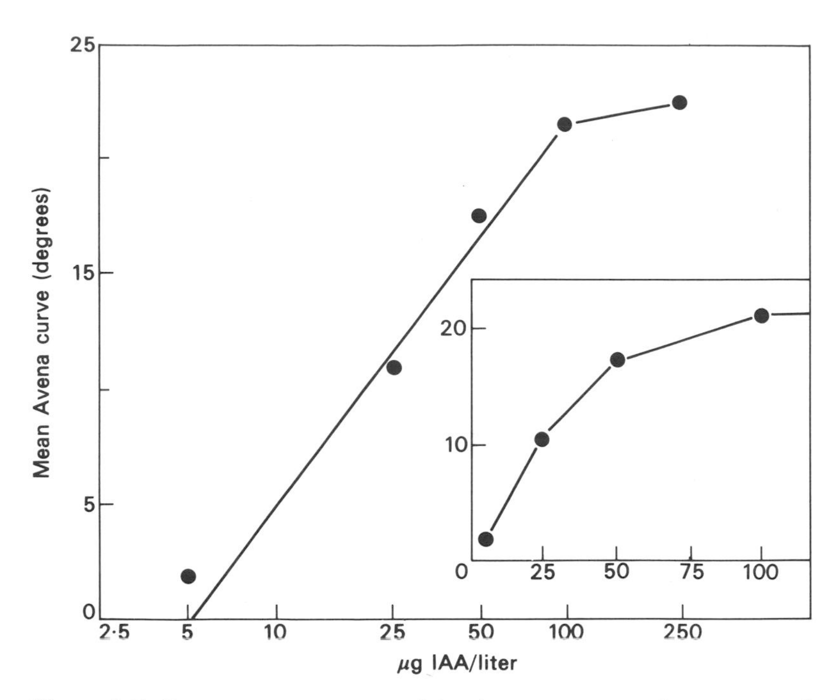

Figure 1-7. Dose-response curve of the *Avena* curvature bioassay, with the *X*-axis representing the concentration of IAA added in calibration blocklets and the *Y*-axis showing the average coleoptile curvature resulting (Scott & Jacobs, unpublished; pooled data from several experiments).

The *Avena* coleoptile section assay

The section assay requires less skill and simpler equipment than the curvature bioassay. For many botanists these advantages outweigh its disadvantage of being not as specific as the curvature test. The basic plan of the section assay is to cut one section of specific length (usually about 5 mm long) from near the tip of each coleoptile, float these sections on solutions in Petri dishes, incubate at 25°C, then measure the length of the sections after 24 hours (Bonner 1933, 1949). The measurements are made with an ocular micrometer and a dissecting microscope.

The section assay shows some response to sucrose alone, so 2% sucrose is routinely added to the solutions for both IAA calibration and the unknowns. Typical results from a section bioassay are shown to the left in Figure 3-5.

The only sizeable disadvantage to the *Avena* section bioassay is the relative lack of specificity. If IAA and sucrose are present, then arginine, methionine, or glutamic acid give still more section growth, and so will succinic, malic, fumaric, pyruvic, or α-ketoglutaric acids (Bonner 1949). Thiamine, nicotinic acid, adenine, or adenosine also increased elongation over IAA plus sucrose. Kinetin (see Chapter 6) greatly increases the elongation resulting from IAA and is, in fact, active by itself (Hemberg & Larsson 1972). That is quite a depressingly long list of substances that can give auxinlike responses in the section bioassay. Another disadvantage of this assay is the likelihood of bacterial contamination and resulting complications from incubating nonsterilized tissues in a 2% sucrose solution for 24 hours. Ease is not the sole desideratum.

2

The action of light in phototropism

In the chain of reactions that constitute phototropism what are the links between exposure to unilateral light and the asymmetry in auxin levels that eventually occurs? This chapter discusses two links. One involves experiments to discover if the auxin asymmetry was solely the result of lateral transport (as Went interpreted his data) or of auxin destruction on the illuminated side. The second concerns the pigment that absorbs phototropically active light. The section on techniques then describes methods of determining valid action and absorption spectra as well as a formal system for evaluating experimental evidence.

Attempts to determine the pigment of phototropism

A basic law of photochemistry states that light must be absorbed to be active in a photochemical reaction. An absorbing chemical is, by definition, a pigment. Identifying the pigments involved in various light-controlled processes in living organisms has been a major activity of photobiologists, who have devised ingenious methods to aid in the identification. One of the most valuable of these methods is to compare the action spectrum of a process with the absorption spectrum of various endogenous pigments in the hope of finding a close match between one of the pigments and the action spectrum of the process. (Refer to the section on techniques of this chapter if you are not familiar with methods of determining action and absorption spectra.)

Blaauw (1909, 1918) was the pioneer in the quantitative study of light in phototropism. He discovered that blue light was the most effective part of the visible spectrum in causing phototropism and that, within the low range of energies that he investigated, the phototropic curvature was proportional to the total energy received. Using broad-band filters, van Overbeek (1933) confirmed the rela-

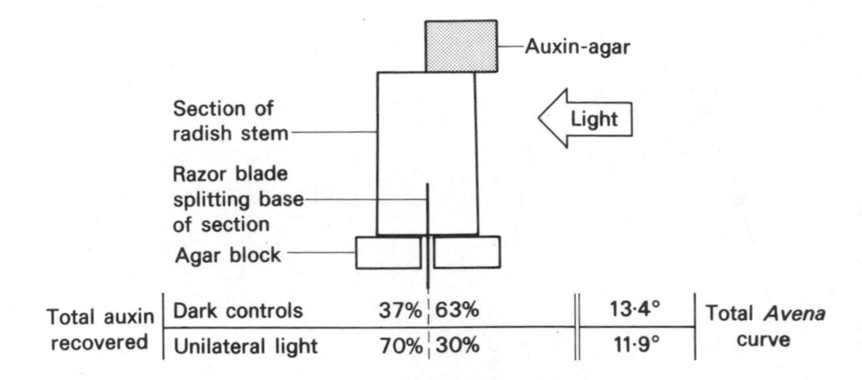

Figure 2-1. One of van Overbeek's experiments (1933), showing that phototropically active, unilateral light causes a shift to the shaded side of auxin added to the apical cut of an excised section of radish hypocotyl.

tive effectiveness of blue light. He added the valuable observation that blue light was also the most effective portion of the visible spectrum in causing lateral shift of exogenous auxin to the receivers under the shaded side of excised sections (Figure 2-1). The unilateral light decreased the total auxin little, if at all, judging by the summed average curvatures from the *Avena* bioassay of the receivers. (This thesis of van Overbeek on radish hypocotyls was one of the few early but thorough auxin studies that investigated an organ other than the coleoptile.)

Johnston (1934) refined the techniques used by earlier men and the first indication of the action spectrum of phototropism in coleoptiles began to take shape (Figure 2-2). Within the blue end of the spectrum, there were two peaks (a maximum at approximately 440 nm and a secondary peak at about 475 nm), with a valley between. Galston and Baker (1949) reported a similar action spectrum; Shropshire and Withrow (1958) and Thimann and Curry (1960), whose action spectra extended the wavelengths investigated, also confirmed Johnston's results within the visible spectrum.

The photoreceptor could not be auxin itself because auxin does not absorb visible light. As early as 1933, Voerkel suggested that carotenoids, which absorb heavily in the blue end of the spectrum, were the pigments involved in phototropism. Wald and du Buy (1936) bolstered this suggestion with evidence that carotenoids were present in extracts of coleoptiles, even of those grown in the dark. This was particularly pleasing to Wald, whose research had been primarily on the visual pigments of the mammalian retina, because it added one more case to the list of animals and plants in

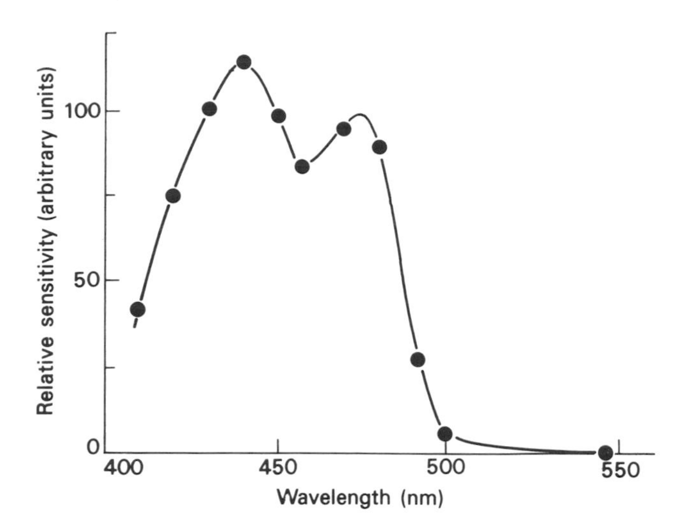

Figure 2-2. Action spectrum for phototropism of the *Avena* coleoptile in the blue end of the visible spectrum (after Johnston 1934).

which the presence of carotenoids was correlated with sensitivity to light. For visible light the absorption spectrum of carotenoids quantitatively matched the action spectrum of phototropism (Figure 2-3): the absorption spectrum showed two peaks within the blue end of the spectrum, with the maximum at 440 nm. The exactness of the match – closer than that between the spectrum for chlorophyll and photosynthesis as shown later – was convincing evidence that carotenoids were the pigments absorbing light for phototropism. Studies on the distribution of carotenoids within the coleoptile were even more convincing: histochemical tests revealed more and more carotenoid as sections were cut closer to the tip, with only the distal 0.2 mm showing a decline (Bünning 1937, 1955; Brauner 1955). Bünning pointed out the nice correlation in different levels of the coleoptiles between the amount of carotenoid from his data and the sensitivity to phototropic illumination from the data of Lange (1927).

Phototropism and the destruction of auxin

The preceding reports, as well as other early papers mentioned by Went and Thimann (1937), seemed to explain adequately various aspects of phototropism. The carotenoids, heavily concentrated in the tip of the coleoptile, absorbed unilateral light and somehow mediated the lateral transfer of more auxin to the shaded side.

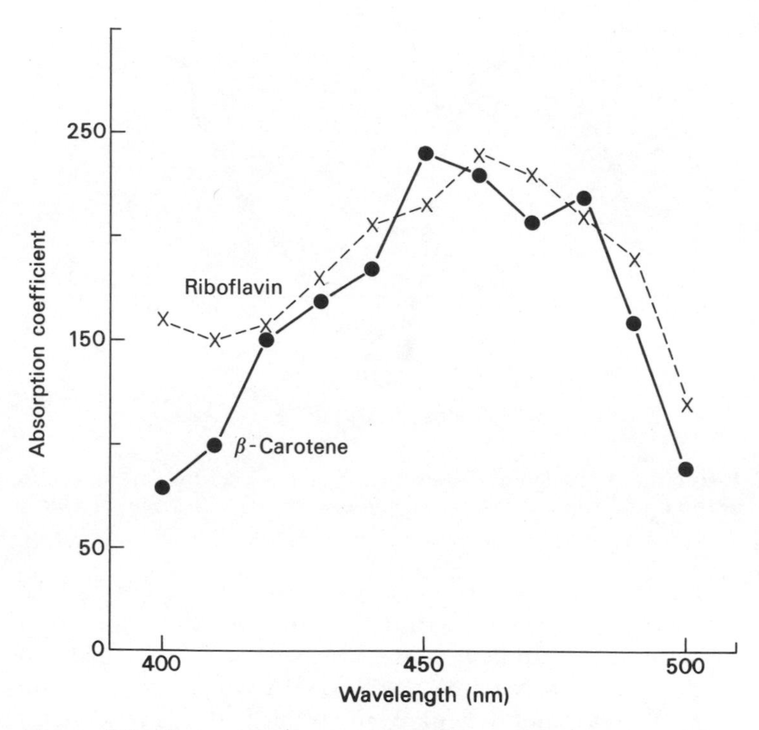

Figure 2-3. Absorption spectra of riboflavin and β-carotene (after Galston & Baker 1949).

Riboflavin and auxin destruction

However, Galston (1949b) discovered that if he added riboflavin (vitamin B_2) to an in vitro solution of the synthetic auxin, indole-3-acetic acid (IAA), riboflavin destroyed IAA if the mixture was illuminated, but not if it was left in the dark. He suggested that riboflavin also destroyed IAA (or whatever the endogenous auxin was) in the intact coleoptile during phototropism and, therefore, that riboflavin rather than a carotenoid was the photoreceptor for phototropism. Riboflavin was present in the coleoptile, as Galston and Baker (1949) soon demonstrated with a bioassay. There are two obvious consequences of this hypothesis: (1) IAA must be destroyed during phototropism, rather than merely shunted to the shaded side; (2) the action spectrum for phototropism must match the absorption spectrum of riboflavin at least as closely as that of carotene. Galston was a persuasive advocate of his hypothesis. But the absorption spectrum of riboflavin did not match the action spectrum as

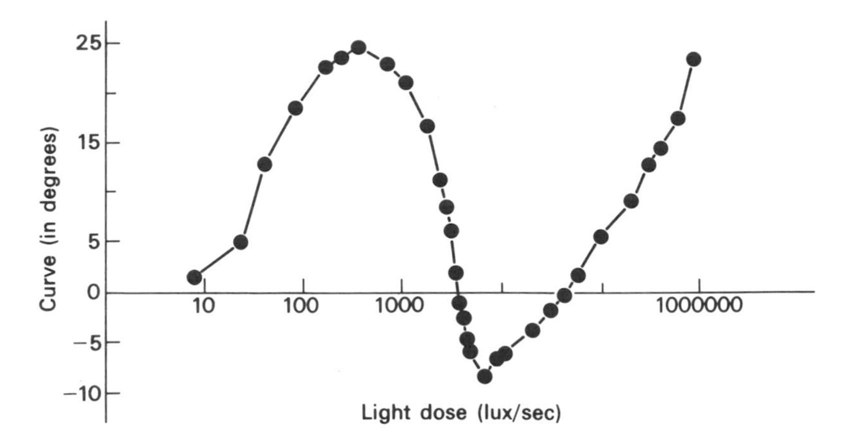

Figure 2-4. Relation between a unilateral light dose and the strength and direction of the phototropic curve in *Avena* coleoptiles (after Steyer 1967).

closely as did the carotenoids' absorption spectra (Figure 2-3). Riboflavin absorbs light heavily within the blue end of the spectrum but it shows only one broad peak in the blue instead of two (Galston & Baker 1949; Thimann & Curry 1960). Is IAA destroyed during phototropism? The original data of Went (1928) had been inconclusive, as pointed out in Chapter 1, and in addition gave rise to the suspicions aroused by results described only as percentages (i.e., with no original data). Stimulated by Galston's hypothesis, several physiologists carefully investigated the question.

Before we discuss whether phototropically active light destroys auxin, we need to be aware that the phototropic curve does not respond to increasing energy levels in the simple way one would intuitively expect (du Buy & Nuernbergk 1934). It is true, as Blaauw reported, that if one illuminates coleoptile tips for a short time with a very weak light and compares the average phototropic curve resulting as the energy is increased, the average curvature increases with increasing energy. However, beyond a surprisingly low level of energy, the curve not only does not increase further but decreases, reaching zero or close to it. (*Avena* coleoptiles, an apparently special case [Steyer 1967], even curve away from the light in this restricted energy range.) At still higher energy levels of unilateral light the coleoptile again grows toward the light – a growth pattern that is called the second positive curvature (Figure 2-4). The complexity of the responses is well illustrated in a recent three-dimensional graph (Figure 2-5 from Blaauw & Blaauw-Jansen

Figure 2-5. Three-dimensional graph of the effect on the phototropic curvature of *Avena* coleoptiles (vertical axis) of increasing exposure (left to right) and decreasing intensity (front to rear) (from Blaauw & Blaauw-Jansen 1970).

1970). Obviously, one needs to keep careful track of the intensity of one's unilateral light; preferably one should gain extra precision by illuminating with a narrow band of wavelengths in the blue end of the spectrum. If white light is used, the response from the phototropic receptor is apt to be blurred by absorption at the other wavelengths. A further complication is that the Bunsen-Roscoe reciprocity law, described in the section on techniques, is only valid for energies in the first positive curvature range.

Bioassay measurements

Briggs (1963b) examined in detail the fate of endogenous auxin in *Zea* coleoptiles illuminated with white light. Coleoptile tips from *Zea* are larger in diameter and produce three or more times as much diffusible auxin as do *Avena* coleoptile tips (Dolk 1930; Briggs 1963a); therefore, they were a reasonable choice for the investigations. Briggs used the *Avena* curve bioassay to measure the auxin in the receiver blocks, but apparently he did not routinely measure serial dilutions (as ideal bioassay techniques require). Several values in his Fig. 2 he cites as above the proportionality range. Unilateral white light at energy levels of 1000 meter-candle-seconds

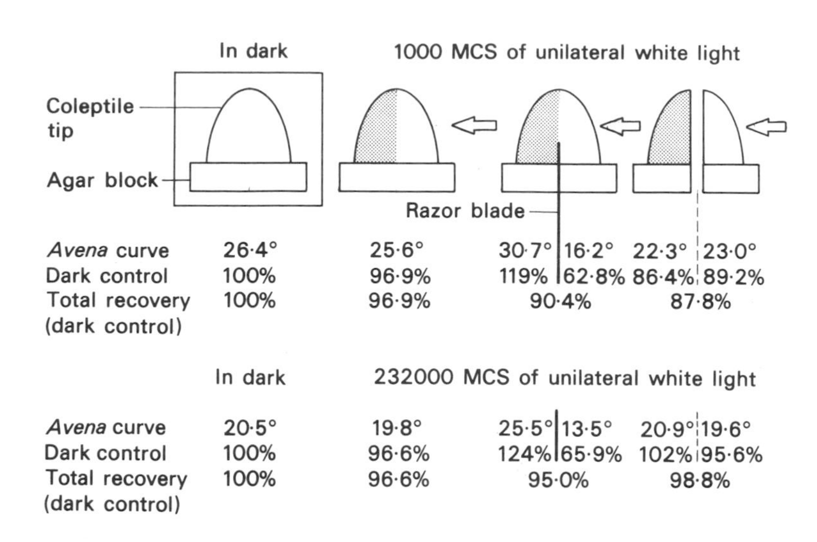

Figure 2-6. (above). Summarization of pooled data of Briggs's Table V (1963b) on the effect of unilateral light on the redistribution of endogenous diffusible auxin and on total recoveries. The 1000 MCS of white light was expected to give maximal first positive curvatures. The average curves were based on three intact tips per receiver or six half-tips (for tips partially or completely split); hence, if there was no auxin destruction or no lateral transport, all the *Avena* curves would be expected to be equal. Bottom portion represents same with the much higher light intensity of 232,000 MCS (expected to provide phototropic curves in the second positive curve range; based on data in Table VI of Briggs 1963b).

(MCS), which produced maximal first positive curve in these *Zea* coleoptiles (Briggs 1963a), did not decrease the total auxin diffusing from coleoptile tips but altered the proportions diffusing from the shaded and illuminated sides. Compared to a control group of tips kept in the dark, the illuminated tips diffused more auxin from the shaded side and less from the illuminated side[1] (Figure 2-6, top). Both conclusions were confirmed by statistical tests of the data, using the t-test (Snedecor 1956). Thus, for this particular energy level, Went's dubious 1928 conclusion was finally, convincingly, confirmed and Galston's riboflavin-IAA destruction hypothesis was correspondingly weakened. Energy levels high enough to give sec-

[1] The pooled results of five experiments showed 96.9% as much auxin coming from illuminated tips as from tips in the dark. (Briggs's calculation shows 99.3% as much, but that includes two extra experiments run only in the dark with no illuminated counterparts, a questionable procedure in view of the variability among experiments, which Briggs mentions.)

ond positive curves (232,000 MCS) similarly caused an increase in auxin diffusing from the shaded side, a decrease in auxin from the illuminated side, but no decrease in total auxin from the tip (Figure 2-6, bottom). (These differences apparently were not analyzed statistically, and fewer experiments were run at this energy level; hence, the results are not so secure.)

At the highest energy level used (1,000,000 MCS), Briggs's conclusions and his data do not match so beautifully. He states, for example, "under these rather drastic conditions of illumination, a small amount of inactivation may occur." But if one focuses on his actual data there is strong evidence that the total amount of auxin decreases when the coleoptiles are exposed to light of such high energy levels. For the 6-mm tips of his Table IV, the decrease in average *Avena* curvature averages 20%.[2] Twelve of the 14 pairs in his Table IV show less diffusible auxin collected from the illuminated tips.

To summarize Briggs's data, illumination did not destroy a significant amount of auxin at the two lower energy ranges but did destroy auxin at the highest energy level tested.

The fate of radioactively labeled auxin

IAA with radioactive ^{14}C in the carboxyl groups became available in the early 1960s and Pickard and Thimann (1964) added it to the very tip of *Zea* coleoptiles illuminated unilaterally with white light providing energy in the first or second positive curvature ranges. Applied symmetrically to the unwounded tip of coleoptiles 6.5 mm in length, the supplemental dose of labeled IAA shifted predominantly to the shaded side. Receiver blocks of agar placed under the shaded or illuminated halves collected average counts of 48 cpm and 16 cpm, respectively. Thus, the ratio of counts from the shaded side to those from the illuminated side was 75 : 25 for the first four experiments run with light that gave first positive curvatures.[3] Pickard and Thimann assumed that the ^{14}C remained with the IAA, although they presented no evidence for this view. They ran no

[2] Briggs apparently used the t-test for multiple comparisons of data from his Table IV (instead of restricting use of the test to pairs), a procedure that can lead to spurious "significance."

[3] I have not averaged all nine experiments of Pickard and Thimann's Table 1 in which 2.2 μM IAA were added in the donor blocks because the data show an inexplicable, consistent increase in cpm (from 28 to 170 cpm) in the shaded receivers as one goes from experiments 1 to 8. The last five of the experiments show a 66 : 34 ratio of counts in the receiver blocks.

statistical tests of significance. However, all 13 experiments of their Table I show higher counts in the shaded receivers.

Tissue taken from the shaded half of the coleoptile showed higher counts of ^{14}C than tissue taken from the illuminated half, although the distribution of 65 : 35 showed a slighter difference than Pickard and Thimann had observed for the receivers beneath the tissue halves. All eight of the experiments in their Table III show higher counts in the shaded tissue. Pickard and Thimann explain the smaller ratio of ^{14}C in the tissues as probably the result of difficulties in exactly splitting the coleoptiles; it seems more likely to result from the conversion of ^{14}C-IAA to nonmobile compounds that still have the radioactive label.

Addressing the question of whether or not phototropically active light destroys any auxin, Pickard and Thimann presented data for subapical sections (rather than for coleoptile tips) exposed to 5.4×10^6 MCS, energy levels in the second positive range. All the counts added in the donor blocks were recovered in the tissue, receivers, and final donors. If these counts entirely represented ^{14}C attached to unchanged IAA – a point on which no direct evidence was presented – they would indicate no destruction of the added ^{14}C-IAA. (Galston had reported in 1949 that the in vitro destruction of IAA by riboflavin and light involved release of carbon dioxide; hence, Pickard and Thimann were assuming that, if their ^{14}C-IAA were destroyed by light, CO_2 would be removed from the radioactively labeled carboxyl group, thereby causing a loss in total counts recovered in agar blocks and tissue.)

In general, Pickard and Thimann's results fit the view that phototropism results from lateral transport rather than unilateral destruction of auxin.

Although, broadly speaking, they ran the same experiment as did Pickard and Thimann, Shen-Miller and Gordon (1966) added several important improvements in technique. First, they illuminated with blue light rather than with white, thereby increasing the probability that any results of illumination would be associated with the blue-absorbing photoreceptor of phototropism rather than with some other process mediated by a pigment absorbing wavelengths elsewhere in the visible spectrum. Second, they increased greatly the probability that they were actually counting ^{14}C-IAA by extracting the tissue and receiver blocks, partitioning between acid and base, separating components of the acid fraction on paper chromatograms, eluting the zone of the paper to which indole-3-acetic acid ran, and then counting the eluate in a liquid scintillation counter.[4] (Pickard and Thimann, by contrast, had directly measured the counts coming from tissue pulverized on planchets.) Third, they provided evidence from statistical tests of significance that the differences they discussed were real and not merely sampling errors. Shen-Miller and Gordon painted onto the top of 5-mm-long coleoptile tips of *Zea* a tiny drop of ^{14}C-IAA in solution. (The radioactive label was in the methylene group of the acetic acid side-chain, rather than in the carboxyl group.) Immediately after adding the ^{14}C-IAA they illuminated from one side with a dose of broad-band blue light sufficient to give a first positive curvature. Blocks of agar collected material coming from the shaded and illuminated halves (as in Figure 1-4). After 2.5 hours they found significantly more counts in the receivers from the shaded side, thus confirming Pickard and Thimann. Essentially all the ^{14}C in the receivers was still with IAA, judging by the fact that 90–95% of the counts on paper chromatograms of extracts were at the Rf typical of IAA. However, extracts of the tissue, unlike those of the receivers, gave a different picture: there were fewer counts from the shaded tissue than from the illuminated. Adding the lower counts in the shaded tissue to the higher counts in the receivers on the shaded side gave total counts that were not significantly different from the total counts of the illuminated tissue and receivers. Similar results were obtained when the endogenous auxin was followed after illuminating the *Zea* tips with the same blue light and using the *Avena* curvature bioassay to measure the auxin. More endogenous auxin was in the shaded-side receivers (although the difference was only at the 9% level of probability, rather than at the 5% level conventionally considered statistically significant), and significantly less endogenous auxin was extracted from the tissue of the shaded half. Shen-Miller and Gordon concluded that there was no evidence from the total assemblies of any lateral movement of auxin; the first positive curve of phototropism must result from decreased basipetal transport of auxin through tissue on the illuminated side.

Further evidence that illumination decreases basipetal movement of both endogenous auxin and added ^{14}C-IAA was soon provided from the same laboratory (Naqvi & Gordon 1967). Reverting to white light but still using energy levels said to cause phototropic curvatures in the first positive range, they illuminated *Zea* coleoptile tips from two sides. The total endogenous auxin diffusing from the tips decreased by about 40%, a highly significant effect. If ^{14}C-

[4] See Chapters 3 and 9 for critical discussions of paper chromatography and radioisotope techniques, respectively.

IAA was added to tips that had just been illuminated, significantly more counts remained in the tissue near the applied IAA.

The most impressive evidence that phototropism involves inhibition of auxin transport was in a paper by Shen-Miller et al. (1969) on phototropism in *Avena* coleoptiles. They compared the dose-response curves of ^{14}C-IAA transport and phototropism using three different light sources; as they shifted from white light to broad-band blue to monochromatic blue light and at the same time increased their sample size from 3 to 20 to 30, statistical tests of significance changed from not significant to significant at the 5% level to highly significant at the 1% level. The action spectrum for phototropism also agreed quite well with the action spectrum for inhibition of ^{14}C-IAA movement.

Hager and Schmidt (1968) confirmed that light decreases auxin transport; and Menschick et al. (1977) confirmed that light energies sufficient to cause phototropism did not cause any detectable destruction of IAA through decarboxylation. In experiments reminiscent of van Overbeek's early work on radish (Figure 2-1), but using carboxy-labeled IAA, dela Fuente and Leopold (1968) demonstrated that IAA added to one side of *Zea* coleoptile sections was shifted by blue light to the shaded side. Their data gave no indication of IAA destruction by the light.

Summary

It seems clear that first positive phototropic curves elicited in coleoptiles do not involve any destruction of auxin. The phototropic curve results from relatively more auxin moving down the shaded side of the coleoptile, probably as a result of decreased movement on the illuminated side. The interrelations of transport inhibition and lateral transport are still obscure.

However, auxin relations for the first positive curve – relatively detailed though our information is about them – tell us little about phototropic curves as they occur in nature. The latter typically are elicited by energies at the high end of the second positive curvature range. At these higher radiant densities the few publications report that auxin destruction does occur (Briggs 1963b; Zenk 1968), although Zenk calculated that the amount of destruction was too small to account for the observed differences in growth rate.

As for the phototropism of organs other than the coleoptiles of *Avena* and *Zea*, our ignorance is as extensive as we should like our understanding to be.

Techniques

The PESIGS rules for organizing and evaluating evidence, and their application to auxin and phototropism

The PESIGS rules are a convenient system for assessing the extent and strength of evidence that a given chemical controls a given process in an organism (Jacobs 1959). The capital letters are a memorization device for the six rules and represent Parallel variation, Excision, Substitution, Isolation, Generality, and Specificity. The more of these rules that are satisfied by the available evidence, the more credence we give to the hypothesis. The more quantitative that evidence, the greater still our credence.

The usual scientific assumption is made at the start that probably only a single chemical is limiting a given process. This approach stems from the principle of logic called Occam's Razor (after the medieval philosopher William of Occam), which tells us to select the simplest of the various hypotheses that could explain the observed data (Russell 1945). (I suspect that the widespread use of Occam's Razor is more in deference to the limitations of the human mind than it is in tribute to its validity in the real world, but it is nevertheless a convenient starting point for fallible humans.)

The first of the PESIGS rules asks that the chemical in question be present in the organism and that parallel variation be demonstrated between the amount of the chemical and the relative activation of the process. For the positive phototropism of coleoptiles, auxin has been shown to be present both by bioassay and by more specific chemical identification (Chapter 3). Both auxin and phototropism vary in parallel fashion: the top millimeter of the coleoptile is the locus of greatest sensitivity to phototropically active light and is also the site of greatest auxin production (Went 1928; Briggs 1963b). Also, the phototropic curve shows strict basipetal polarity in its movement (Rothert 1894; van der Wolk 1912), and so does auxin moving through excised coleoptilar sections. The speed with which the phototropic curve moves basipetally (about 6 mm/hour) is close to the speed at which auxin moves through coleoptilar sections (9–15 mm/hour, from results of various authors, discussed in Chapter 9). As a final case of parallel variation, the wavelengths with maximal effectiveness in causing phototropism (the blue end of the spectrum) are the ones with maximal effectiveness in changing auxin transport (van Overbeek 1933; Shen-Miller et al. 1969).

The excision rule of PESIGS refers to the practice of removing the organ, tissue, or organelle that is known to be the source of the chemical under investigation and demonstrating a subsequent cessation of the process. The removal is often by excision, as the name of the rule suggests, but can be by genetic mutation or by adding chemicals believed to inhibit specifically the chemical in question. The logical difficulties with this rule are that actual excision usually removes many chemicals other than the one under investigation, that mutants may have pleiotropic effects, and that chemical inhibitors are rarely known to be as specific in their effects as the users hope. If one can show that a graded excision results in a graded reduction in the process, the evidence is, of course, thereby strengthened. As applied to phototropism, actual excision of the tip (the portion producing the auxin) removes almost all the phototropic sensitivity.

Substitution of pure chemical for the excised normal source should lead to restoration of the process. The ideal application of this rule would be to add a number of concentrations of the pure chemical, with some of them *exactly* replacing the amount provided by the natural source, and to show that quantitatively increasing restoration of the process ensued, with the exact replacement of chemical providing exactly normal activation of the process. Substituting auxin agar for the excised coleoptile tip can partially restore the phototropic curve, but to my knowledge no one has even tried exact replacement of the normal auxin supply.

Although the ideal application of the substitution rule involves replacing the excised natural source of the chemical with the pure chemical, researchers are often compelled, through ignorance of the chmical's identity, to try less direct substitutions. These may consist of substituting an active extract of the source, or macerated pieces of source tissue, or merely replacing the excised source on the cut in the hope that the chemical will cross the diffusion gap and show activity. The latter procedure was tried with coleoptiles before IAA was identified as "the" (or, at least, "a") native auxin, and a sizeable percentage of the phototropic curvature was restored.

The isolation rule states that one should isolate as much of the reacting system as is feasible and demonstrate that the chemical's effect is the same as in the less isolated system. If one has isolated the system such evidence reduces the probability that the chemical is primarily acting on another part of the organism, and only secondarily affecting the process in question. Progressively more extreme spatial isolations could consist of demonstrating the effect in an excised organ (as in a leaf cutting, for example), a piece of organ or tissue, an individual cell in aseptic culture, or preparations of subcellular organelles tested in vitro. From this special viewpoint, the part of biochemistry that deals with in vitro reactions of extracted chemical systems represents the extreme application of the isolation rule of PESIGS. In studying phototropism, physiologists have applied the isolation rule by physically isolating the tip of the coleoptile and showing that auxin production and lateral asymmetry in diffusible auxin can result from unilateral irradiation, if only the distal 0.5 mm of the tip is still unsplit (Briggs 1963b). Galston's efforts to show that riboflavin was the photoreceptor in phototropism and therefore that auxin was destroyed derived from his in vitro experiments with IAA and riboflavin, which in a PESIGS analysis would be considered extreme isolation.

A second form of isolation, which was not included in the original formulation (Jacobs 1959), is isolation in time. This is particularly useful in trying to discover the primary effect of a hormone. The effect that occurs most quickly after a chemical is added is considered most likely to represent the primary action. Such an application of the isolation rule has been used creatively during the last decade in trying to pinpoint the biochemical basis of auxin action (see Chapter 5).

Demonstrating the generality of the results is usually of great scientific importance. Most biologists hope to discover general principles, not merely results that apply, for instance, only to the Winesap variety of apples growing on the shores of Lake Cayuga. Hence, ideally one should show that the other PESIGS rules apply to genera from many different families, as well as to various developmental stages and various organs of any one species. In practice, this rule is slighted more than any other. American physiologists seem to be particularly guilty. We may be acquainted with the hormonal physiology of *Avena* and *Zea* coleoptiles, or pea and bean, or tobacco, or *Coleus*, or tomato stems while blithely, if silently, assuming that the other 170,000 or so species of angiosperms have the same hormonal relations. It is not a very wide base for generalizations. With regard to phototropism and auxin, only the grass coleoptiles have been thoroughly investigated.

The last PESIGS rule concerns tests of the specificity of the given chemical. Particular attention is paid to show that other naturally occurring chemicals do not fit the PESIGS rules for the same process. It is useful also to test analogues or presumed precursors of the chemical believed to control the process. Such a procedure often provides clues as to the biosynthetic pathways leading to the major controlling chemical and clues as to the particular features of the molecule that result in activity. Studies of phototropism have barely investigated chemical specificity.

Action spectra and absorption spectra

As we know from elementary physics, light visible to humans is only a small portion of the electromagnetic spectrum, the latter extending from the very short wavelengths of cosmic rays through X-rays, ultraviolet radiation, visible (e.g., blue, green, and red) to the still longer wavelengths of infrared radiation and radio waves. A specific location on the electromagnetic spectrum is denoted by wavelength in nanometers (nm), what we call *blue light* covering a broad zone in the 400–450 nm area, *red light* being in the 600–700 nm area. Pigments do not absorb all wavelengths to the same degree (which is why chlorophyll, the pigment that absorbs light for photosynthesis in leaves, looks green to us; it absorbs heavily in the red and blue ends of the visible spectrum and reflects the nonabsorbed green wavelengths). A quantitative measure of how much light is absorbed by a given pigment at each wavelength is called the *absorption spectrum* of that pigment. Figure 2-7 shows the absorption spectrum of chlorophyll a (Curtis & Clark 1950). Although the chemical structure of the pigment is the major determinant of the location and height of the absorption peaks, the solvent in which the pigment is dissolved can also affect the exact location of the peaks; hence, the solvent should be cited with each absorption spectrum.

Before the 1940s the absorption spectra were laborious to determine. The investigator had to set each wavelength on the spectrometer by hand, then read and write down the absorption at that wavelength from the swinging needle of the meter. All these individually determined absorbancies would then be plotted against wavelengths to give the finished absorption spectrum. Since the 1940s spectrometers have been developed that do all this automatically, including plotting the final absorption spectrum. Of equal importance, they are commercially available for about $4000, which puts them within reach of the average grant-holding U.S. researcher.

An action spectrum (which always refers to a process rather than a pigment) is a quantitative measure of the effectiveness (i.e., action) of various wavelengths in causing a light-controlled process. For instance, the action spectrum of photosynthesis would be determined ideally by illuminating the green leaf with light of many selected wavelengths and measuring the amount of photosynthesis that occurred at each of the wavelengths. Plotting the action of each such wavelength against the wavelength provides the action spectrum of photosynthesis (Figure 2-7).

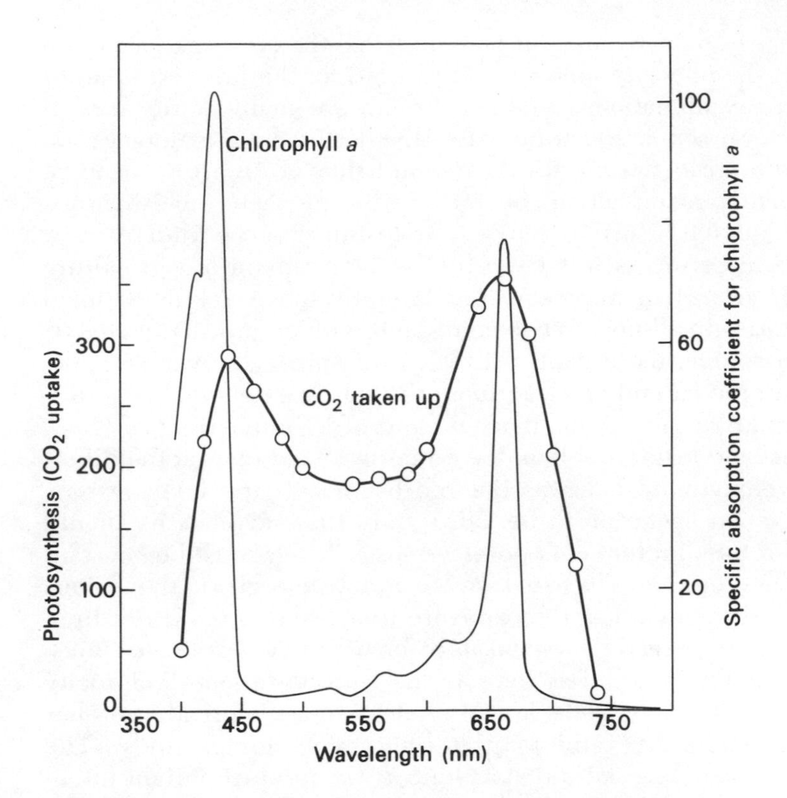

Figure 2-7. Comparison of the action spectrum of photosynthesis with the absorption spectrum of chlorophyll *a* (after Curtis & Clark 1950).

Unfortunately, the determination of action spectra is still not even equivalent to the pre-1940 era of absorption spectroscopy, as far as technology is concerned. There is not commercially available apparatus that will automatically determine the action spectrum of a process with precision like that of the absorption spectrometers. The two major problems are (1) to use very narrow bands of wavelengths – as many and as narrow as possible – so that the action spectrum can reveal fine detail; (2) to equate the energies at the different wavelengths. Before 1940 it was not feasible to determine action spectra with many narrow bands; a very few narrow bands could be obtained (by electric discharge through gas) or a larger number of very broad bands could be obtained (by using colored filters of gelatin or glass, such as the Wratten filters made by Kodak). Even though several broad-band filters could be combined to provide a fairly narrow range of wavelengths, the combinations often lowered the intensity below the threshold for the process; in addition, such combinations could not provide the many points needed for a precise action spectrum. After the 1940s the development of the interference filter, with its transmission of a narrow band of wavelengths, made action spectra feasible for many investigators. The alternative to using many separate interference filters was to use a giant prism, as was done by the Department of Agriculture research groups in Maryland and later by the Argonne National Laboratories in Illinois. When combined with an unusually powerful light source, these giant prisms could spread the visible spectrum over the far end of a long room to such an extent that an entire plant could be placed in a narrow band of wavelengths.

The second major problem, the equating of energies at the different wavelengths tested, was ignored by many early investigators. For an action spectrum to be valid, only the wavelengths should change at the different data points: energy levels must be constant across the spectrum. The usual way to equate energies at the various wavelengths is to adjust the exposure times so that the lower light intensities are exactly compensated for by longer exposure times. This procedure is valid as long as the Bunsen-Roscoe reciprocity law – also referred to as the law of photochemical equivalence – has been shown to be valid for the light effect under study. The Bunsen-Roscoe law states that as long as the product of light intensity and exposure time remains constant, the photochemical effect also remains constant. Of course, this law, if it is shown to be valid for a given process, is only valid within set limits: one cannot extrapolate to exposure times longer than those tested and assume that reciprocity still holds.

The hope of an investigator who determines the action spectrum of a process occurring in some organ is that a pigment found in that organ will have an absorption spectrum exactly matching the action spectrum. Such an exact match would be considered evidence the pigment was absorbing light for that process. For photosynthesis and chlorophyll the actual data of Figure 2-7 do not show the ideal, exact match. Although the peaks of the two spectra are in the same general location, their relative heights are reversed, and green light is more effective than expected from chlorophyll's absorption. The lack of an exact match can be explained away on the grounds that the absorption spectrum is determined in a pure solution of pigment in an optically polished glass vessel, whereas light for the action spectrum had to pass through the waxy cuticle layer on the leaf's surface, the walls and cytoplasm of the epidermal cells, the outer wall and cytoplasm of the palisade parenchyma cells, and the membranes of the chloroplasts before actually reaching the chlorophyll pigment ensconced therein. All these surfaces and structures would provide reflective and absorptive surfaces for the impinging light and could be expected to blur the exact action of the various wavelengths on photosynthesis itself. This explanation is strongly supported by the much more exact match found between photosynthetic action spectra and pigment absorption when seaweeds with simpler histology have been investigated (e.g., Haxo & Blinks 1950). Another obvious explanation for the lack of an exact match between an action spectrum and the absorption spectrum of a given pigment could be the involvement of other pigments in the action spectrum.

3

The chemical nature of endogenous auxin

The chemical identification of a hormone is a prerequisite for fast progress in understanding the physiology of its action. Hormones are typically present in extremely minute concentrations in plant and animal tissue (just a few micrograms per kilogram), so that isolating sizeable amounts of an endogenous hormone is difficult. Most university laboratories, in fact, are not equipped to handle the huge volumes of solvent and tissue such isolations require. The result is that a hormone isolated from a natural source is apt to be rare and expensive. Once a hormone has been identified, chemical synthesis has often been found to be much cheaper than further isolation. This has been particularly true for the currently known plant hormones, which are all of relatively small molecular weight. Knowledge of the chemical structure also allows investigation of the effects of modifications of the structure in the hope of finding an "artificial hormone" with a valuable activity related to the endogenous one. This has been preeminently successful in several cases, ranging from oral contraceptives to the million-dollar-a-year sales of 2,4-D (2,4-dichlorophenoxyacetic acid), a weed killer that apparently has chemical properties close enough to those of endogenous auxin to stimulate growth but is chemically different enough to escape the endogenous auxin regulating systems. In addition to these practical values, the synthesis and testing of hormonal analogues helps physiologists to determine what properties of the endogenous chemical's structure provide the hormonal activity – a point of obvious value in understanding the hormone's physiological role. Another major asset of chemical identification is that it opens the door to fast chemical measurements as contrasted to the slow bioassays that otherwise must be relied on.

In this chapter I shall describe the convoluted history of the attempts to identify endogenous auxin chemically, the effect of personalities and techniques on progress, and the current state of knowledge, unresolved though some of the problems still remain.

Auxin A and B or the fungal auxin, indole-3-acetic acid?

Went (1928) had diffused coleoptile tips onto gelatin blocks, which then caused curvatures when bioassayed in the *Avena* curvature test (see Chapter 1). The assumption was made that this biological activity was due to only one substance, auxin, coming from the coleoptile tips. No bioactivity was lost from such gelatin-auxin blocks by heating to 90°C or by illuminating the blocks with various doses of white light. More specific evidence about auxin's chemical nature was that its molecular weight was estimated to be 376, based on a calculation of the diffusion coefficient as described in Chapter 1. The calculation assumed that there was only one molecule involved, that the active molecule was not dissociated, and that the molecular weight was between 50 and 500. None of these assumptions was known to be true. A few years after Went's thesis, a chemist at Utrecht reported the isolation of auxin A and auxin B (Kögl et al. 1933). The structures of these molecules were quite complex; they were reported to autoinactivate, and only very small amounts were extractable. Kögl declared the molecular weight of auxin A to be 328, not too far from the calculated molecular weight of 376 reported in Went's thesis. Botanists accepted these two compounds as the naturally occurring auxins even though they were primarily obtained from human urine. For a number of years Kögl supplied auxin physiologists with "pure auxin". On request, one would receive a small vial of solution from Kögl that showed the expected auxin activity in the *Avena* curvature bioassay.

In the next 2 years a substance was isolated of very different structure, but with high activity in the *Avena* curvature bioassay for auxin. First reported as an auxin from human urine (Kögl et al. 1934), then from yeast (Kögl & Kostermans 1934), it was also isolated from the medium in which the fungus *Rhizopus* had grown (Thimann 1935a). This substance was indole-3-acetic acid (IAA) (Figure 3-1), a compound known previously but not as an auxin. Its molecular weight was 175, clearly out of the range of the molecular weight of the presumed native auxin of *Avena* coleoptiles. Because it had not been found in higher plants and because its molecular weight differed from those of the "native" auxins, it was called *heteroauxin* ("different auxin") and was not thought to occur in higher plants. However, its availability in synthetic form made it more accessible than auxin A, which was available only from Kögl.

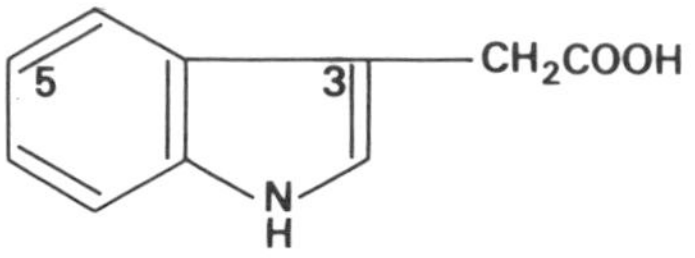

Indole-3-acetic acid (IAA)

Figure 3-1. The chemical structure of indole-3-acetic acid (with position 3 and 5 marked on the indole ring).

What in hindsight seems a depressing number of papers published in the 1930s compared the activity of IAA to the activity of auxin A.

As is inevitable in any field where a number of people are working, eventually scientists attempted to repeat the isolation of auxin A and auxin B. All failed, including Haagen-Smit, who had helped with the original work in Kögl's lab. Some fruitlessly tested human urine once again as a source of auxin A, but with no luck. Botanists speculated that the original discovery of auxin in urine was the result of some peculiarity in the Dutch diet, but extractions of Dutch urine yielded no auxin A or B. Meanwhile, the structure of biotin as it was reported by Kögl's laboratory was shown to be inaccurate (György 1954) and a bit of suspicion spread through the auxin world about the validity of the auxin A isolations, although, of course, these suspicions remained unpublished. Thimann mentioned at national meetings (in an unpublished statement) that he had obtained a solution of auxin A from Kögl and had found it active as an auxin by bioassay, as expected, but when he tested this solution with a color reaction specific for an indole ring it gave a positive response. (No indole ring was present in auxin A as described by Kögl.) In the past four decades no one has ever found auxin A or auxin B again, but IAA has been isolated from higher plants as well as from urine (Haagen-Smit et al. 1941; 1942; 1946). The work of Berger and Avery (1944) confirmed Haagen-Smit's isolation of IAA from *Zea* seeds.

Human urine is scarcely the most satisfactory source for isolating plant hormones and the isolation of IAA from kernels of *Zea* does not constitute convincing evidence that IAA is the sole hormone of all organs in all plants. Nevertheless, there have been surprisingly few further isolations and chemical identifications of endogenous auxin. However, several investigators, using the diffusion coefficient, have calculated that the endogenous auxin of various plants had a molecular weight much closer to that of IAA than to that of the presumed auxin A. These findings increased the belief that IAA was probably the endogenous auxin of most plants. For instance, the auxin activity extractable by ether from guayule showed a molecular weight calculated as 166; IAA in parallel tests gave a calculated molecular weight of 168 (Smith 1945). Similarly, the auxin activity obtained from tomato stems, whether by ether extraction or by diffusion, showed a molecular weight calculated to be 201 to 203 (Kramer & Went 1949). The most thorough of this class of paper was published by Gordon and Sanchez Nieva (1949). The auxin extractable from pineapples by ether was probably IAA, judging by the calculated molecular weight of 181, the pH sensitivity, and the dilution activity in the *Avena* curvature bioassay. Wildman and Bonner (1948) partially resolved the mystery of the calculated molecular weight of 378 for auxin from oat coleoptiles, a value that Heyn had confirmed fairly closely (1935). Wildman and Bonner reported a calculated molecular weight of 306 for auxin activity diffusing into agar blocks from *Avena* coleoptiles. They made an interesting calculation using a different preparation. If they extracted the diffusate blocks with ether and added the reduced extract to agar blocks that had not been in contact themselves with plant tissue, the calculated molecular weight from diffusion under these conditions was only 206, much closer to the molecular weight of IAA than to the 328 of auxin A as described by Kögl. In addition, they reported that the amount of IAA, as estimated by the Salkowski color test, agreed reasonably with the amount of IAA present as determined by activity in the *Avena* curvature bioassay. (There was 1.8 μg of IAA by Salkowski color test and 2.3 μg by the bioassay.)

Note that this evidence supports the view that IAA is present in the extracts of *Avena* coleoptile tips but does not provide any evidence that it is the sole endogenous auxin of that organ. Despite the paucity of convincing chemical evidence, by the end of the 1940s most auxin physiologists felt that IAA was undoubtedly the native endogenous auxin of vascular plants and that auxin A and B did not exist. But during the next decade the development of paper chromatography revived the controversy and stimulated a great proliferation of papers on growth stimulators and growth inhibitors in extracts of various plant parts.

The new technique of paper chromatography

Description of the technique

Paper chromatography is a technique for separating chemicals in solution. As a simple illustration, a small drop of the solution is

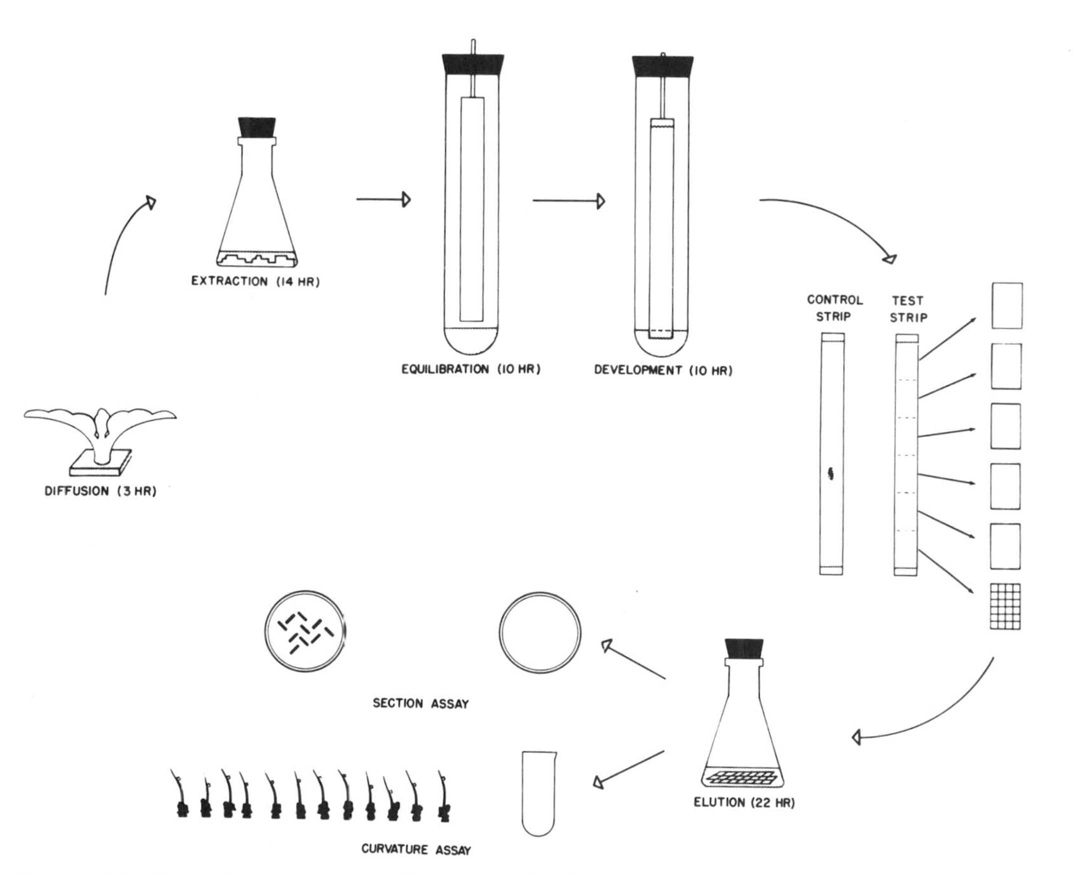

Figure 3-2. Flow diagram of procedures used by Scott and Jacobs (1964) to identify the diffusible auxin of *Coleus* shoot tips.

pipetted onto a pencil dot near one end of a 20- to 30-cm strip of filter paper (Figure 3-2). After the drop has dried, that end of the paper strip is lowered into a small volume of solvent, with care being taken not to submerge the pencil dot. Solvent will move upward in the filter paper, migrating over and past the spot containing the chemicals. After 12 or so hours the front of the migrating solvent will be about 20 cm from the application spot. The paper strip is then removed from the closed container and the front quickly marked with pencil. If the solvent is a judicious mix of liquids, the chemicals that were spotted onto the pencil spot (called the *origin*) will now be separated from each other and will be located at specific locations on the paper strip. A specific location is denoted by Rf – the *r*atio of that distance from the origin to the distance of the *f*ront from the origin. Thus, Rf = 0.5 denotes a location halfway between the origin and the front, Rf = 1.0 denotes the front itself. The Rf of a given chemical differs according to the particular mix of solvents into which the strip is dipped (with which the chromatogram is developed), but it is reasonably constant for a given solvent combination. For precise localization, however, the Rf of known amounts of pure chemical (the marker spots) must be determined in control chromatograms that are developed at the same time. The judicious mix of solvents for separating given chemicals had to be determined empirically for the most part (Smith 1960).

Paper chromatography is simple, effective, and cheap. The 12 or more hours needed for development of the chromatogram made overnight runs convenient. Not surprisingly, in view of this attractive set of qualities, the technique was adopted quickly by researchers in many fields.

Yamaki introduced paper chromatography to auxin physiologists in 1950. There followed a spate of papers reporting the localization on chromatograms of bioassay activity, indole color, fluorescence under ultraviolet light (UV), or various combinations of these, after extracts had been spotted and run on the paper. Most researchers tested extracts of plant tissue, rather than of agar diffusates, presumably because the former involved much less work. The same criterion apparently promoted the use of the coleoptile section growth bioassay, despite the fact that the *Avena* curvature bioassay is more specific. (The section assay does, however, have the advantages of more easily revealing zones of inhibition.)

One of the best and most thorough of the early papers reporting paper chromatography of plant extracts was that of Kefford (1955; preliminary account in Bennet-Clark et al. 1952). Figure 3-3 illustrates both the method and what has become the conventional way

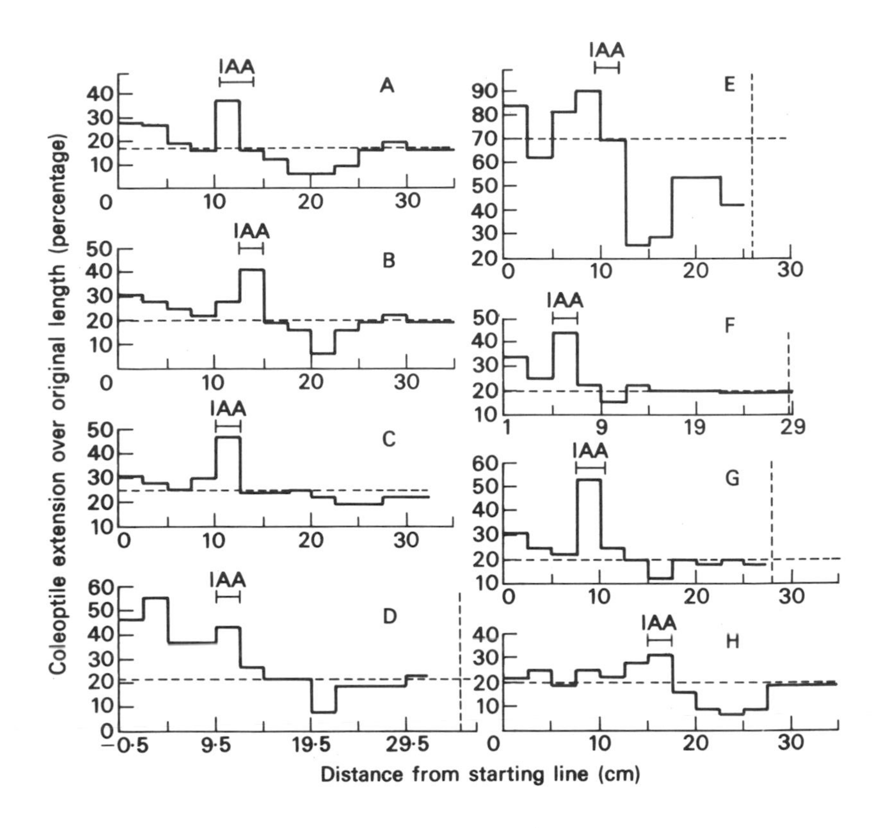

Figure 3-3. Amount of promotion or inhibition of growth (in the coleoptile section bioassay) from various zones of paper chromatograms of extracts of *Vicia* shoot (*A*) and root (*B*), *Pisum* shoot (*C*) and root (*D*), etiolated *Helianthus* shoot (*E*), *Zea* root (*F*), etiolated potato shoot (*G*) and "skin" of potato tubers (*H*). The location to which IAA ran on control chromatograms is shown by the line marked "IAA" above each histogram (figure from Kefford 1955).

to present results from paper chromatograms. Along the horizontal axis is plotted either the absolute distance from the origin (as in Figure 3-3) or the distance in Rf values. Along the vertical axis is plotted the activity of various Rf zones in a bioassay. The chromatogram is dried and then cut into 10 or more zones. The dried zones can be added directly to the medium of the coleoptile section growth assay or chemicals can be eluted from the dried zone with solvent, with the eluate subsequently bioassayed. The response of the controls is indicated by a line (the dashed lines of Figure 3-3) or preferably by a horizontal band providing some specified measure of the variability of the control response (such as the 95% confidence intervals). The most useful representation, adopted by some later investigators, is to represent the controls with a horizontal line, but to fill in with stippling or solid black those zones whose responses were significantly different from the controls (e.g., Figure 3-5).

Early results using paper chromatography

Kefford's results with chromatographed extracts containing acidic substances from roots or shoots of several genera usually revealed three zones of bioassay activity (Figure 3-3). One zone of stimulation (and hence, by definition, of auxin activity) was at the Rf zone to which synthetic IAA ran (designated by the line labeled "IAA" above each graph). A second zone of auxin activity was usually found nearer the origin and a zone of inhibition (called β-inhibitor by Kefford and his successors) was usually found running ahead of the ^{Rf}IAA^{Rf} zone with his particular mix of solvents. Kefford suggested that these three zones were physiologically significant in the control of lateral bud growth because extracts of lateral buds showed more β-inhibitor and less auxin activity than did extracts of the main (elongating) shoot.

More and more people ran chromatograms of plant extracts. Several new compounds were reported from such chromatograms; when these new compounds were used subsequently as control markers on a separate chromatogram run in the same tube as the chromatogram of an extract from another genus, activity would often appear on the experimental chromatogram at these same Rfs. The practitioners of this technique jumped to the conclusion that the zones of coleoptile inhibition and stimulation that they had found on their chromatograms represented the various inhibitors and auxins present in the original plant. In the other direction, a number of workers reported no sign of IAA on their chromatograms.

As is often the case when a new technique is introduced into an experimental field, critical evaluation of paper chromatography of plant extracts trailed behind the advent of the technique. The ethyl ester of IAA, which is active in auxin bioassays, was reported to have been isolated from *Zea* seeds (Redemann et al. 1951), and it was often thereafter added to control chromatograms when examining extracts of other materials. Quite often activity at the Rf of the ethyl ester would then be found (e.g., Nitsch 1956). But ethanol had been used as the extracting solvent in the original isolation, and after Henbest et al. (1953) pointed out the possibility, Fukui et al.

(1957) retracted the original report, deciding that the ethyl ester was an artefact of isolation. Indoleacetonitrile, also active as an auxin, was reported isolated from cabbage leaves (Henbest et al. 1953), but Gmelin and Virtanen (1961) showed that indoleacetonitrile was also an artefact of isolation, formed from glucobrassicin by enzyme action during the extraction (Gmelin 1964; Kutáček & Procházka 1964). Before indoleacetonitrile was recognized as an artefact, it was sometimes used as a marker on control chromatograms and was also reported to show up on the experimental chromatograms (Nitsch 1956). However, if the control chromatogram was run in the same tube as the "unknown" paper strip, substances could cross over from one strip to the other through the air – a quite unforeseen source of error (Raadts & Söding 1957; confirmed, apparently independently, by Nitsch & Nitsch 1960). Hence, an unknown number of the papers reporting the occurrence of IAA, its ethyl ester, or IAN as endogenous auxins were presumably merely reporting the results of such an artefactual transfer from the marker chromatogram.

Investigators invalidated the blithe and tacit assumption that everything appearing on the developed chromatogram was also present in the original extract and endogenous to the intact plant. Much of the IAA disappeared from the chromatogram if the paper was not tested soon after drying (Fig. 3 of Kefford 1955). The developing solvents could themselves produce new chemicals. Jepson (1958) showed that ammoniacal solvents could produce indoleacetamide, as an artefact, from indoleacetylglucosiduronic acid. Zenk (1961) reported a similar effect on the glucose ester of IAA: ammoniacal isopropanol destroyed 57% of it, with 13% hydrolyzed to indoleacetamide. Despite these demonstrated artefacts, isopropanol-ammonia-water has continued to be the most popular solvent combination for paper chromatograms: controls are seldom run to measure the extent of breakdown of sugar esters.

In addition to chemical transformations and losses occurring on the paper, the solvent itself can inhibit the section bioassay, as can material in the paper (Hancock & Barlow 1953; Farrar 1962; Scott & Jacobs 1964). Some of the early conclusions that IAA was *not* the endogenous auxin of various plants (see the list in Bentley 1958) undoubtedly resulted from the failure to realize that the color tests require more IAA (as Booth & Wareing 1958 and Scott & Jacobs 1964 demonstrated). In addition, IAA chromatographed with a plant extract can give a different color from what it gives when it is tested by itself (Housley & Griffiths 1962; Scott & Jacobs 1964); it also may run to a somewhat different position on the chromatogram (Cutler & Vlitos 1962; Terpstra 1953).

Despite these problems with paper chromatography, it is obviously beneficial to separate whatever chemicals are in the raw extract. But if one runs a one-dimensional chromatographic strip and then cuts the strip into 10 subdivisions for testing, it seems obvious, though apparently unappreciated, that *everything* added to the origin has to terminate in one of these 10 zones. The Rf typical of IAA for a given solvent is the Rf typical of many other compounds, too. Water extracts, run on paper and then bioassayed with the section growth test, can show auxin activity and IAA color at the Rf of IAA, but this can result entirely from sugars (Booth 1958). (Adding optimal levels of sugar to the medium before adding the test strips will prevent one's being misled in this way.) Bandurski and Schulze (1974) warn that p-coumaric acid can easily be confused with IAA by typical Rf, UV absorption peak, and Ehrlich color reaction. Inhibitory chemicals can run to the IAA zone and block auxin activity in the bioassay, as Pavillard and Beauchamp (1957) showed for scopoletin and Kaldewey (1964) confirmed for his unidentified inhibitors. Malonyl tryptophan runs to the same Rf in several solvents as does indoleacetylaspartate, and gives similar color reactions when treated with both Erhlich and Salkowski reagents (Good & Andreae 1957).

Despite the growing number of reported artefacts and false identification, faith in the conclusions persisted among the practitioners of paper chromatography. Such faith is understandable; no one would want to spend a decade exploring a hypothesis that is itself the artefact of an inexact technique. The subconscious bias would strongly favor the belief generated by paper chromatography that the "auxin," estimated by direct bioassay before the days of paper chromatography, was really a mixture of several different promoters and inhibitors.

The most serious weakness of the chromatographic work, however, was the general lack of evidence quantitatively relating the zones of activity found on the chromatograms to the normal physiology of the plant.

Diffusible auxin compared to extractable auxin

Two papers at the 1963 Gif Conference clarified the situation (Kaldewey 1964; Scott & Jacobs 1964). The latter concentrated on the nature of diffusible auxin in *Coleus*, as contrasted to the extractable auxin examined by most chromatographers. (Diffusible auxin from 200 to 1000 coleoptile tips of dark-grown *Avena* seedlings had been collected and chromatographed by Terpstra (1953) and Shibaoka

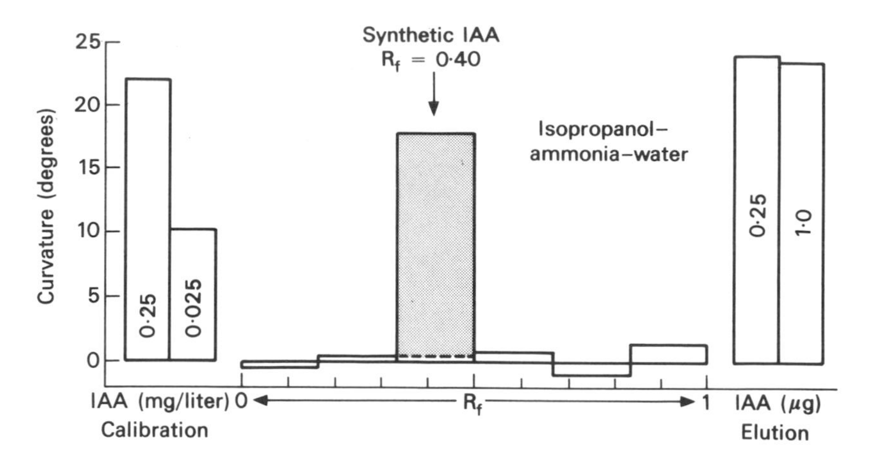

Figure 3-4. Auxin activity, as measured by the *Avena* curve assay, of various zones of paper chromatograms of the diffusible auxin collected from *Coleus* shoot tips. All the diffusible auxin ran to the same Rf as synthetic IAA (from Scott & Jacobs 1964).

and Yamaki [1959]. They had found auxin activity, by the *Avena* curve assay, only at the Rf of IAA. Ether extracts of the tips gave the same result [Terpstra].) In green plants of *Coleus*, diffusible auxin, estimated by the direct bioassay of diffusate agar blocks, had been quantitatively related to various physiological processes. (See Chapter 4 on xylem differentiation and Chapter 8 on abscission.) The evidence was particularly detailed for xylem regeneration. Jacobs (1956) determined the amount of synthetic IAA that would produce in the *Avena* curvature bioassay a curvature equal to that produced by the diffusible auxin coming from a given number of young *Coleus* leaves. He found that the amount of auxin thus determined also exactly replaced those leaves in their controlling effect on the number of xylem strands regenerated in the subjacent stem. Was the diffusible auxin activity, which could be exactly reproduced by pure synthetic IAA, actually the result of an unknown mixture of promoters and inhibitors, as the adherents of chromatographed extracts suggested?

To answer this question Scott and Jacobs selected the extraction method used most often by the plant chromatographers. The rationale was to avoid, in case the diffusible auxin proved to be solely IAA, the argument that "you *would* have found several auxins and several inhibitors if you had only used *our* extraction methods." Accordingly, overnight extraction in cold diethyl-ether was fol-

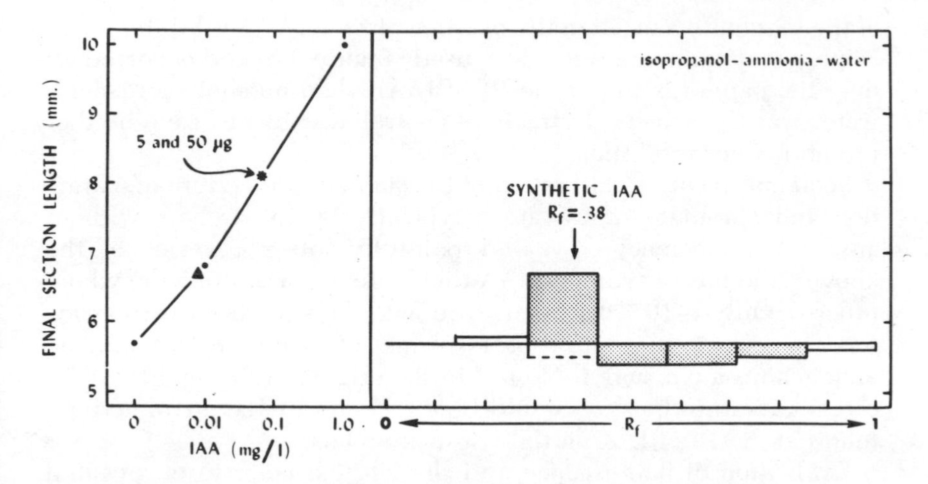

Figure 3-5. Auxin activity and inhibitory activity from chromatographed extracts of "diffusible auxin" receivers of *Coleus*, measured by the *Avena* section assay (Scott & Jacobs 1964).

lowed by one-dimensional paper chromatography using isopropanol-ammonia-water, with activity measured in the *Avena* straight-growth assay and compounds localized with Ehrlich's color reagent (Figure 3-2). To supplement these standard methods they also used the *Avena* curvature bioassay (a more specific test than the straight-growth assay), an acid solvent combination for developing the chromatograms, and short-wave UV for localizing fluorescent compounds such as IAA on the paper. Quantitative methods were introduced in two ways. First, they calibrated the efficiency of the whole standard process by adding known amounts of synthetic IAA and seeing how much was finally recovered (such calibration was rarely reported in the literature; Hamilton et al. 1961, who reported 15–50% recovery, were notable exceptions). Second, they tracked all the auxin estimated to be present by direct bioassay in the diffusion blocks of agar (Jacobs et al. 1959) through the standard process to see if the recovered IAA by itself could account for the original auxin activity. Three-hour diffusions of 80 shoot tips of *Coleus* were calculated to give enough IAA for detection by bioassay – *if* all the auxin activity was due to IAA alone.

By both the curvature and section bioassays, there was one and only one auxin present and that ran to the Rf typical of IAA (Figures 3-4 and 3-5). Both solvent combinations gave this result (see Fig. 3 of Scott & Jacobs 1964). Statistically significant inhibition of growth of the sections occurred in the position of Kefford's β-inhibitor when isopropanol-ammonia-water was used (Figure 3-5) and occurred all over the paper (except at the Rf of IAA) when butanol-acetic-acid-water was the solvent. Extracts of control agar blocks gave neither promotion nor inhibition.

Scott and Jacobs's calibration of the standard procedure of extraction and chromatography – used typically by researchers without any stated efficiency – revealed painfully low recoveries of the known amounts of synthetic IAA that were taken through the whole process. Only 1–10% was recovered with the straight-growth assay.

After correction for the observed loss of synthetic IAA, all the endogenous auxin activity found in the original diffusion blocks by direct curvature bioassay could be explained by the auxin activity found at the Rf of IAA on the chromatograms.

Evaluation of fluorescence and the Ehrlich color tests revealed how misleading these tests could be if they were used without parallel bioassays. (Lack of typical IAA color on a chromatogram had been interpreted by some earlier workers as showing that IAA was not present in the extracted plant.) The diffusible auxin from 80 shoot tips, recovered from the paper chromatograms, was more than enough to give a significant response in the bioassays but remained undetectable by the Ehrlich color test. Jacobs and Scott increased the number of shoot tips to 200 so that the minimum amount of IAA would be present on the chromatograms for detection by Ehrlich's reagent; again the assumption was that all the auxin activity was due solely to IAA. The only spot on the resulting sprayed chromatogram from the extract of the diffusible auxin was at the Rf of IAA. But the color was not typical of pure IAA. However, pure IAA, added to such an extract before chromatography, resulted in an intensified spot at the Rf of IAA, a spot of the same aberrant color.

Scott and Jacobs concluded that the diffusible auxin of *Coleus* shoots was IAA and only IAA. They also argued for the view that diffusible auxin was more meaningful physiologically than was extractable auxin. "If one wished to measure endogenous hormones (which, by definition, move from one location to another within the organism), it seemed reasonable to use 'diffusion' – a technique which collected those materials which were being actively transported and diffusing out of the organs which served as endogenous sources of hormone. Extraction of those organs, on the other hand, would be expected to provide not only some of the hormone(s), but also precursors, breakdown products, plus the grab-bag of compounds which happened to be soluble in the solvent used." Extractable auxin was, in addition, not a single entity; drastically different amounts of auxin activity had been obtained using different solvents and extraction conditions. "Although some extraction method could presumably be worked out" which would give the same result obtained with "diffusible auxin", Scott and Jacobs championed the efficiency of letting the "living plant do a major portion of the purification." Their conclusions were supported by later work on *Coleus*. Böttger (1970b) confirmed the presence of $^{Rf}IAA^{Rf}$ in diffusates, using the straight-growth bioassay, whereas Hemberg (1972) extracted *Coleus* leaves and found both $^{Rf}IAA^{Rf}$ and a second zone of auxin activity (his Fig. 2A) in the extracts after electrophoresis.

By a quite different route Kaldewey (1964) independently came to the conclusion that diffusible auxin was only IAA. Continuing his studies of *Fritillaria* fruit stalks and introducing thin-layer chromatography into the auxin field, he also found the only auxin activity from extracts of diffusion blocks ran to the Rf of IAA and gave expected color reactions. Such activity was found in extracts of sections, too. But if the sections were diffused first and then extracted, the auxin activity at the Rf of IAA was gone. Inhibitory zones were present in extracts of sections but were seldom found in diffusates. Hence, unlike the $^{Rf}IAA^{Rf}$, the inhibitors did not seem to be functioning as hormones.

These two extensive studies on light-grown plants, coupled with earlier, less detailed ones on coleoptiles or shoots kept in the dark (Terpstra 1953; Shibaoka & Yamaki 1959; Shibaoka 1961), and the later work of Ohwaki (1970) supported the hypothesis that the "hormonal," diffusible auxin is IAA and only IAA – whatever may be the multiplicity of other zones found in extracts of tissue sections.

More recent views and the introduction of mass spectrometry

The Kaldewey and Scott and Jacobs papers seemed finally to convince most plant physiologists that the hormonal auxin – the auxin that actually moved around the plant – was IAA only and that "diffusion" was the easiest way to measure it.

The reports of other auxins dwindled. For a while, for instance, the endogenous auxin of citrus fruits was said to be nonindolic (Khalifah et al. 1963, 1966). However, although there has been no published retraction so far, Lewis (personal communication) feels that the reports were incorrect and that IAA is actually the citrus auxin, a view confirmed by Goldschmidt and Monselise (1968). As people carried out more thorough separations – using several solvent systems, two or more bioassays, and extra color tests – they tended to find only ^{Rf}IAA^{Rf} even in extracts (Housley & Griffiths 1962; Clark & Bonga 1963). And earlier reports that IAA was not present in extracts of tobacco (Vlitos et al. 1956) were contradicted by Kefford (1962) and Knegt and Bruinsma (1973).

However, it was not until 1971 that mass spectrometry provided evidence more convincing to chemists (as contrasted to the more easily persuaded physiologists) that IAA was present in plant extracts. In that year Igoshi et al. used mass spectrometry to identify IAA in extracts of citrus fruits. In 1972 Greenwood et al. reported that IAA could be identified by mass spectrometry in the diffusate from *Zea* coleoptile tips (although lack of evidence of 100% recovery of the original auxin activity prevented the conclusion that IAA was the only auxin). Bandurski and Schulze (1974) used mass spectrometry to show that IAA was present in extracts of *Avena* "shoots" (apparently coleoptiles and enclosed leaf, judging by their descriptions), calculating 16 μg/kg as the upper limit for free IAA. Again there was no evidence that IAA was the only auxin present. Similar evidence from mass spectrometry identified IAA in extracts of *Zea* roots and cotton ovules (Elliot & Greenwood 1974; Shindy & Smith 1975, respectively).

Is IAA the only endogenous auxin? Except for the research already discussed on green shoots of *Coleus* and dark-grown coleoptile of *Avena*, there is little evidence available on this point. No one has demonstrated with actual chemical isolations that all the auxin activity of an organ can be accounted for quantitatively by the auxin(s) isolated. My guess is that when this is done auxins other than IAA will be found, even though IAA will probably be the major auxin of shoots. In support of this guess I can cite both the murky situation in roots (see Chapter 10) and the recent evidence of Wightman (1977) that phenylacetic acid, a substance with weak auxin activity, is present (along with IAA) in extracts of the shoots of various species. Aside from the question of identifying all the auxins in a given species, it is likely that some families of angiosperms have evolved slightly modified auxins. For instance, 4-chloroindole-3-acetic acid, or its methyl ester, has been reported in extracts of immature seeds of *Pisum* (Gandar & Nitsch 1967; Marumo et al. 1968; Engvild 1975). Indole-3-acrylic acid has been reported in extracts of roots of several legumes (Hofinger et al. 1970). Indole-3-acrylic acid is active in some auxin assays and has an Rf close to that of IAA in some commonly used solvents.

The latest, and presumably last, word on auxin A and auxin B

What about the mysterious auxin A and auxin B? After Kögl's death Vliegenthart and Vliegenthart (1966) found authentic samples of "auxin A and auxin B" and their derivatives in his laboratory. They determined the chemical structure of these samples by mass spectrometry. Not one of the samples had the structure originally proposed. The bottle labeled "auxin A" was really chloric acid, that labeled "auxin B" was thiosemicarbazide. The Vliegentharts concluded that auxin A and auxin B are nonexistent, a conclusion that certainly fits the physiological evidence from chromatography. Soon after his thesis work with Kögl, Haagen-Smit had moved to the California Institute of Technology in California, where he made a brilliant career exploiting one of the great "natural" resources of the Los Angeles area by becoming an expert on the chemistry of smog – work for which he was awarded a National Medal of Science in 1973. His recollection of the events of the early 1930s (personal communication) is that Dr. Erxleben, as Prof. Kögl's skillful and trusted co-worker, carried out the elaborate final steps leading to crystalline material for all the isolations in the laboratory. "Kögl trusted her implicity and used her work to check on the workers in the laboratory," but apparently he did not use anyone to check on her. No one else in the laboratory at that time doubted Erxleben's integrity, and Haagen-Smit can only guess at what happened: "It is possible that the initial mistake was to advertise the purity of auxin A prematurely. Prof. Kögl's eagerness to publish and his dictatorial behaviour possibly made it very difficult for Miss Erxleben to retract her error, although this could have been done quite readily in the early period. . . . It was Erxleben's persistence in covering up which led to the unwitting involvement of many of her associates." As one of those so involved, Haagen-Smit (with his students) not only published the isolations of IAA that I have already cited from the 1940s but tried again with the improved methods of the 1960s to isolate auxins A and B from *Zea* (unpublished). He concluded that the activity was due to IAA.

Lessons to be learned by the student

What can we learn from this strange history? First, do not trust anyone, including yourself. If you, as the head of a laboratory, like Kögl, are going to put out a paper with your name on it, not only should you insist that the experiment be run twice (or preferably thrice) with essentially the same results but you should run through the entire procedure yourself for one of those repeats. Of course, this will slow down your publication rate, but it will also decrease the chances that you will be famous mainly as the senior author of one of the biggest scientific errors of the century – and your name will mean what it should mean on a paper: that you personally vouch for the adequacy of the contents. Your experiments should be designed to eliminate at all possible points your own subconscious biases: do readings from coded material (so that you do not know, for instance, which spot on the chromatogram is supposed to be brighter according to your working hypothesis), and ask others in the laboratory – without giving them any clue – how many spots they see and which spot they consider brighter. Calibrate the people as well as the machines: regularly feed "unknowns," the contents of which you actually know, to laboratory personnel who are doing routine analyses to see how dependable their work is. (A friend, while a postdoctoral researcher in a well-known and sizeable laboratory group, discovered by such calibration that the assistant who for years had been doing the laboratory's nitrogen analyses would slide the "results" either up or down according to the views expressed by the scientist. If he was surprised they were so low, on the next run they would be higher. I am sorry to say that my friend, reluctant to be the bearer of such bad news, did not tell the professor in charge of the laboratory of his discovery.) Ask the laboratory personnel regularly to feed you, among your unknowns, some that are "blind repeats" (i.e., you have actually measured them before) to calibrate the reproducibility of your measurements. It is much more work, but the data will be likely to have real meaning.

The brouhaha about paper chromatography and plant hormones teaches a general lesson about the history of a new technique: researchers are apt to swoop down on an easy new method, use it experimentally with uncritical enthusiasm, and rush papers out to the journals. Several years typically pass before the artefacts produced by the technique begin to surface. I can think of no way instantly to discover the artefacts. But one can take the following precautions:

1. Calibrate the procedures when possible (e.g., when we did this for percentage of IAA recovered, using the most widely used extraction methods in the literature, we found that most workers had been using techniques that gave the appallingly low values of 1–10% recovery [Scott & Jacobs 1964]).

2. Relate what you find to the intact organism, preferably in a

quantitative way. For instance, agar blocks containing diffusate from young leaves of *Coleus* showed auxin activity when bioassayed directly in the *Avena* curvature test (Jacobs 1952, 1956). Extracts of such agar blocks showed evidence of only one auxin, $^{\mathrm{Rf}}\mathrm{IAA}^{\mathrm{Rf}}$; all the auxin activity in the original diffusate blocks was explainable by the amount of $^{\mathrm{Rf}}\mathrm{IAA}^{\mathrm{Rf}}$ recovered from the chromatograms, when correction was made for the loss in the extraction and chromatography (Scott & Jacobs 1964). But with the section bioassay there was a small but statistically significant amount of *inhibition* from the zone of the chromatogram in the position of Kefford's β-inhibitor. Was this inhibition significant physiologically? Considering just the data resulting from the chromatograms developed in ammoniacal isopropanol and bioassayed with coleoptile sections (Figure 3-5), one might argue that the inhibitor was a meaningful component of diffusible auxin (as assayed directly). An apparent 20% of the original diffusible auxin was recovered, but only 1–10% of synthetic IAA was recovered. This apparently greater recovery of the diffusible auxin might be due to the fact that on the paper (unlike in the diffusion block) the β-inhibitor has been separated from the $^{\mathrm{Rf}}\mathrm{IAA}^{\mathrm{Rf}}$ so that the true, higher level of IAA can at last be seen. But if one relates these results to the organism, this interpretation is unlikely. Calculating the activity found in *Coleus* leaf diffusion blocks by direct curvature bioassay as if it were all due to IAA (instead of the hypothetical mixture of β-inhibitor plus *more* IAA), and then substituting for the leaf an amount of synthetic IAA that exactly replaces by bioassay the leaves' diffusible auxin gives exact replacement of the leaves' action in causing the regeneration of xylem cells (Figure 4-10). In excised internodes of *Coleus* synthetic IAA alone was later shown to be able to replace quantitatively both the rest of the shoot and the whole root system in causing regeneration of xylem and sieve tubes (Fig. 1 of Thompson & Jacobs 1966).

On a less complicated level, there is no discernible virtue in looking for hormones in extracts of organs that have not been proven hormonally active on the rest of the plant. It is the young, fast-growing leaves that produce the most diffusible auxin and that have significant effects on xylem differentiation and prevention of abscission (Jacobs 1952; Jacobs & Morrow 1957; Jacobs et al. 1959), and it is such leaves that show larger amounts of $^{\mathrm{Rf}}\mathrm{IAA}^{\mathrm{Rf}}$ in extracts (Hancock & Barlow 1953; Linser et al. 1954; Pegg & Selman 1959), yet researchers will extract fully grown leaves to report that no $^{\mathrm{Rf}}\mathrm{IAA}^{\mathrm{Rf}}$ is present (Luckwill 1957; Vendrig 1960).

Summary

Current evidence points to IAA as the endogenous auxin of the higher plants. Heteroauxin has become *The Auxin*. But actual chemical identification has been painfully scanty. There is still not a single case of *full* identification of the auxin of green shoots (as contrasted to etiolated seedling organs). As usual in biology, the sampling of various families of organisms has been minimal. Some family of plants may prove to have evolved an auxin different from IAA. As always in an actively investigated field, there are many loose ends; only in elementary textbooks is science beautifully clear-cut.

4

Other developmental effects of auxin

As soon as Went's thesis work on auxin and phototropism became known, feverish activity began on the reinvestigation of developmental phenomena that were apt to be controlled by a polarly moving substance like auxin. Geotropism of coleoptiles, the polar regeneration of roots on cuttings, apical dominance, and – much later – the polar differentiation of vascular cells were all shown to involve auxin. And after the commercially available indole-3-acetic acid (IAA) became known as an auxin in the mid-1930s, it was added to the organs of many genera whether or not polarity was involved. So many effects of auxin have been reported in the last 50 years that merely listing them all would take an inordinate amount of space, much less discussing them with the detail given to phototropism in the preceding chapters. Therefore, in this chapter I select several of auxin's major developmental effects for detailed discussion, briefly describe some others, and list still more along with references to the literature where more detailed treatment can be found.

The role of auxin in geotropism, abscission, flowering, and fruit development

Geotropism was an obvious phenomenon to investigate after Went's success with phototropism, and auxin was presented as the controlling substance in several fine early papers from the Utrecht laboratories (e.g., Dolk 1930, 1936 on coleoptiles; Dijkman 1934 on *Lupinus* hypocotyls). The approach and methods paralleled those of Went and are well described in Went and Thimann's book (1937). In roots the quality of the evidence was much lower but the unifying power of the Cholodny–Went theory of tropisms was so great that it has been only in the last decade that the role, if any, of auxin in root geotropism has started to be critically reevaluated (see Chapter 10).

Auxin is the hormone, coming from leaves and fruits, that prevents abscission ("leaf fall" or "fruit drop"). Chapter 8 describes the experimental evidence for this conclusion. Evidence about the role of auxin (and some other hormones) in fruit development is summarized in Nitsch's encyclopedia article (1965), and the possible involvement of auxin in the control of flowering is covered in Chapter 7.

The role of auxin in apical dominance

Two papers in 1933 reported that auxin could inhibit lateral buds if substituted for the apical bud of the main shoot of *Vicia*. The interpretations differed, however. Laibach noted swelling on the shoot stump to which he applied his auxin source (the pollinia of an orchid) and considered that the auxin was causing local growth stimulation and thus was only indirectly inhibiting the lateral buds. Thimann and Skoog (1933, 1934) considered the inhibiting action of auxin to be directly on the lateral buds; they noticed no extra elongation of the shoot stump from adding auxin (impure, from medium in which the fungus *Rhizopus* had grown). One of the key experiments from Thimann and Skoog's thorough 1934 paper is summarized in Figure 4-1. It shows that auxin substituted for the apical bud fully replaces it in slowing down elongation of the lateral bud. The apical bud was shown by *Avena* curve assay to produce much diffusible auxin.

Many people have also obtained results like those in Figure 4-1 when adding pure IAA to *Vicia* or to other members of the Leguminosae family. For plants in some other families of Angiospermae, however, apical dominance does not seem to be controlled by IAA from the apical bud. For instance, when IAA was substituted for *Coleus* shoot tips in amounts that gave exactly the normal amount of auxin judging by the *Avena* curve bioassay, xylem regeneration bioassay, and an abscission-speeding bioassay, that amount of IAA had no inhibiting effect whatever on the growth of the lateral buds (Jacobs et al. 1959). Even in legumes, where Libbert (1964) demonstrated that the level of added IAA that gave essentially the normal amount of auxin (by the *Avena* curve assay) also substituted quantitatively for 2 days for the apical bud in inhibiting a lateral bud, a second hormone from the apical bud is apparently involved. If gibberellic acid (see Chapter 7) is substituted along with IAA for the apical bud in legumes, lateral inhibition lasts longer than if only IAA is substituted (Jacobs & Case 1965; Scott et al. 1967; Phillips 1971a). Ruddat and Pharis (1966) independently and by a different

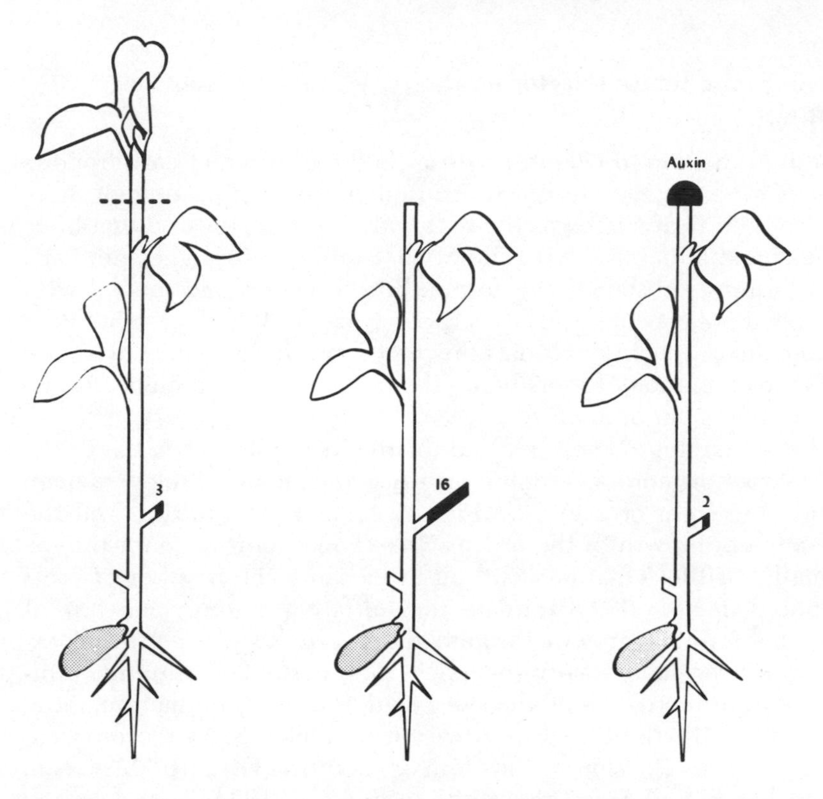

Figure 4-1. Summarization of the experiment of Thimann and Skoog (1934), demonstrating that auxin substituted for the main shoot tip of *Vicia* inhibits the elongation of a lateral bud as strongly as does the main shoot tip itself (figure from Jacobs et al. 1959).

logical route arrived at the conclusion that gibberellins were involved in apical dominance in both soybeans and redwood trees.

Some sense of the range of apical dominance, and of existing responses to decapitation, comes from Champagnat's review in the *Encyclopedia of Plant Physiology* (1965). Reading his review helps dispel the notion that all angiosperms act like bean, pea, or *Coleus*. To my mind the major unresolved question about apical dominance is the original one raised by the divergent interpretations of Laibach and Thimann and Skoog, namely, when auxin from the apical bud inhibits the outgrowth of lateral buds does the auxin act by stimulating growth in the main shoot, thereby diverting materials from the lateral buds, or does the auxin act by moving down to the laterals and directly inhibiting their growth?

Auxin as the limiting factor for the regeneration of roots on cuttings

The introduction to Chapter 1 described the frequent early reports that if roots regenerate on an excised piece of stem or root, they typically form at what was the root end of the cutting and that buds or leaves stimulate such regeneration. Sachs (1880) suggested that a root-forming substance was formed in the leaves and moved with basipetal polarity to induce this root formation. It took over 50 years before auxin was identified as this root-forming substance, with the polar regeneration of roots being the presumed consequence of the polar movement of auxin.

It is important to keep in mind the distinction between the initiation of root primordia, which typically occurs many cell layers deep inside the stem or root (see Fig. 17.11 of Esau [1953]), and the subsequent growth of the organ. One cannot merely count the externally visible roots on a cutting to measure the number of roots initiated; factors that stimulate the initiation of roots can inhibit their subsequent growth through the overlying tissues to the exterior. Furthermore, many species in their normal development initiate root primordia along the stem, with these primordia remaining dormant until stimulated to grow out by a change in the environment, such as excision of the stem as a cutting (Esau p. 528) (e.g., many but not all *Salix* species [Van der Lek 1924; Carlson 1938, 1950]; *Populus* and *Ribes* [Van der Lek 1924]; *Acer, Ulmus, Malus, Thuja* [Borthwick 1905]). Roots also may develop such dormant root primordia (e.g., *Convolvulus* [bindweed], a prolific regenerator judging by research on cultured segments by Bonnett and Torrey 1965b).

The view generally held at present (e.g., Phillips 1971b, pp. 23, 70) is that the endogenous auxin in a freshly excised stem cutting moves basipetally because of the polar transport of auxin and collects at the "root-tip" end of the cutting, where it causes the initiation of roots, just as synthetic auxin added from outside does in so many species when auxin is used as a rooting hormone in horticultural practice. One of the most frequently confirmed observations in the hormone literature is that auxin applied to the base of a stem cutting increases the rooting. Books on horticulture or practical aspects of hormone research typically list dozens of such species (e.g., Audus 1953). Dipping the root end of a cutting in Rootone (the trade name for a mixture of talcum powder and the synthetic auxin indole-3-butyric acid) is routine propagating practice. Those species that root profusely even without the addition of exogenous auxin are presumed to contain high levels of auxin. Let us review briefly the quantitative evidence for this interpretation.

Parallel variation between movement of endogenous auxin and regeneration of roots

Regeneration in root cuttings. Soon after a cutting is excised, does auxin increase at the base[1] (root end) of the cutting, as one would expect if root regeneration results from locally increased auxin? Warmke and Warmke (1950) provided evidence that within a few days auxin decreased at the shoot end and increased at the root end of a 15-mm segment cut from the fleshy taproot of chicory. Ether-extractable auxin, bioassayed with the *Avena* curve test, was found to increase at the root end and decrease at the shoot end on the fourth day after regeneration started; there was no difference in auxin levels at the two ends at 2 or 8 days, although the averages were less than at time zero (their Fig. 4). The auxin activity obtained with a much longer ether extraction (overnight) was separated into acid and neutral fractions. As early as the second day after excising, auxin activity in such an acid fraction showed a bigger increase in slices from the root end than in slices from the shoot end, and it stayed higher in the 4-day collections. (No statistical evidence was given for the significance of any differences.) Warmke and Warmke interpreted their results as showing that the "type of regeneration – whether shoots or roots differentiate from *Taraxacum* and *Cichorium* root cuttings – depends on auxin level." This view was bolstered by the fact that treatment with added auxin (IAA, naphthaleneacetic acid, or indolebutyric acid) caused roots to regenerate from the treated end, whether that was the shoot or root end of the cutting. Unfortunately, the statement in their discussion (and illustrated in their often reproduced diagrammatic Fig. 1) that the cuttings showed a "strong tendency . . . to produce leaves from the proximal ends and roots from the distal ends in normal regeneration" is not true for the root regeneration: the distal (root-tip) ends were described repeatedly in the results as forming callus only, with rootlets developing "only occasionally." If auxin is the endogenous controller of root regeneration, why does the increased

[1] "Root end" indicates the end of a cutting, whether of a stem or root, that was nearer the root tip of the intact plant; "shoot end" indicates the end nearer the shoot tip of the intact plant. (This fits the hormone transport in the intact plant and this terminology is less confusing than "distal" and "proximal," in which the root end is distal for a root cutting but proximal for a stem cutting.)

auxin level at the root-tip end of the cutting not cause ample root regeneration there – particularly since, as pointed out by Warmke and Warmke, the maximal levels of endogenous auxin were about 10 times higher than those previously extracted from other plants using the same methods?

Camus's extensive thesis (1949), also on regeneration of pieces of *Cichorium* roots but using aseptic, nutrient culture, was apparently unknown to Warmke and Warmke. He placed pieces, 20 mm long, cut from chicory root, horizontally on the nutrient medium and followed organ regeneration and extractable auxin at the root and shoot end of the pieces. (Cold-ether extracts for short periods were assayed with the *Avena* curve bioassay, similar to the methods of Warmke and Warmke.) In the first few days some preexisting root primordia grew out but no new root primordia formed. Strong polarity of callus development was evident by 6–8 days, with much more callus on the root end of the cultured pieces. It was only during this first week that Camus also found sizeable polarity in the amount of auxin extractable from the two ends (his Fig. 83, p. 163). It declined at both ends compared to zero-day extractions, but auxin levels from the shoot end declined faster. Hence, both 2- and 4-day collections showed more auxin at the root end than at the shoot end. By 7 days auxin from the root end had declined to approximately the same low level as that in the shoot end. At 9, 11, and 14 days, as the ether-extractable auxin started to increase, Camus felt there were no meaningful differences between the two ends. (Throughout there were no tests of statistical significance, nor any variance data.) Buds were first visible in microscope sections at 9 days, and were macroscopically visible by 11 with no apparent polarity of initiation. By 14 days, however, the buds at the shoot end were much further developed even though, as Camus emphasized, the levels of extractable auxin were not that much different at the two ends by that time. Even in this nutrient medium, it was not until 28–30 days, after the buds at the shoot end had grown out into leafy shoots, that the first *new* roots appeared. Their regeneration was strictly polar, at the root end.

Hence, neither of these papers on chicory roots fits the detailed expectations from the standard hypothesis. Camus found no evidence that auxin was "piling up" at the root end (there was, instead, a steady decrease in auxin level through the first week), and although both papers reported higher averages of extractable auxin at the root end than at the shoot end (as expected), this relatively higher level was not followed by initiation of roots to a significant degree.

Vardjan and Nitsch (1961) took up the chicory problem, using improved techniques of lyophilization and paper chromatography. Several zones of growth activity were found using Nitsch's mesocotyl bioassay, after running the cold methanol extracts on the chromatograms. One zone was present only if they cut up the tissue at room temperature rather than in a cold room and was, therefore, presumed to be an artefact. (Perhaps, they suggested, it was the neutral auxin of the Warmkes.) Substance C, the zone that caused the most growth in their bioassay, showed higher levels in extracts of the root end than the shoot end of a 100-mm regenerate, although the levels declined steadily in both regions, from 0 to 3 to 6 days. (By 12 days buds had regenerated at the shoot end.) A zone that regularly appeared but showed much less activity in the mesocotyl bioassay was dubbed Substance B. This zone did increase in activity from 0 to 3 to 6 days, judging by the average values in their Fig. 4. Hence, although it declined again at 9 and 12 days, for the first 6 days Substance B seems to pile up at the root end of the regenerate as the general hypothesis would lead us to expect. Although Vardjan and Nitsch refer to the activity as due to auxins, the mesocotyl assay is not specific for auxins: gibberellins (see Chapter 7) and helminthosporol are known to show activity too, and many other substances have not been tested for activity. And Substance B showed Rf values and color reactions typical of tryptophan rather than of IAA.

Regeneration in shoot cuttings. For regenerating pieces of shoot, information on auxin levels is even more sparse. From a paper concerned primarily with other points, we can glean that *Nicotiana* (tobacco) stem pieces 10 mm long, stripped of the extracambial tissues and cultured on White's nutrient medium, showed essentially the same amount of auxin in the shoot-end half as in the root-end half (48% and 52%) when tested immediately after excision (Niedergang-Kamien & Skoog 1956). After 3 hours 89% of the total auxin was in the root-end half. However, if we compare total amounts of auxin extracted, we find that there was a decline in absolute levels in both halves. So the endogenous auxin was not "piling up at the root end" to a level higher than normal; it was merely not declining there so quickly. (Cold ether was used for extraction, and auxin was measured with the *Avena* curve bioassay.) If these stem pieces are allowed to develop on the culture medium they form callus only at the root end, where the endogenous auxin level is relatively higher. If IAA is added to the medium, the callus spreads over the whole surface. Surprisingly, if buds regenerated,

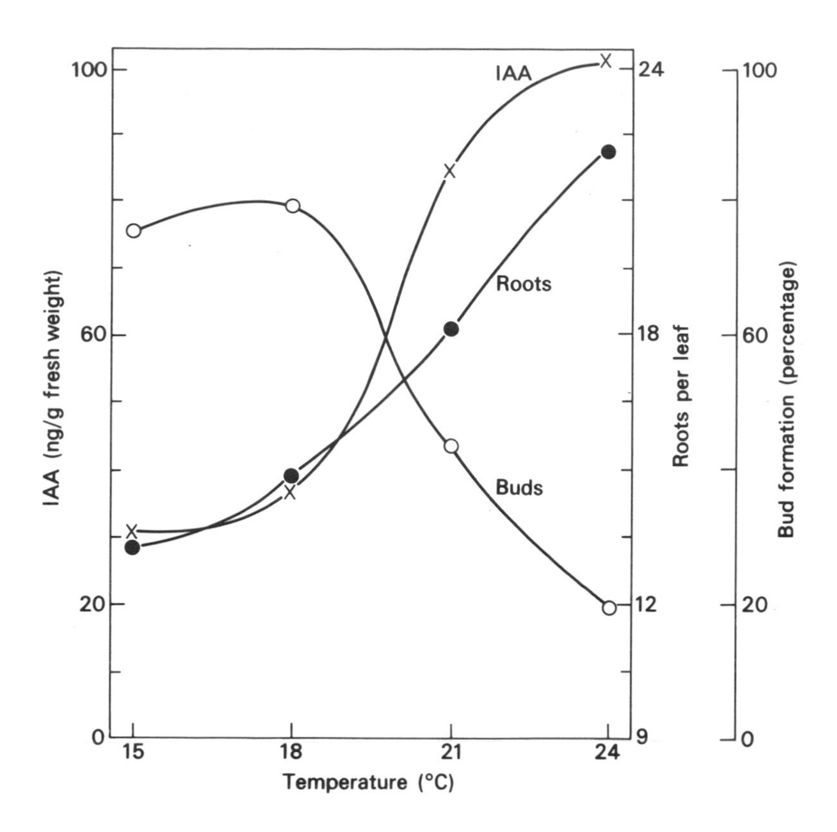

Figure 4-2. Changes with temperature in root and bud regeneration from *Begonia* leaves, as related to changes in level of endogenous [Rf]IAA[Rf] (after Heide 1972).

they formed only on the basal (i.e., root) end of stem pieces cultured root-end down in the nutrient medium! (By searching through the older literature, I found that Plett [1921] had found that only three of the 32 species he tested regenerated shoots exclusively on the *root* end of shoot cuttings and that *Nicotiana* was one of the three.) As usual with tissue culture work, it is difficult to judge whether some substance normally limiting in regeneration has been made nonlimiting in the cultured piece by virtue of being supplied as one of the many chemicals in the culture medium. However, even with this caution in mind, the cultured stem piece seems to act like the root pieces investigated by others: auxin declines throughout, but less so in the root-end half. Paralleling this polar distribution of endogenous auxin is a polar formation of callus, apparently evoked by the auxin.

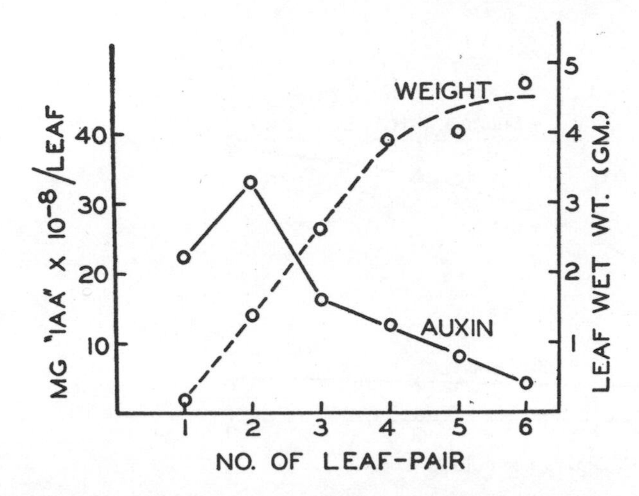

Figure 4-3. The production of diffusible auxin by *Coleus* leaves of various ages and weights as measured by the *Avena* curve assay. Pair 1 is nearest the apical bud, and pair 6 is the oldest pair on the plant (from Jacobs 1952).

Although questions of polarity within the cutting were not involved, a clear case of parallel variation between the amount of rooting and the levels of endogenous auxin has been worked out with *Begonia* leaves (Heide 1967, 1968). Either high temperatures or continuous light produced more rooting of the leaf cuttings than did lower temperatures (15°C) or short days (9 hours of light per 24-hour period). The higher temperature or continuous light also increased the endogenous auxin (which was IAA, judging by extraction with cold ether, Rf values after paper chromatography in each of five solvent systems, and bioassay with the *Avena* curve test) (Figure 4-2). As further evidence that this relation between auxin and rooting was causal, Heide (1965) demonstrated that NAA added from outside stimulated rooting.

Substitution of chemicals for leaves in root regeneration

An old and frequently confirmed observation is that the presence of leaves and buds on a cutting typically stimulates rooting, with young leaves being particularly effective (see Dore 1965, p. 52, for references). Young leaves were shown to be the major sources of auxin in shoots in most cases where endogenous levels have been measured (e.g., Goodwin 1937; Delisle 1938; Myers 1940; Jacobs 1952) (Figure 4-3). Therefore, it was reasonable to guess that the rooting effect of young leaves and buds is due to their auxin produc-

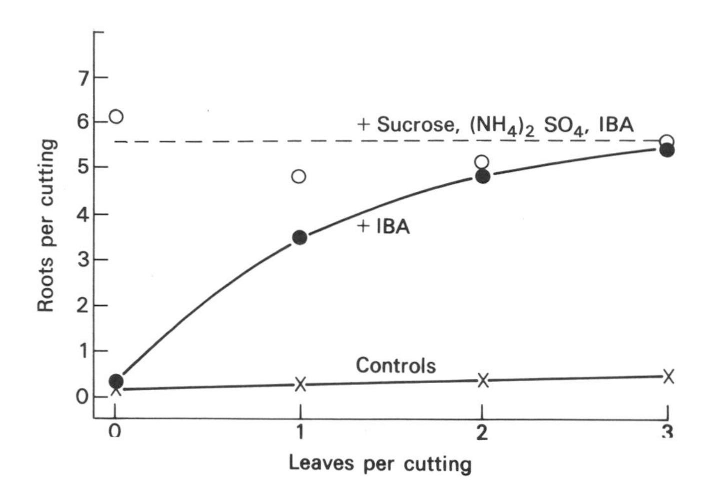

Figure 4-4. The effects of indolebutyric acid (IBA), the number of mature leaves, and sucrose plus ammonium sulfate on the number of roots regenerated per stem cutting of *Hibiscus* (after van Overbeek et al. 1946).

tion. Before the discovery of auxin production by leaves, the effectiveness of leaves was attributed to photosynthate ("nutritive factors produced by the leaves"); indeed, even Bouillenne and Went (1933), who proposed the term "rhizocaline" for special root-formative substances other than nutritive ones, reported that 1.5% sucrose or glucose alone could replace almost half the root-forming effect provided by intact cotyledons to *Impatiens* hypocotyls. (Seven roots/hypocotyl were obtained with the added sugar, 17 with cotyledons, 1 in the control with no cotyledon.)

A satisfyingly quantitative analysis of nonauxin leaf effects on rooting in *Hibiscus* was carried out by van Overbeek and Gregory (1945) and van Overbeek et al. (1946). Stem cuttings of common red hibiscus rooted profusely if some leaves were left on the stem and if the base was also treated with indole-3-butyric acid (IBA). IBA without leaves was ineffective and even as many as three leaves without IBA were ineffective. Presumably, therefore, the leaves were providing something other than auxin to stimulate rooting. The full rooting effect of three leaves combined with IBA treatment could be obtained by substituting for the leaves 4% sucrose plus 10 ppm arginine, or 4% sucrose plus 0.1% ammonium sulfate (Figure 4-4). (Some other nitrogen-containing compounds gave partial replacement when added with the sucrose and IBA, but many had no effect.) Sucrose plus ammonium sulfate had no added effect when three leaves remained on cuttings treated with IBA, indicating that the rhizogenic effect of the leaves was due to their production of sugar and nitrogenous material – an interpretation bolstered by the fact that, when the leaves remained on the cuttings, sugars and "soluble nitrogen" increased at the base of the cuttings. The authors concluded the "main function of leaves in the process of root initiation is to supply the cutting with sugars and nitrogenous substances." This is misleading with regard to the role of auxin. They always had to add auxin (IBA) to the cuttings, along with leaves or the nutrients, to get rooting. One might expect auxin to be provided endogenously by the leaves. However, a careful reading shows that they had removed the terminal and axillary buds; the leaves remaining were "mature." Thus, the young, fast-growing leaves – the organs expected to be the major sources of endogenous auxin – were always excised; the IBA was substituting for them.

In summary, regeneration of roots on stem cuttings seems to require auxin and nutrients, the latter not being surprising when one considers the metabolic work obviously involved in initiating the numerous meristems inside the stem tissue, differentiating them into root apices, then providing the energy for the elongation of the young roots through the stem tissue to the outside. If one provides enough nutrient in the stem cutting (as by using longer pieces) only auxin need be added to get rooting (suggested by Went for his results with *Acalypha* cuttings [Bouillenne & Went 1933] and demonstrated by van Overbeek et al. for red hibiscus).

Specificity of chemicals for root regeneration

The specificity of added chemicals for stimulating rooting has been tested in many genera. When nutrient levels are high enough to be not limiting, various auxins are active (IAA, IBA, naphthaleneacetic acid, etc.). (See lists in Audus [1953] referred to earlier.) Rather surprisingly, IBA is often more effective than IAA (e.g., Fig. 4-5). The literature contains various guesses, some rather irrational, to explain IBA's potency. Mullins (1972) recently provided specific evidence on this point: both IAA and IBA stimulated rooting in mung bean, but both substances also stimulated ethylene production. However, IBA caused the production of only about one-third as much ethylene as did IAA, and added ethylene inhibited rooting. Mullins's conclusion was that IBA provides more total rooting than does IAA because IAA counters it own rooting stimulation with its subsequent greater stimulation of ethylene production.

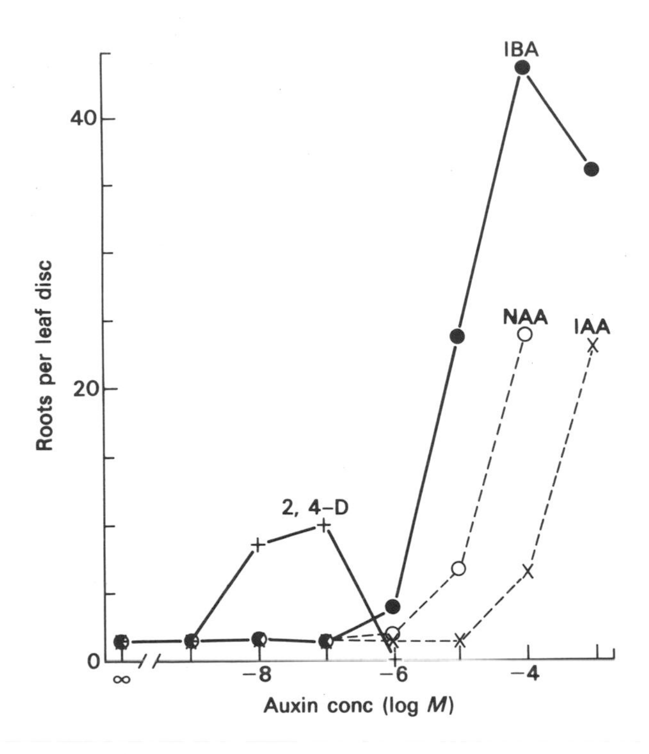

Figure 4-5. Relation of chemical structure to activity of various auxins in causing roots to regenerate on discs cut from *Streptocarpus* leaves (after Heide 1972).

The surprisingly high concentrations of auxin needed to cause the maximal number of roots to regenerate is a point of importance that has been neglected. As Figure 4-5 shows for rooting of *Streptocarpus* leaf discs, even the highly effective IBA requires 10^{-4} M to show its maximal rooting effect, with IAA requiring even higher concentrations. Such data were published even back in the 1930s (e.g., the optimal concentration of IAA for rooting in citrus cuttings was more than 10^{-3} M [Biale & Halma 1937]). Bottelier (1959), in an interesting comparison of various response curves to a wide range of IAA concentrations, pointed out that when *Ageratum* petioles were used in a rooting bioassay one had to add 7–10 times as much IAA (ca. 10^{-3} M) to get maximal response as when one used those petioles for an epinasty bioassay. Furthermore, by interpolating the "zero IAA added" value on the two response curves, he estimated the endogenous levels of IAA in the petiole. The untreated petioles gave an epinastic response equivalent to that evoked by 7×10^{-9} M of added IAA; but their rooting response was equivalent to that evoked by 25×10^{-7} M of added IAA. Bottelier pointed out that this big difference in presumed endogenous IAA concentrations might be a result of IAA being active at completely different positions for the two different processes. But why does initiation of root primordia require such a high level of auxin, whether added or endogenous? In studies of tracheary cell regeneration (see next section of this chapter), calculations made on the basis of several explicit assumptions indicated that 14 times more IAA was required to initiate a regenerated file of tracheary cells than to differentiate a file from procambium during normal development (Jacobs & Morrow 1957). The same argument used there can be applied to root regeneration, namely, it is reasonable to require for regeneration larger amounts of a hormone than is needed for normal development; otherwise tracheary cells and roots would be regenerating all over the plant. Presumably, in both cases described, cutting the normal path of longitudinal movement causes the hormone to rise to more than the normal level and thus initiates regeneration.

Auxin and tracheary differentiation during regeneration

The role of auxin as a limiting factor for cell differentiation was not established until much later than the establishment of auxin's effects on shoot growth and root regeneration. The reason for the delay lies mainly in the relative ease of the techniques. It is easy to measure lengths or to weigh, but it is more difficult to follow individual cells and to count them. A subsidiary reason was that auxin was initially thought of as a growth hormone; only after the multiplicity of its effects started to be known did it occur to botanists that auxin might be limiting cell differentiation also. But which cell type was auxin likely to be controlling? As a clue, Jacobs (1952) seized on the property, generally considered at that time to be unique to auxin (Söding 1952), of moving with strict basipetal polarity in shoots. He searched anatomical literature for a report of some cell type that differentiated with a similar strict basipetal polarity. Such a cell type existed in the xylem. When xylem strands in a young stem or leaf were cut, tracheary cells were reported to regenerate with strict basipetal polarity (Simon 1908; Freundlich 1908; Kaan Albest 1934; Sinnott & Bloch 1945). Jacobs ran various experiments that confirmed the hypothesis that the auxin, IAA, was the normal limiting factor for tracheary differentiation during the regeneration of a vascular strand (1952, 1954, 1956).

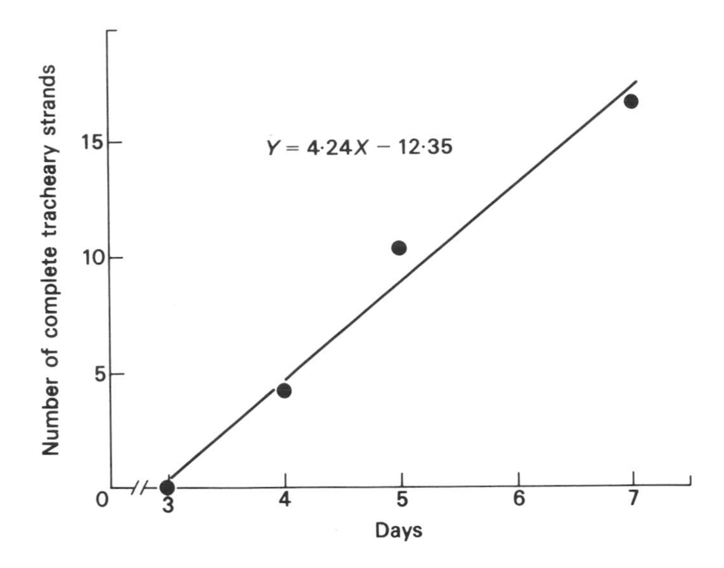

Figure 4-6. The time course of regeneration of tracheary strands around a wound in a young internode of *Coleus* (data points from Aloni & Jacobs 1977a, linear regression and figure unpublished).

Using a genetically identical sample population of *Coleus*, at a uniform developmental stage, Jacobs bioassayed and compared the amount of auxin produced by each of the leaves, as well as the amount produced within the stem itself. The fastest-growing leaf pair on the stem (leaf pair 2, the second pair below the apical bud) produced more diffusible auxin than any of the other leaves tested, judging by a direct assay of the collecting blocks of agar (Jacobs 1952). About 350–400 pg IAA were produced in 3.5 hours. There was a progressive decrease in the amounts of diffusible auxin from leaves progressively older and lower on the stem (Fig. 4-3). Internode 2 (just below leaf pair 2), when excised and diffused separately, gave off much smaller amounts of auxin than leaf 2. Hence, this internode was chosen as the optimal site for the wound to initiate xylem regeneration, on the assumption that it would show the greatest response to the excision of leaf 2.

To measure quantitatively and efficiently the tracheary cells that regenerated, Jacobs counted the number of joining strands that regenerated around the sides and back of the <-shaped wound. Changes in the total amount of tracheary regeneration were said to be reflected more in changes in the number of such strands than in

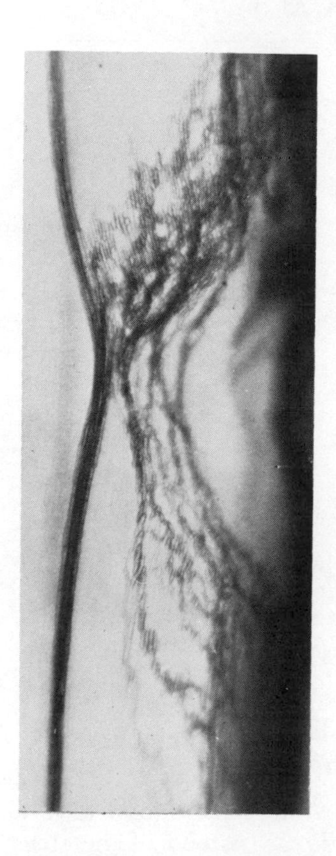

Figure 4-7. Photograph of a cleared whole mount of tracheary regeneration around one side of a wound (*right middle*) in a young *Coleus* internode. Tracheary cells have been stained with basic fuchsin (Jacobs 1952).

the number of tracheary cells in a transverse section of individual strands. The technique that made it feasible for him to study cell differentiation on a quantitative basis with statistically adequate sample sizes was the use of transparent *whole-mounts* of the regenerating area, eschewing the thousands of microscopically thin transverse slices that earlier investigators had used. With these methods, 10–20 joining strands regenerated in a week in *Coleus* plants that were intact except for the wound in internode 2. The range in actual mean values resulted from seasonal changes in the growing conditions of the greenhouse. Typical results from a recent repeat of one such experiment are shown in the time course of Figure 4-6. Figure 4-7 provides an example of a transparent whole-mount of the regenerated joining strands.

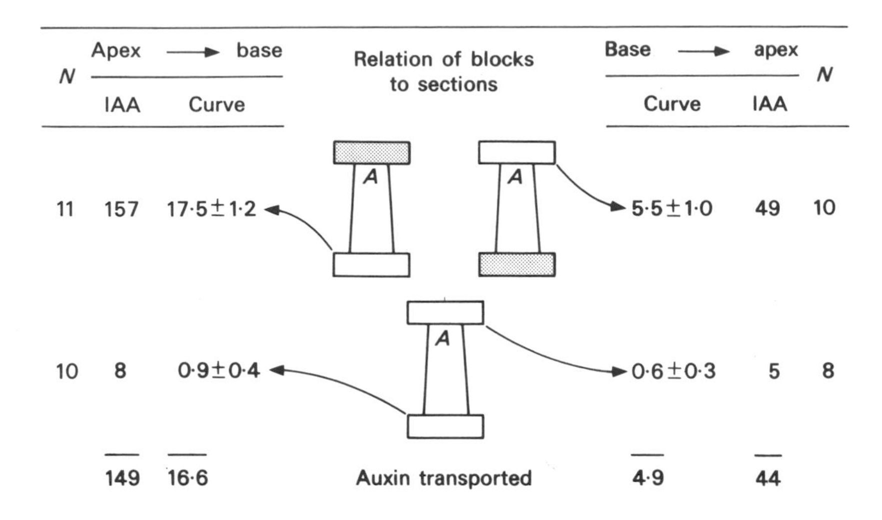

Figure 4-8. Amounts of auxin, assayed by the *Avena* curve test and calculated as if IAA, collected in 3-hour diffusions and transports from apical (A) and basal ends of sections cut from young internodes of *Coleus*. *N* designates number in each sample (Jacobs 1952). IAA as mg $\times$ 10^{-8}.

Parallel variation between the polarity of IAA transport and that of tracheary regeneration

Surprising results were obtained when Went-type transport tests were run on excised sections of internode 2. Instead of the strict basipetal polarity of IAA movement expected from the earlier literature on coleoptiles, there was a significant amount of acropetal movement (Figure 4-8). Almost one-third as much IAA was transported acropetally as basipetally when the donor blocks contained 2 mg IAA/liter (Jacobs 1952, 1954). These results, based on the *Avena* curve assay for auxin, were confirmed on a different variety of *Coleus*, using ^{14}C-labeled IAA, paper chromatography, and the liquid scintillation counter to measure amounts of IAA (Naqvi & Gordon 1965).

This discovery seemed to shatter the parallel between strictly basipetal auxin transport and tracheary regeneration that was the working hypothesis of the research project, particularly since *Coleus* had been used by the earlier workers on xylem regeneration. However, collections of tracheary regeneration in matched internodes 2, using large samples and collecting throughout the regenerating period, revealed what had not been reported by the

Table 4-1. The effect of excision of distal organs on the regeneration of xylem in *Coleus* internode No. 2

Controls	Leaf pairs no. 1 and 2 excised	No. of plants per treatment	Controls	All distal organs excised	No. of plants per treatment
11.2 ± 1.0	8.9 ± 0.6	9	17.1 ± 1.0	4.9 ± 0.9	8
t = 1.90	$t_{0.05}$ = 2.12		t = 9.24	$t_{0.01}$ = 2.98	

Source: Jacobs 1952.

earlier investigators: although tracheary regeneration was predominantly basipetal, strands of tracheae *did* regenerate in continuity with the apical cut end (the cut end below the wound). "These regenerated xylem cells formed short projections, at the most ten cells long, usually only three to four cells long" (Jacobs 1952). In other words, the parallel between auxin transport and xylem regeneration was reestablished, but with the additional observation that acropetal IAA movement was paralleled by acropetal xylem regeneration.

Excision of auxin sources

If auxin is the internal factor that normally controls the regeneration of tracheary cells, then excising the leaves, which are the main sources of endogenous auxin in *Coleus*, should result in a decrease in the amount of xylem regenerated. The data in Table 4-1 confirm this. If the two unfolded leaf pairs just above the wound were excised (1 and 2), there was a decrease in the average number of regenerated tracheary strands that did not quite reach the 5% level of probability by the t-test with the sample size used. After pairs 1 and 2 were excised, there still remained on the plant as auxin sources the five small leaf pairs in the apical bud (Jacobs & Morrow 1957) and the four to six pairs below internode 2. If all the distal leaves were excised (the apical bud in addition to pairs 1 and 2), the decrease in number of regenerated tracheary strands was beyond the 1% level of probability (Table 4-1). The five or six strands regenerated when only proximal leaves remained (Tables 4-1 and 4-2) would be expected, since the proximal leaves do produce auxin

Table 4-2. The effect of 2 mg/liter IAA substitued for excised distal organs on the number of xylem strands regenerated in young *Coleus* internodes

	Control plants (all leaves and buds left on)	Distal organs excised and 2 mg/liter IAA substituted for them	Distal organs excised	Number of plants per treatment
Average number of xylem strands ± S.E.	15.0 ± 1.3	11.5 ± 1.2	6.3 ± 1.4	11
	$t = 1.98$		$t = 2.87$**	
		$t_{0.05} = 2.09$; $t_{0.01} = 2.84$		

Source: Jacobs 1952.

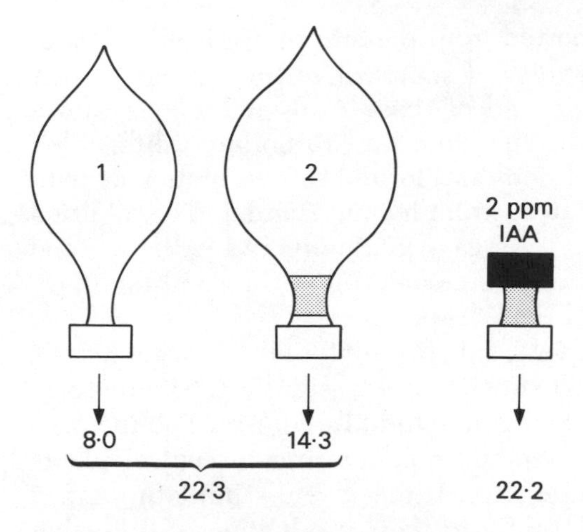

Figure 4-9. The average amounts of auxin obtained per hour from *Coleus* leaves 1 and 2 and from petiole 2 sections capped distally with 2 mg/liter of synthetic IAA. The endogenous auxin, assayed by the *Avena* curve test, is given as equivalent "IAA" units $\times 10^{-8}$ mg (Jacobs 1956).

(Fig. 4-3) and the young stem can transport auxin upward (Figure 4-8).

So long as all the distal leaves were left on, excision of *all* the proximal leaves did not reduce the total number of xylem strands regenerated in 1 week (Table 3 of Jacobs 1952), even though the excised proximal leaves represented 96% of the total unfolded leaf area on the main shoot.

Substitution of synthetic auxin for the leaves

If the demonstrated xylogenic action of the distal leaves was due solely to their production of auxin, one should be able to substitute synthetic auxin for the leaves and replace their xylogenic effect. Exact substitution should give exact replacement.

Pooling the results of all three experiments, the amounts of diffusible auxin produced per hour by the leaves just above the wounded internode 2 are shown in Figure 4-9 (Jacobs 1956).

Leaf 2, immediately above the wound, produced an average of 14.3×10^{-8} mg of IAA/hour; the next leaf above that (leaf 1) produces 8.0 units/hour. The method shown to the right of Figure 4-9 was devised to determine the physiologically significant amount of added IAA that was transported from the site of application.[2] Sections were cut from the petioles of standardized no. 2 leaves, IAA was added to the leaf-blade end of those excised sections, and a block of agar was placed on the stem end to collect "diffusible" auxin, under the same conditions and for the same duration as with the collection of diffusible auxin from leaves 1 and 2. The addition of 2 mg IAA/liter to the distal end gave 22.2 units of IAA/hour at the proximal end (Jacobs 1952), the same amount as the combined production of leaves 1 and 2 (Figure 4-9).

This amount of synthetic IAA, substituted for all the distal leaves, produced a marked and statistically highly significant increase in the number of xylem strands regenerated (Table 4-2). The increase was 35% of the number of strands regenerating in control plants with all leaves intact (Figure 4-10). If the xylem-stimulating effect of leaves 1 and 2 was due solely to their production of diffusible auxin and if the *Avena* curvature bioassay adequately measured that auxin, then excising leaves 1 and 2 with their 22 units of auxin should give a decrease of 35% in the number of xylem strands. Figure 4-10 shows that this was the case. The expected relations were found also in other experiments where only leaf 2 was removed (with its 14.3 units of auxin) or where the 8.0 units from leaf 1 were removed (Jacobs 1956). (The fact that the results of Figure 4-10 were given only as percentages of strands regenerating, rather than as absolute values, should automatically arouse the reader's suspicions. Jacobs said that the decrease in the number of strands regenerated in intact greenhouse plants progressively later in the spring made the use of percentages the only valid way to compare experiments done during different months of the year.)

Thus, the entire effect of the distal leaves in causing tracheary regeneration was quantitatively explained in terms of their production of diffusible auxin, if one equates auxin with pure indoleacetic acid. A further implication of the results was that no other substance

[2] The literature of auxin physiology was replete with arguments as to whether so-and-so's results should be discounted because the auxin was added in "unphysiologically high" concentrations. In none of these cases did the disputants actually check how much auxin was getting into the plant. To measure how much auxin activity had disappeared from the source of applied auxin (as was done by Thimann & Skoog 1934, their Fig. 1) is not sufficient, since it gives no information as to how much of the auxin was destroyed at the cut surface, how much was sequestered and metabolized in the adjoining tissue, or finally (and the only information pertinent to auxin's correlative effects), how much passed through the petiole to act as a hormone elsewhere (Jacobs et al. 1959).

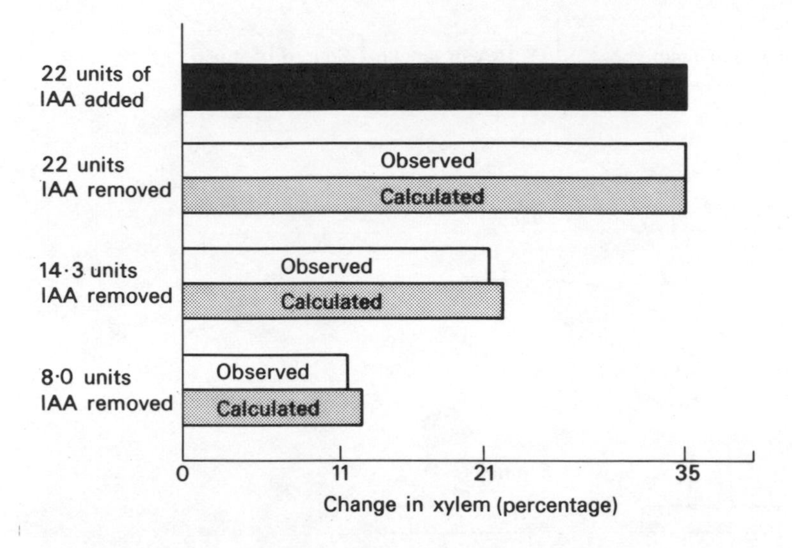

Figure 4-10. The effects on the number of tracheary strands regenerating per week of adding 22 units of synthetic IAA, or of excising leaves 2 and/or leaves 1 (producing 14.3 and 8.0 units of "IAA" per hour, respectively). The data are given as the percentage of strands that regenerated in intact control plants during the same period (Jacobs 1956).

that is necessary for basipetal auxin transport normally and exclusively came from these distal leaves.

The IAA used in the substitution experiments was added in water. The use of lanolin as an auxin carrier is well established in the literature (Laibach 1933). To get physiological effects equivalent to those from IAA in water or agar, much larger amounts of IAA need to be added to lanolin (Jost & Reiss 1936), but the effects of one application last longer. Apparently, IAA is taken up by tissues much more slowly from a lanolin–IAA mixture than from an IAA–water solution.

IAA in lanolin was tested for its effects on tracheary regeneration. It gave results similar to those from IAA in water when added through petiole 2 (Jacobs 1952). When added to the stem stump of plants decapitated through internode 1, all three experiments agreed in showing that, as assayed by *Avena* curves, IAA at 1% in lanolin gave the same amount of auxin through node 2 as did the endogenous supply of diffusible auxin coming from the top of the shoot (Figure 4-11). (This was applying the same technique as that

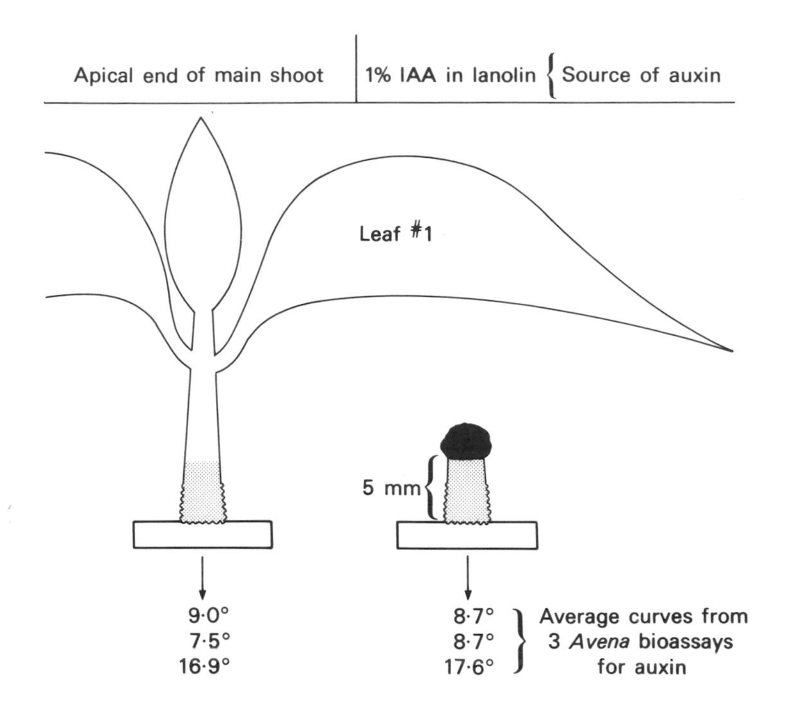

Figure 4-11. Evidence that 1% IAA in lanolin provides the same amount of hormonal auxin through *Coleus* node 2 as does the tip of the main shoot (Jacobs et al. 1959).

described for Figure 4-9.) One percent IAA in lanolin also completely replaced the top of the shoot in causing tracheary regeneration in internode 2 (Table 3 in Jacobs et al. 1959).

Relation of acropetal auxin transport to tracheary regeneration

During the same time and under the same conditions that internode 2 moved 65×10^{-8} mg of IAA basipetally, matching internodes moved 20×10^{-8} mg acropetally (average of all six experiments, Jacobs 1954). Is this acropetally moving auxin, observed in isolated sections, of significance in the more intact plant?

Quantitative evidence for the significance of the acropetally moving auxin is shown in Figure 4-12. If all the leaves above the wound were left intact, an average of 16.0 tracheary strands regenerated in a week. If, on the contrary, the leaves above the wound were excised while all the leaves below were left intact (thus providing

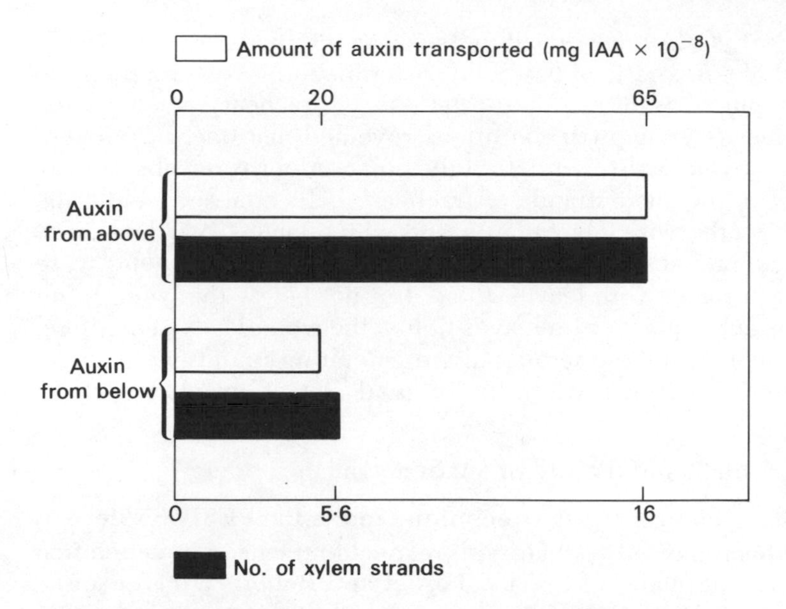

Figure 4-12. Relations between the amounts of auxin moved through excised sections of young *Coleus* stems and the number of tracheary strands regenerated when leaves above or below are excised. The "auxin" transported into receiver blocks is calculated as "IAA" $\times 10^{-8}$ mg from *Avena* curve bioassays; 2 mg IAA per liter was added in the donor blocks at either the apical (auxin from above) or basal (auxin from below) ends of the sections (Jacobs 1956).

sources for acropetally moving auxin), then 5.6 strands regenerated (Jacobs 1954, 1956). By comparing the ratio of IAA movement in opposite directions through sections with the ratio of the amounts of tracheary regeneration obtained when endogenous auxin was available from opposite directions, we see that there was a close parallel (Figure 4-12). The observed value of 5.6 strands for acropetally moving auxin was only 0.6 strand higher than the value calculated on the basis of perfect proportionality. There was also good agreement in xylogenic activity between the 5.6 strands formed in this experiment, presumably with 20 acropetally moving units of endogenous auxin, and the 5.2 strands formed with 22 basipetally moving units of exogenous IAA added through petiole 2 (Table 4-2).

In short, there was quantitative parallel variation between the movement of auxin and the regeneration of tracheary cells when auxin was available either above or below internode 2.

Under greenhouse conditions giving much faster growth, this same clone of *Coleus* recently showed almost absolute basipetal polarity of ^{14}C-IAA movement in its young internodes (Jacobs 1977). Instead of a 3 : 1 ratio of basipetal to acropetal movement, the ratio was as much as 85 : 1. Investigations of tracheary regeneration under these new growth conditions revealed that tracheary regeneration was now also very strongly polar (Aloni & Jacobs 1977a). (However, the short strands of tracheary cells attached below the wound in otherwise intact plants, and which Jacobs 1952 had interpreted as probably resulting from acropetal IAA movement, were present in plants with leaves above but not below the wound, but were absent in plants with leaves below the wound only. Therefore, it is unlikely that these acropetally regenerating strands were due to acropetally moving auxin in the original "intact" plants.)

Isolation of the site of auxin action

Isolation as an investigative technique under the PESIGS rules has not yet been pushed very far with respect to tracheary regeneration in young internodes of *Coleus*. Fosket and Roberts (1964) excised internode 2, surface-sterilized it, placed it base down on 2% sucrose-agar, and applied sucrose and 1 mg IAA/liter to the apical end. They said that the 7-day tracheary regeneration around a mid-internode wound was similar in pattern to that in intact control plants. Isolating still smaller units, they cultured thin transverse slices of internode 2 on 2% sucrose in agar and added various compounds. IAA at 0.05 mg/liter, when added with sucrose, gave a big and statistically significant increase in the number of tracheary cells regenerated throughout the slice in 7 days, resulting in 1286 tracheary cells. Sucrose alone resulted in 483 cells, more than 20 times as many as in the plain agar controls. Because the cells counted in these slices were not regenerating around a wound, there was no direct way to compare the numbers with those in intact controls. IAA concentrations of 0.5 mg/liter and higher inhibited tracheary cell number compared to sucrose-agar controls. IAA added to agar not containing sucrose was said to reduce further the number of tracheary cells, but unfortunately they did not report results from 0.05 mg/liter IAA, the sole (and low) concentration that stimulated when sucrose was present. Kinetin (see Chapter 6) inhibited at the four concentrations tested. Fosket and Roberts made the sizeable improvement of using a multiple comparison statistical test, estimating the significance of differences with Duncan's multiple-range test.

Quantitative data for the excised young internode were provided by Thompson and Jacobs (1966), who checked regeneration in iso-

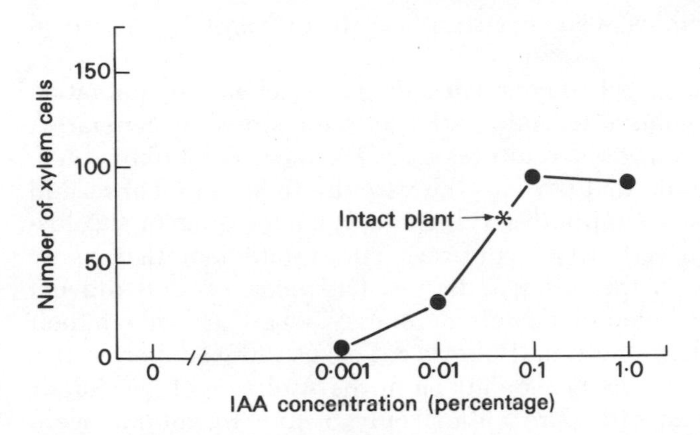

Figure 4-13. The number of tracheary cells regenerating around a slit wound in excised young *Coleus* internodes in relation to increasing concentrations of IAA in lanolin added to the apical cut surface. The value for the number regenerated around such a wound in otherwise intact control plants is shown by the asterisk (Thompson & Jacobs 1966).

lated internodes 2, which had regenerated for a week in glass dishes in either the greenhouse or a growth chamber. They counted the number of cells that regenerated around a wound in the small vascular strand in the middle of one of the flat sides of the internode. (Sinnott and Bloch had previously studied the regeneration around a wound in the large corner bundle and Jacobs had counted the "complete strands" but not cells.) With no IAA added to the apical end of the isolated internode, no tracheary regeneration occurred. With increasing amounts of IAA added, increasing numbers of tracheary cells regenerated until a plateau was reached with 0.1% IAA in lanolin (Figure 4-13). (The plateau confirmed the saturation of basipetal IAA transport in this young internode, observed earlier with direct transport tests [see Chapter 9, Jacobs 1961, and Scott & Jacobs 1963].) By interpolating on this dose-response curve the values from control plants that were intact except for the slit wound in the vascular bundle, Thompson and Jacobs determined that 0.05% IAA in lanolin exactly replaced the rest of the plant in its effect on tracheary regeneration in this young internode (asterisk in Figure 4-13). Assuming that the IAA relations are the same for the corner wound used by Jacobs (1952) and the side wound used here, results of this more detailed examination suggest that the average of 11.9 strands resulting from 1% IAA in Jacobs et al. (1959) was really larger than the average of 7.7 strands in the "intact" control even though the samples were not significantly different by statistical test.

Fosket and Roberts had contrasted the fast tracheary regeneration in the young *Coleus* internode with the much slower regeneration reported for various tissue cultures (e.g., Wetmore & Sorokin 1955), and they suggested that perhaps the vascular tissue in Jacobs's and their experiments supplied a xylogenic factor missing in the histologically deprived tissue cultures. Their statement that tissue slices from which the vascular tissues had been excised did not regenerate any "wound vessel members" when grown on their usual sucrose–IAA medium (Roberts & Fosket 1966) supported this suggestion. Earle (1968) gave data on further isolation of the *Coleus* system: blocks of pith parenchyma only 6 mm^3 in volume were cultured in the dark on a much more complete tissue culture medium, which she had used previously for studies of *Convolvulus*. She found that IAA at 5×10^{-5} M caused sizeable numbers of tracheary cells to differentiate in a few weeks in such pith blocklets, whether the pith was from internode 2 or 5 (approximately 3 weeks older). Without IAA only a few tracheary cells differentiated; without sucrose but with IAA, none did. Earle made the important point that the large increases in numbers of tracheary cells regenerated that she obtained with added IAA were not associated with any statistically significant increases in fresh weight. The necessity of the many other chemicals in her basic medium was not tested, but she guessed that their presence might explain why she obtained tracheary cells in pith cultured with IAA plus sucrose, contrary to the stated result of Roberts and Fosket using their unfortified IAA–sucrose agar to grow pith slices. Although the difference reported might, as so often is the case, also be due to varietal differences (Earle having used the Princeton clone), it seems more likely that Roberts and Fosket did not test a wide enough range of IAA concentrations; Comer (1972) confirmed Earle's positive results but added that only a narrow range of IAA concentrations was effective when added to plain sucrose agar.

We can conclude that these isolation experiments rule out the interpretations that IAA acts primarily on preexisting vascular tissue, with the new tracheary cells being a secondary effect of that primary action. As Comer pointed out, essentially the same concentration of IAA that exactly replaced leaves 1 and 2 in causing tracheary regeneration in the otherwise intact *Coleus* plant (the 2 mg/liter of Jacobs 1952, 1956) gave the maximal number of tracheary cells in cultured pieces of pith (the 1 mg/liter of Comer's Table 1) as long as sucrose was present too. There is no indication, therefore, that the vascular tissue of *Coleus* provides anything for tracheary regeneration except possibly sucrose and IAA.

Autoradiographs have been used by several authors to try to isolate the site of IAA action. The radioactive label of choice for high-resolution autoradiographs is ^{3}H; the low energy of its β-radiation allows better resolution than does ^{14}C or other radioisotopes with even more penetrating radiation. The basic problem with all attempts to make hormonal autoradiographs is how to obtain radioactively labeled hormone of such high specific activity that one can add a low, normal hormonal dose and still have enough radioactivity to show up in the autoradiographs. As soon as such ^{3}H-IAA became available, it was added to the *Coleus* internode regeneration system already so well known, and ultrathin sections were cut to improve further the resolution. The time course of labeling was followed: after only 3 hours the label was found specifically localized in the cell walls of the youngest tracheary cells, and in those cells the label was restricted to the portions of the wall that showed secondary thickenings (Sabnis et al. 1969). Such localization fitted in with the fact the main criterion used in the earlier work for deciding that a tracheary cell had differentiated was the presence of the distinctive secondary wall. Although the authors did not determine what chemical ^{3}H was attached to in the tracheary wall, the previous evidence relating IAA to tracheary differentiation suggested that the tritium might still be with IAA or a close derivative of it. Later works, using ^{14}C-IAA and thick sections, also reported localization of radioactivity in the tracheary walls of *Coleus* (Wangermann 1970; Gee 1972).

Generality of results

To what extent are these results true only for the Princeton clone of *Coleus blumei* and for its young internode 2? For excised internode 5, approximately 3 weeks older and typically in its first week of cambial growth, IAA at the same interpolated 0.05% "concentration" in lanolin provides the same exact replacement of the whole rest of the plant in causing tracheary regeneration around a transverse wound (Fig. 1 of Thompson & Jacobs 1966). The polarity of this IAA effect also matches the polarity of IAA movement: IAA moves only basipetally through excised cylinders of tissue cut from internode 5 (Jacobs & McCready 1967) and IAA stimulates tracheary regeneration when added at the apical end of the excised internode but not when added at the basal end (Table 3 of Thompson & Jacobs 1966). Although no other genus, species, or organ has been investigated in such detail, enough has been done to show that IAA can control tracheary regeneration in many other cases. Excised internodes of the Golden Bedder variety of *Coleus* showed similar complete restoration of tracheary regeneration by apical, but not by basal, IAA application, and *Lycopersicon esculentum* (tomato) reacted similarly (Thompson & Jacobs 1966, Table 3). (Skoog had shown long ago, by direct bioassay, that tomato stem pieces show strictly polar basipetal transport of added IAA [1938].) Isolated internodes of peanut and okra showed similar polar stimulation of tracheary regeneration by added IAA (Thompson 1968, 1970). Isolated plugs of tobacco pith that were cultured on a non-growth medium for many weeks finally showed haphazard differentiation of tracheary cells in a zone of recently divided cells near the tip of a pipette that was feeding in 0.1 to 1.0 mg IAA/liter (Clutter 1960; Sussex et al. 1972).

Tracheary regeneration in the leaves of many genera was investigated by Freundlich (1908); active regeneration that was typically polar was found in leaves of *Ginkgo* and various dicotyledenous plants. (He found little or no regeneration in leaves of ferns or monocotyledons.) Jost (1942) reinvestigated leaves and discovered that IAA could substitute, qualitatively at least, for the xylem-regenerating effect of the distal portion of a *Fittonia* or *Elatostemma* leaf. Because he only got the regeneration by adding very high concentrations of IAA to intact leaves, Jost concluded that a wound was indispensable along with auxin for the tracheary stimulation. This conclusion was called unnecessarily limited by Jacobs (1952), who felt it resulted from Jost not appreciating that much less IAA penetrates an intact epidermis than a cut surface.

Specificity of chemical

Evidence that it is the auxin properties of IAA that provide its xylogenic power came from quantitative studies in isolated internode 5 of *Coleus* (Table 9 of Thompson 1965). The synthetic auxin naphthaleneacetic acid (NAA) or the auxin-type herbicides, 2,4-D and 2,4,5-trichlorophenoxyacetic acid (2,4,5-T), all gave as many regenerated tracheary cells as did the "intact" control plants, although 2,4-D gave many more cells than the other auxins did. Tryptophan, thought by many to be a precursor of IAA, also replaced the rest of the plant when given at 10 times the level used for the auxins. Various nonauxins were without effect at the one concentration tested (thiamine, nicotinic acid, and ascorbic acid). Gibberellic acid was also without effect when added by itself; when added with IAA, it usually increased the mean number of tracheary cells regenerated, although not to a statistically significant extent (Thompson 1965).

Jacobs (1952) pointed out how many developmental effects brought about by auxin required an adequate supply of sugar, and he hypothesized that the unusual pattern of normal differentiation of tracheary cells in the shoot apex resulted from limitation by both auxin and sugar. And although no exogenous sugar, but only IAA, was needed to restore completely the normal level of tracheary regeneration in the brightly illuminated excised internodes 2 of Thompson and Jacobs (1966), other workers who kept their excised internodes for longer periods in dimmer light (e.g., the 24 days of Beslow & Rier 1969) found that without adding sucrose the tracheary regeneration from added IAA was negligible. Even with no exogenous IAA, however, increasing concentrations of sucrose gave increasing numbers of tracheary cells in *Coleus*, reaching a maximum of 289 cells with 7.5% sucrose (Beslow & Rier). The addition of IAA at 2 mg/liter increased the general level of response to sucrose, reaching a maximum of about 450 cells at 6.5% sucrose. Fosket and Roberts (1964), using their 2-mm transverse slices from internode 2, also found essentially no tracheary "regeneration" unless sucrose was added. The effectiveness of sucrose alone was puzzling. Fosket and Roberts suggested that perhaps auxin was synthesized at the cut surfaces, which of course constituted a relatively large area on their thin slices of tissue. However, perusal of the methods reveals that they removed all leaves and buds for 2 days before excising the slices ("to remove endogenous auxin sources"). It seems reasonable to interpret the need for sucrose in the cultured slices as a result of this 2-day starvation period, coupled with the relative paucity of stored food material in the smaller pieces of tissue. And in contrast to Thompson's work, the tissue was not grown in the full light of a greenhouse or growth chamber.

Auxin and tracheary differentiation in normal development

After collecting the preceding evidence that IAA was the normal limiting factor for the regeneration of tracheary cells in the stem, researchers turned to the differentiation of tracheary cells from procambium during normal shoot development. As with regeneration, although with much less total evidence, quantitative data have been obtained that IAA is also the limiting factor for normal differentiation.

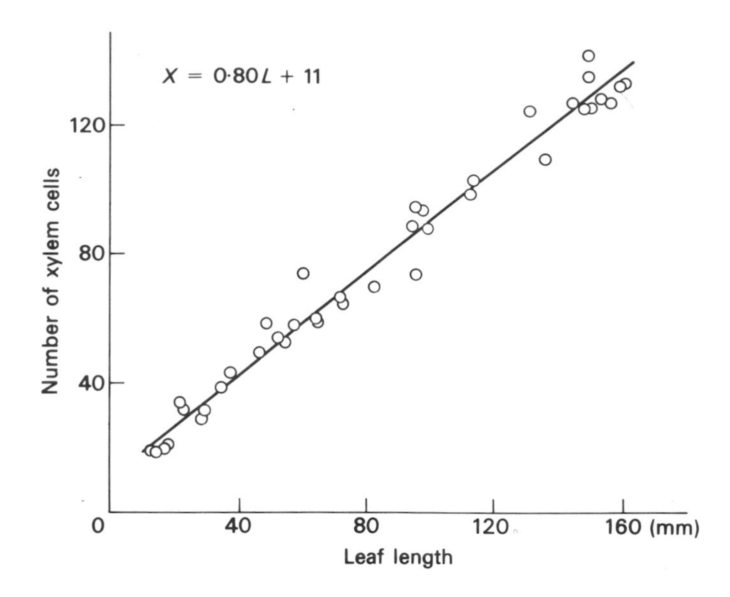

Figure 4-14. Leaf length of *Coleus blumei* plotted against the number of tracheary cells in a transverse section of the petiolar base. The calculated linear regression line and its equation are also shown (Jacobs & Morrow 1957). (The counts and regression calculations were by P. Green & J. Daube.)

Parallel variation

The amounts of diffusible auxin produced by *Coleus* leaves had already been determined (Figure 4-3), so the question was, "Does normal tracheary regeneration show a quantitative, parallel variation with the normal auxin production?" The numbers of tracheary cells observed in transverse sections from the base of petioles were plotted against leaf length (Figure 4-14); a straight line fitted the data well (Jacobs & Morrow 1957). No statistically significant improvement in fit over the linear regression shown was achieved by fitting the third and fourth polynomial (unpublished). Hence, the linear regression equation was used, along with information on rates of leaf growth, to convert these data to rates of production of tracheary cells, so they would be comparable to the data on rate of production of diffusible auxin from leaves of the same clone (Table 4 of Jacobs & Morrow 1957). Converting both sets of data to relative rates of production (with 100% taken as the summed production of

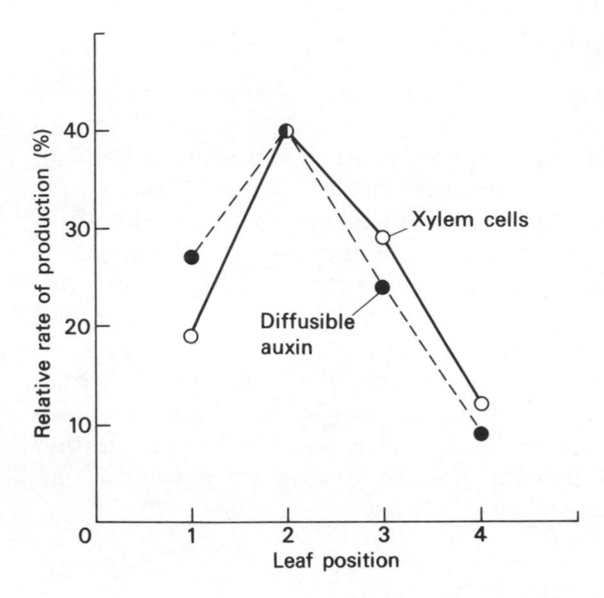

Figure 4-15. The relative rate of production by leaves of diffusible auxin and tracheary cells in relation to leaf position on the stem of *Coleus* plants. The production by leaves 1–4 inclusive is taken as 100% (Jacobs & Morrow 1957).

leaves 1–4) revealed a close, quantitative parallel variation (Figure 4-15, Jacobs & Morrow 1957).

By comparing the number of normal tracheary cells differentiated per week at the base of petiole 2 with the number regenerated around a wound under the influence of leaf 2, it was calculated that 14 times as much IAA was needed to regenerate a tracheary cell from parenchyma as to differentiate one normally from procambium. This differential sensitivity seemed to explain nicely why the pith parenchyma did not normally differentiate into tracheary cells: it was only after cutting the vascular strand that the local IAA level would rise to the "14×- normal" level that would induce the parenchyma cells to differentiate (Jacobs & Morrow 1957). The authors felt that too many assumptions were involved in calculating the 14× factor for the exact value to be of great importance, but they hoped that the order of magnitude was correct. (The assumptions included that diffusible auxin continued to be produced by leaf 2 for a whole week at the same average rate as for 3 hours and that there was no decrease in diffusible auxin from petiole base to mid-internode.)

Excision and substitution

By excising young leaves and checking the changes in tracheary differentiation in the subjacent traces, Jost (1893) concluded that a stimulus for differentiation moved down the leaf trace from the leaf. In 1940 he resumed these experiments and concluded from a series of poorly designed, qualitative experiments that IAA was involved.

Wangermann (1967) provided excellent quantitative data: excising a young *Coleus* leaf reduced the number of tracheary cells differentiated in the subjacent leaf trace. "Auxin" (chemical nature not given) at 0.1 M in lanolin, when substituted for the leaf, gave full, or more than full, replacement of the effect of the leaf on normal tracheary differentiation. She noted a strict basipetal polarity of effect of leaf excisions on tracheary differentiation, lending further credence to the view that the polarly moving auxin was the endogenous regulator.

Generality and specificity

Evidence that auxin was involved in normal tracheary differentiation even in a plant as phylogenetically far removed as a fern was provided by Steeves and Briggs (1960). Diffusible auxin, assayed with *Avena* curves, was found moving basipetally from *Osmunda* leaves. Excision of the pinnae from such leaves decreased both elongation and tracheary differentiation within the leaf. Substituting IAA-lanolin at 0.5% restored some or all of the elongation and gave tracheary differentiation that looked normal in transverse sections (their Fig. 4-7).

An interesting variant was the report that the free-living little gametophyte of the fern, *Todea,* which DeMaggio (1972) said did not form any tracheary cells in nature, could be induced to form such cells within 63 days in almost half the gametophytes if IAA at 0.01 mg/liter was added to the mineral salts and agar on which the plants were grown. However, sucrose at 2–5% was even more effective than IAA, and both chemicals added together seemed to be less effective than either one alone. NAA, gibberellic acid, or kinetin when added at 0.01 ppm each increased somewhat the average percent of plants that differentiated tracheary cells. The higher concentrations of NAA and gibberellic acid inhibited tracheary differentiation. Judging by the graphs, control gametophytes gave about nine tracheary cells per plant and about 6% of plants with tracheary cells. (Data were all given as averages, with no statistical tests of significance.) These experiments on fern gametophytes differ from all the other experiments mentioned in that there was no excision of gametophyte parts; all the chemicals were being added to the intact organism.

Torrey (1953) grew 5- to 10-mm root tips of *Pisum* in sterile culture to see the effect of added IAA on the root tips' development. (We are faced with the difficult decisions, as usual with "tissue culture" results, of whether the development observed is normal [as the tissue culturists usually suggest] or regenerative [as one would expect from the fact that an excised little piece of plant is involved]. For convenience we mention the results here under "normal development.") IAA at 1 mg/liter gave the expected inhibition of elongation of the excised root but also resulted in tracheary cells differentiating closer to the root tip. By a nice comparison of elongation rate per day with the linear differentiation rate per day of the tracheary cells, Torrey concluded that, during the first day of culture at any rate, the added IAA must have increased the differentiation of tracheary cells: the file of tracheary cells was 330 μm longer than in the controls from that first 24 hours of IAA treatment.

Auxin as the controlling factor for the differentiation of sieve cells

The main transporting cells of the major land plants are the tracheary cells and the sieve cells (the sieve cell reaching a stage of further differentiation in angiosperms that merits the modified name "sieve-tube element"). The sieve cells have long been believed to be as important as the channel of movement of organic materials throughout the plant as the tracheary cells are as the channel for moving water and inorganic salts. After some of the preceding evidence had been amassed that IAA was the factor normally limiting both the regeneration and normal differentiation of tracheary cells, it was natural to investigate sieve cells in a similar way and to attempt to determine the limiting factor for sieve-cell differentiation.

There were special difficulties involved in such a study, however. Sieve cells are notoriously difficult to identify: they are usually much smaller in diameter than tracheary cells, and do not have the latter's easily recognizable secondary wall with its characteristic patterns and staining properties. In addition, the pattern and timing of sieve-cell differentiation, whether in normal development or in regeneration, was reported to be different from that for tracheary cells, making it unlikely that IAA would be limiting the differentiation of sieve cells as well as tracheary cells. Hence, when quantitative studies of sieve-cell regeneration were started, LaMotte and

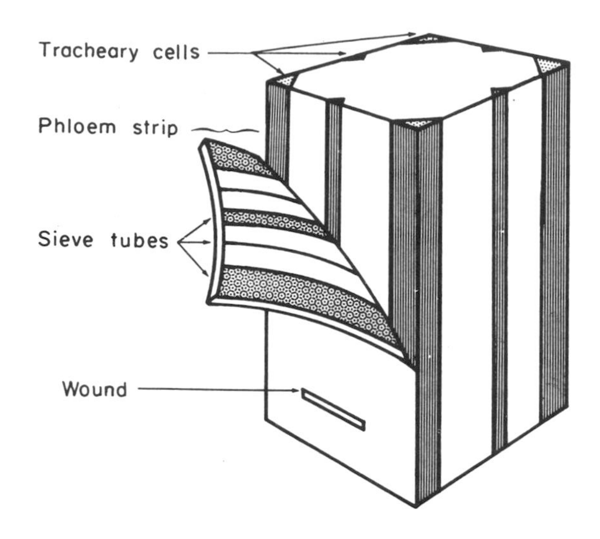

Figure 4-16. Diagram of the technique described by LaMotte and Jacobs (1962) for removing a phloem strip from one side of a *Coleus* stem (the figure is from Jacobs 1970). Note that the smallest strands of sieve tubes do not yet have tracheary counterparts differentiated on the other side of the cambium.

Jacobs (1962) had no inkling as to what the limiting substance might be – a sharp contrast to the earlier situation with tracheary regeneration where the hormone was selected first (for its polar movement) and then a polarly regenerating cell type was found to go with the hormone. Accordingly, the technical demands took precedence. LaMotte and Jacobs modified the clearing and staining techniques used earlier for the tracheary studies so that sieve cells would be heavily stained with aniline blue, switched from the <-shaped corner wound to a slit wound in the middle of one of the flat sides of the *Coleus* stem (so that the sieve cells would regenerate in essentially one plane where they could be seen and counted more easily than in the complex three-dimensional pattern typical of the corner wound), and peeled off the flat strip of phloem and tissues external to it from the xylem-pith remnant (Figure 4-16) (so that the tiny sieve cells could be accurately identified in permanent mounts through the covering layer of thin cells). The phloem strip did not separate easily and cleanly from the interior tissues unless an older internode was used. Internode 5, about 3 weeks older than internode 2, was selected because it was the youngest internode that gave easy separation (correlated with the fact that it was just starting cambial activity).

After devising a reproducible system for counting regenerated sieve-tube strands (with each strand containing an average of 10 sieve cells), LaMotte and Jacobs found that plants intact except for the slit wound in internode 5 would regenerate an average of 14–35 sieve-tube strands in 5 days, the average shifting with the season and other variables (Table 1 of LaMotte & Jacobs 1962).

Excision of organs

Leaves, buds, and branches were cut off *Coleus* plants in varying patterns, both above and below wounded internode 5. The sieve-tube regeneration that occurred in 5 days decreased from 22 strands in the controls to 5 in plants with only the main stem and the root system remaining (Fig. 4-17, LaMotte & Jacobs 1963). This finding was confirmed with similar quantitative data by Thompson (1967), who used the same clone of *Coleus,* and by Houck & LaMotte (1977), who studied regeneration of the smaller, xylemless vascular strands in *Coleus* internode 5. There was a strong polarity in the effectiveness of leaf and bud excision: excising leaves and buds above the wound caused a significant decrease in the number of sieve-tube strands that regenerated, but excising those below the wound had no effect (Figure 4-17). Thompson's data (his Table 2, 1967) show the same phenomenon. (All the differences mentioned were statistically significant [see footnote 3 of LaMotte & Jacobs 1963].)

With the distal organs being so important, one would expect signs of polarity when the time course of regeneration was scrutinized in the area of the wound in otherwise intact plants. This expectation was met. More sieve cells regenerated above the wound than below during the first 3–5 days of regeneration (Table 2 of LaMotte & Jacobs 1963; those data are graphed in Fig. 11 of Jacobs 1970; this polarity of regeneration was confirmed by Aloni & Jacobs 1977b).

These results fit the view that the distal leaves (and perhaps the distal buds) are producing a substance that moves down the stem and is limiting the regeneration of the sieve cells in internode 5.

Substitution of chemicals for the effective organs

LaMotte and Jacobs started out with no clue as to what the limiting substance might be, except that the drastically different patterns and timing reported in the literature for sieve cells and tracheary cells made it unlikely that the substance would be the IAA already known to be limiting tracheary differentiation. Hence, they tested

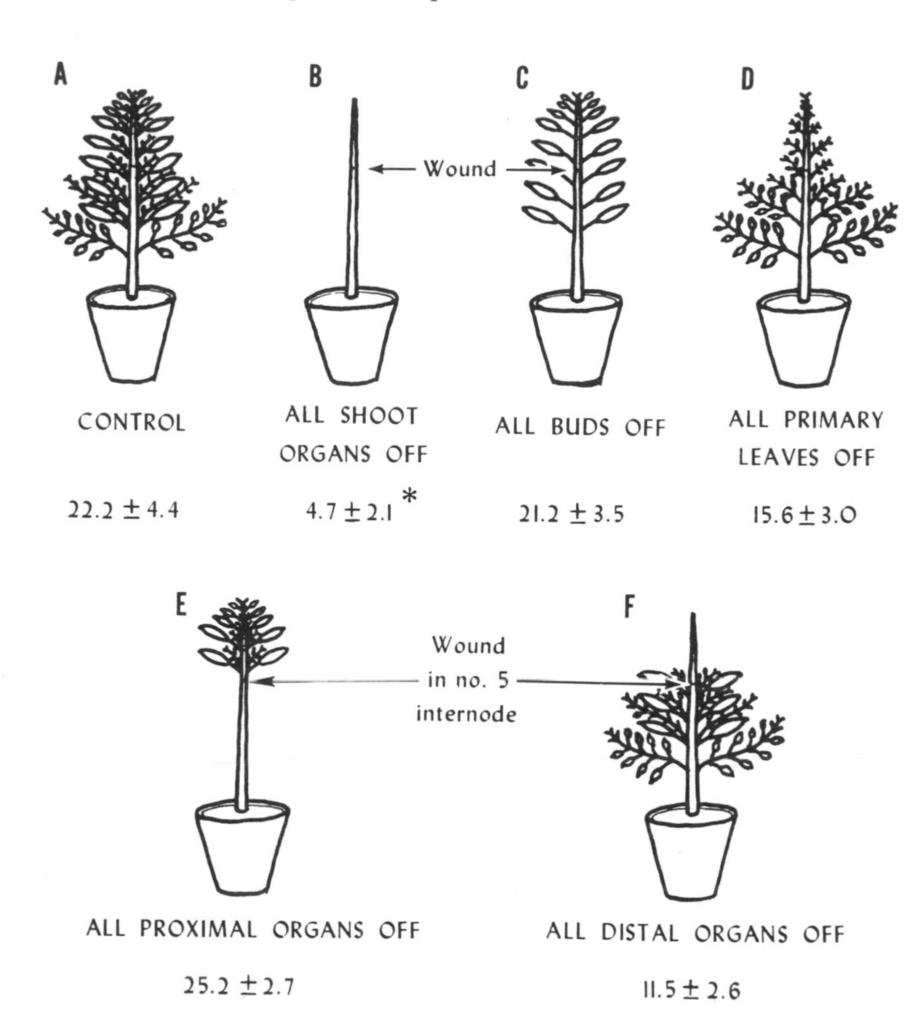

Figure 4-17. Average number of sieve-tube strands regenerated in 5 days around a slit wound in internode 5 of *Coleus* as the result of the various treatments diagrammed (LaMotte & Jacobs 1963).

IAA first, solely to dispose of it quickly. To their surprise, IAA exactly replaced the distal shoot organs, when substituted for them, in their stimulating effect on sieve-cell regeneration (Table 4-3). If, instead of adding IAA in aqueous solution, they added IAA in lanolin, at a 0.1 and 1.0% "concentration," even more sieve-tube strands regenerated than did so under the influence of the distal shoot that it was substituting for (LaMotte & Jacobs 1963). Thompson (1967)

Table 4-3. Effects of IAA and sucrose on sieve-tube regeneration in *Coleus* plants standing 5 days without shoot organs

Treatment	Number of regenerated strands, mean ± S.E.
Intact control plants	23.2 ± 2.2
All shoot organs excised	
Water	7.2 ± 3.4[a]
Sucrose (20 g/liter)	7.6 ± 2.5[a]
IAA (2 mg/liter)	23.9 ± 4.5
IAA + sucrose	20.9 ± 3.9

Source: Table 3 of LaMotte and Jacobs 1963; $n = 4$.
[a] Significantly different (at 5% level) from the values from the intact controls.

confirmed and extended these results, showing that 0.05% IAA in lanolin fully replaced the distal shoot in its stimulating effect in sieve-tube regeneration throughout the 4-day regenerating period. Data from his Table 2 on the time course of sieve-tube regeneration are graphed in Figure 4-18, with the best-fitting straight lines fitted to the data by linear regressions (unpublished). For the smaller vascular strands, consisting only of sieve tubes (and presumably procambium), which Houck and LaMotte (1977) studied, the distal shoot could similarly be replaced completely by IAA.

Isolation of the reacting system

To see whether the root system was necessary in addition to IAA for restoring the normal level of sieve-cell regeneration, regeneration was studied in excised, isolated internodes 5 of *Coleus*. The internodes were placed in glass dishes in the growth chamber, with the light level adjusted so that they were exposed to as many foot-candles as internode 5 normally received in the intact plant. IAA, applied on the apical end at 0.1% in lanolin, gave a highly significant increase in the number of regenerated sieve-cell strands over the lanolin-treated controls (LaMotte & Jacobs 1963; Thompson & Jacobs 1966). The excised internode treated with IAA at either the apical or basal end showed absolute polarity of IAA effectiveness.

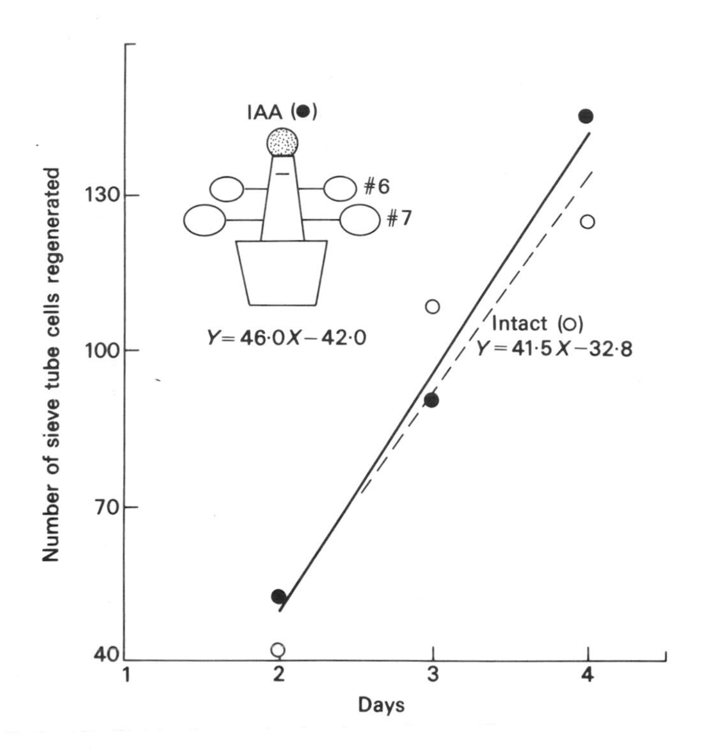

Figure 4-18. The time course of sieve-tube element regeneration around a slit wound in *Coleus* internode 5. (The raw data are from Table 2 of Thompson 1967; the graph and regression calculations are by Jacobs, unpublished.)

IAA applied to the apical end for 7 days could substitute for the whole rest of the plant in causing sieve-cell regeneration; IAA on the basal end had no effect on sieve-cell regeneration occurring just half-way up the excised internode (Thompson & Jacobs 1966). In other words, IAA added to the isolated internode showed the same polarity of effectiveness as did the distal organs compared to the proximal organs in causing sieve-cell regeneration (Figure 4-17), and showed the same complete basipetal polarity that was manifested in direct tests of ^{14}C-IAA movement through cylinders of tissue cut from this internode (Jacobs & McCready 1967). When a graded series of IAA "concentrations" in lanolin was added to the apical end of the excised internodes, it was found by interpolation in the dose-response curve that 0.05% IAA in lanolin gave the same number of sieve-cell strands as were found in the intact controls (Figure 4-19); that is, apically applied 0.05% IAA could quantita-

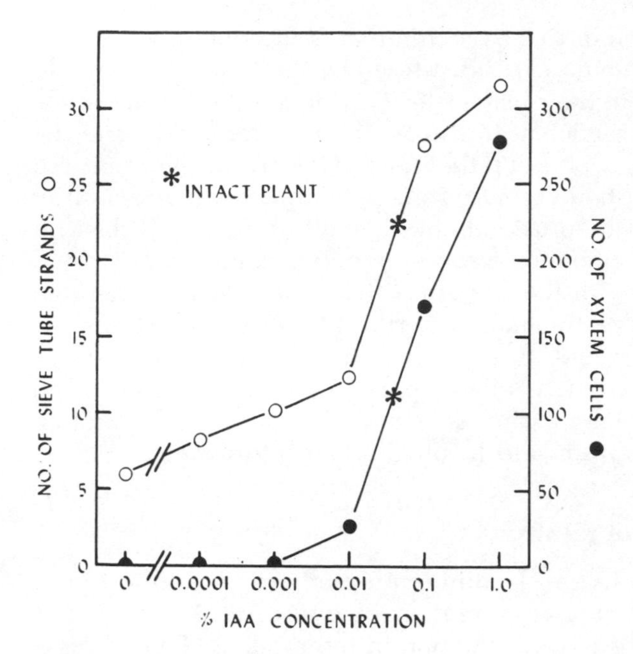

Figure 4-19. The dose-response curve of regeneration of sieve-tube strands (open circles) and tracheary cells (solid circles) in isolated *Coleus* internodes 5 to IAA in lanolin that was added at the apical end of sections. The average values for control plants (intact except for the slit wound) are interpolated on the dose-response curve as asterisks, and demonstrate that IAA at 0.05% in lanolin exactly replaces the excised shoots and roots in their effect on the regeneration of both sieve tubes and tracheary cells (Thompson & Jacobs 1966). Note that the *Y*-coordinates have been selected so that the number of *cells* of the two vascular tissues is being plotted (each sieve-tube strand containing an average of 10 sieve-tube cells according to LaMotte & Jacobs 1962).

tively replace the shoot system and root system that had been cut away (Thompson & Jacobs).

These results confirm the hypothesis that IAA is the substance limiting sieve-cell regeneration in such sieve-tube-xylem vascular strands in *Coleus* internode 5, that IAA normally comes from the leaves and stem above the wound, and that the ineffectiveness of proximal organs (roots, leaves, and stem below the wound) results from the very strong basipetal polarity of IAA movement – a phenomenon also reflected in the polarity of regeneration seen in the time course in "intact" plants.

An interesting extension of this picture came from investigations in slightly younger internodes by Houck and LaMotte (1977) of the smaller vascular strands that have differentiated sieve-tubes but not tracheary cells. Although IAA substituted for the distal organs could fully replace them in their major effect on sieve-tube regeneration, IAA alone was not sufficient to replace both distal organs and the roots and proximal shoot. But if the isolated internodes were treated with IAA on the distal cut and with a cytokinin, such as zeatin or zeatin riboside, on the proximal cut, then all or nearly all the "intact" sieve-tube regeneration was restored. (Cytokinins were tested because of evidence in the literature that root exudate contained cytokinins and that zeatin and zeatin riboside were present in some roots, as Chapters 6 and 10 describe.) It may be that cytokinins needed to be added here along with IAA because Houck and LaMotte were studying earlier stages of vascular strand development than did Thompson and Jacobs in their internodes 5.

Generality of results

Because the 0.05% IAA in lanolin that exactly replaced the rest of the plant for sieve-tube regeneration in internode 5 also exactly replaced it for tracheary regeneration in internode 2 (Thompson & Jacobs 1966), it was obviously necessary also to see if IAA restored full sieve-tube regeneration in internode 2. It did not. Although added IAA did increase the number of sieve-tube strands that regenerated in the excised young internode, no amount of added IAA brought the number up to that found in the "intact" plant (Fig. 5 of Thompson & Jacobs 1966). Apparently in this young, still elongating internode, some factor or factors in addition to IAA are limiting sieve-tube regeneration. The extra factor is probably cytokinin, judging by the results of Houck and LaMotte on a somewhat older internode.

There is little literature on other varieties or species. The Golden Bedder variety of *Coleus blumei* showed similar effectiveness of IAA, applied apically to excised internodes, in restoring sieve-cell regeneration to the level found in the intact plant, and similar ineffectiveness of basally applied IAA (Thompson & Jacobs 1966). Excised internodes of tomato treated apically with IAA also showed sieve-cell regeneration equivalent to the intact plant (but there was no plain lanolin control).

IAA was said to replace the distal leaves of *Cucumis* and *Coleus* in their major effect on sieve-tube regeneration, but no data were provided and the method of estimating sieve-tube regeneration was not described (Benayoun et al. 1975).

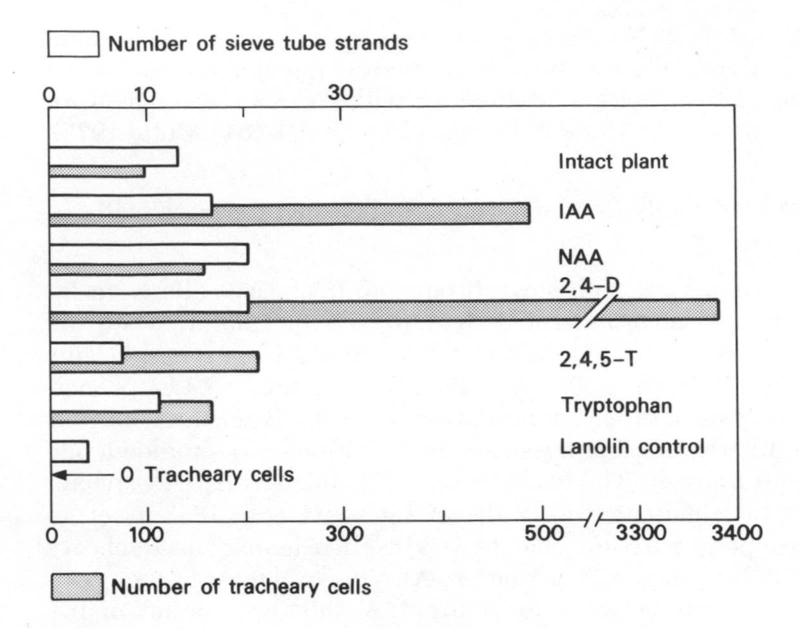

Figure 4-20. The effect of various auxin-type chemicals on the number of sieve-tube and tracheary cells regenerated in excised *Coleus* internode 5 (data from Thompson 1965, figure from Jacobs 1970).

Specificity of chemical

Auxins other than IAA, including the weed killers 2,4-D and 2,4,5-T, were effective in restoring sieve-tube regeneration in excised internodes 5 (Figure 4-20, from Jacobs 1970; data from Thompson 1965). Several nonauxin hormones were without effect (viz., thiamine, ascorbic acid, and nicotinic acid, although each was tested at only one concentration) (Thompson 1965). Gibberellic acid was tested at several concentrations, both alone and with IAA, but had no statistically significant effect on the number of sieve-cell strands regenerating (Thompson 1965). However, in all four experiments gibberellic acid did decrease the average number of sieve-cell strands resulting from IAA added simultaneously, and because the average decrease was 31%, one is left with the impression that the GA effect might reach the standard "significance level" if larger samples were used.

Turning from hormones, sucrose was tested at the usual 2% nutritive concentration because workers on cultures of callus had said that sucrose favored phloem formation in the callus (Wetmore &

Rier 1963; Jeffs & Northcote 1966). However, sucrose was without effect on the number of sieve-tube strands regenerated in *Coleus* internode 5, whether added alone or with IAA as replacement for the shoot organs (LaMotte & Jacobs 1963; Houck & LaMotte 1977).

Does IAA primarily limit sieve-tube rather than tracheary cell differentiation?

Normal numbers of both sieve tubes and tracheary cells were induced by the same amount of IAA (0.05% in lanolin) when the hormone was substituted for the rest of the plant in excised *Coleus* internodes 5 (Figure 4-19). This raises the question, "Is IAA primarily limiting sieve-tube differentiation, with the latter in turn limiting the differentiation of tracheary cells?" From current knowledge this seems unlikely. The localization of ^{3}H added as ^{3}H-IAA specifically in the differentiating walls of tracheary cells (Sabnis et al. 1969) supports the view that IAA is itself reaching the tracheary cells to induce their differentiation. Also the exactness of the quantitative relations between available IAA and the number of tracheary cells would not be expected if the IAA is really acting on sieve-tube differentiation and then some other factor is moving from the sieve tubes across the intervening procambial or cambial cells to cause differentiation of the young tracheary cells. My guess is that these cambial-procambial cells that are on the phloem side of the undifferentiated vascular strand are more sensitive to IAA than those on the xylem side. This is indicated by the dose-response curve for regeneration (Figure 4-19) and fits the fact that in normal shoot development sieve tubes typically differentiate before tracheary cells in the same procambial strand. In regeneration, also, sieve tubes regenerate before (Thompson 1967) or simultaneously with tracheary cells (Aloni & Jacobs 1977b).

Whatever turns out to account for this intimate relationship between the differentiation of sieve tubes and tracheary cells, even the presently available information makes clear that it is now relatively pointless to investigate hormonal effects on tracheary cells without simultaneously studying the sieve tubes (Jacobs 1970).

5

The biochemical basis of auxin action

How auxin acts in the plant cell from a biochemical point of view has been covered by other recent textbooks such as those by Phillips (1971b) and Leshem (1973), as well as by recent review articles (Davies 1973; Ray 1974). There is little point, therefore, in my trying to discuss in detail the biochemical aspects, since I am not an expert in that field. However, the action of auxin from a biochemical point of view and the attempts to find out how it acts do have importance for all of us – trained biochemists or not. In addition, it is reasonable to summarize, however briefly, the discoveries in this field to avoid the gap that would otherwise exist in our mental picture of the chain of causation. And although the biochemical basis of auxin action is still unknown, it is interesting to hear the rationales for the various approaches tried. To one looking at the field of auxin biochemistry, by necessity, from the outside, the approaches seem particularly reasonable as working hypotheses; it is unusually unfortunate that none seems to have provided the key to explain completely auxin action.

In this chapter, therefore, I shall list and discuss briefly the major areas of past or present investigation. One of the difficulties auxin biochemists faced was that auxin seemed to have too many known actions of too much diversity: it controlled shoot phototropism, stimulated shoot growth, inhibited the growth of roots and lateral buds, inhibited leaf and fruit abscission, caused parthenocarpic development of some fruits, caused root primordia to regenerate, and controlled the regeneration of tracheary and sieve-tube cells. Finding a common basis for action among these effects was more difficult than it would have been if the hormone was more limited in its actions.

Auxin and the cell wall

When indole-3-acetic acid (IAA) first became known as a plant auxin in the mid-1930s no animal hormone was known that had similar chemical structure, and IAA itself was reported by several biologists to have no effect on animals. Fungi were also strikingly unresponsive to added IAA (even though IAA was isolated as an auxin from the medium in which the fungus *Rhizopus* was growing). It was the nonfungal plants that responded to auxin – plants with chloroplasts and cell walls based on cellulose (rather than on chitin). It was, therefore, a reasonable guess that auxin was causing growth by affecting the ability of the cellulosic cell wall to expand (Heyn 1931). The early research is summarized by Went and Thimann (1937) and by Heyn (1940).

Pectic substances in the walls were the focus of interest in the 1950s (e.g., Wilson & Skoog 1954; Ordin et al. 1957). Although several biochemists reported increases in the activity of enzymes affecting pectic metabolism, the fundamental problem of demonstrating that this was the *primary* effect of auxin has never been resolved (see Setterfield & Bayley 1961). More recently, with improvements in biochemical techniques of analyzing cell walls and associated enzymes, other aspects of cell wall metabolism as affected by auxin have been investigated (Pohl 1961; Cleland 1971).

Auxin considered as a coenzyme

Many compounds related to IAA have been synthesized and tested for auxin activity. From these studies arose several theories about the chemical structure required for auxin activity (e.g., Thimann 1951, Veldstra 1953). Both the structural specificity and the effectiveness of minute concentrations of auxin, as well as IAA's small molecular weight, supported the view that IAA might be acting as a coenzyme. After World War II a new method for extracting and separating proteins was developed. Wildman and Bonner's basic idea was that by using this new technique they might find a protein fraction that would contain auxin. If that were the case, they hoped that they would be able to show that the auxin-containing protein fraction was associated with an enzyme activity important in plant development. The hypothesis seemed especially intriguing and in the Sixth Growth Symposium Bonner and Wildman (1946) reported that one of their protein fractions from spinach leaves did contain a high level of auxin activity, judging by bioassay of extracts. As is so often the case in a symposium in which one is an invited speaker, and for which there is no outside reviewing of the submitted papers, the authors had not had time to check a point that was raised in the discussion, namely, how they knew that their extraction with hot alkali had not produced indoleacetic acid as an artefact from tryptophan in the protein. Gordon and Wildman (1943) had already shown that heating tryptophan with alkali caused such a conversion. Schocken (1949) demonstrated that the question raised at the growth symposium was pertinent and that there was a direct relation between the amount of tryptophan in various proteins and the amount of auxin activity that was obtained after heating them with alkali, using the Wildman and Bonner technique. Boroughs (1954) further demonstrated that the phosphatase-enzyme activity, which Wildman and Bonner had considered to be part of Fraction I protein, actually could be easily separated from it. Hence, contrary to the first report, there was no reason to think that auxin activity in the original protein fraction was associated with any detectable enzyme activity, the more so since Bonner and Wildman found that adding indoleacetic acid to their protein fraction did not stimulate phosphatase activity as they had expected.

Auxin and "energy release" or water uptake

Another suggested explanation for the multiple actions of auxin was that auxin might be concerned with some basic release of energy, the use to which that energy was put being a function of the site of energy release. This approach was used to see if auxin was involved with the metabolism of four-carbon acids in the Krebs cycle (e.g., Commoner & Thimann 1941; cf. Galston & Purves 1960). A similar approach was used to see if auxin controlled water uptake, one of the striking characteristics of higher plants being that their increase in size is associated with a very large increase in uptake of water that goes into the vacuole of the growing cell (Czaja 1935; Commoner et al. 1943; Bonner et al. 1953). Each of these approaches failed to convince workers in the field that it was the mechanism by which auxin acts on plant cells (see review by Galston & Purves 1960). As recently as 1974 a joint publication from three laboratories (Dowler et al.) reported that no promotion of water exchange by IAA could be obtained, contrary to an earlier report by others.

Auxin and nucleic acids

As is true of any field faced with a recalcitrant problem, as each new fashion swept the world of biochemistry the workers on auxin biochemistry would try to apply the new approach to their problem. Hence, it was not surprising that in the 1960s, the heyday of nucleic acid investigations, many people reported that indoleacetic was involved with nucleic acid metabolism and pointed out that such involvement could explain both the multiple effects and the ability to be active in such minute concentrations. One laboratory reported that IAA acted to stabilize soluble RNA (i.e., to make it more resistant to pancreatic ribonuclease) and suggested that this might be the prime method of auxin action (Bendana et al. 1965). However, these reports were retracted in 1971 (Davies & Galston) when the later authors discovered that the earlier technique was faulty and that there had been inadequate separation of the material in the extracts.

There was no doubt that the addition of IAA to tissues such as the coleoptile, whose growth was limited by auxin, would lead to increases in various constituents – if one waited long enough. Only gradually did investigators awake to the value of isolating auxin's effects in *time* as well as in space. Studying the responses to IAA that occurred the soonest after IAA addition would be more likely to uncover the primary biochemical effect. But it was equally important to see if there was any lag in response after IAA was added, because if IAA acts primarily by causing the synthesis of messenger RNA (which in turn might cause the synthesis of new enzymatic proteins), such RNA synthesis would take at least several minutes. (The lag assumed to be necessitated by RNA synthesis is based mostly on the extensive biochemical literature on microorganisms.) Ray and Ruesink (1962) and Evans and Ray (1969) initiated such studies and reported that a lag period of about 12 minutes occurred before *Avena* coleoptile sections started to elongate after IAA was added. These results led to a spate of papers on rapid actions of IAA (e.g., Nissl & Zenk 1969; Uhrström 1969; dela Fuente & Leopold 1970; see review by Evans 1974). Polevoy (1967) briefly reported data indicating that, if there was any lag at all in the response of his *Zea* coleoptiles, it would have to be less than 5 minutes long. Nissl and Zenk (1969) mounted a full-scale attack on the "necessity for the lag"; they used high temperatures, fast flow of IAA solutions, and higher concentrations of IAA and found that the elongation lag then disappeared. The *Avena* coleoptile sections showed immediate response to the added auxin. Their conclusion was the obvious one that such instantaneous stimulation ruled out the possibility that auxin was acting primarily to cause the synthesis of new messenger RNA, much less new protein. Murayama and Ueda (1973) obtained similar results with *Pisum* stems. Evans's (1974) review covers this and related literature, including the inevitable arguments for not taking Nissl and Zenk's results at face value; it also includes a most useful table of all the auxin effects that occur in 60 minutes or less.

Auxin and binding sites on membranes

Binding sites on membranes or cell organelles are the most recent area of intense interest for those interested in the biochemical basis of auxin action, paralleling the current fashion for studies of membranes and following the lead established by earlier work on steroid hormones in mammals (e.g., Maurer & Chalkley 1967). Tautvydas, who had worked out techniques for isolating nuclei from *Pisum* (1971), joined with Galston (1972) to see if they could get evidence that ^{14}C-IAA bound specifically to the isolated nuclei. Their data indicated that possibility.

Hertel and co-workers initiated the work on membrane fractions of plants. Their first paper reported that 1-*N*-naphthylphthalamic acid (NPA), a substance already known as an inhibitor of IAA transport (Morgan & Söding 1958; Hertel & Leopold 1963), was specifically bound to a cell fraction containing plasma membrane (Lembi et al. 1971). (IAA was tested first, of course, but binding could not be demonstrated [Thomson 1972], so they looked for compounds that did bind but still had some clear relation to auxin action.) The techniques used were as follows: *Zea* coleoptiles were harvested on ice and ground up, with various large particles being (mostly) removed by centrifugation at 0°C. A small amount of NPA labeled with ^{3}H (about 10^{-9} M) was added to the now clear supernatant. If that material was centrifuged at high speed, the resulting pellet would show about 500 cpm by direct counting. If, on the other hand, a much larger concentration (10^{-5} M) of unlabeled NPA were added to the supernatant after the ^{3}H-NPA had been added, the pellet would show fewer counts (about 200 cpm). The interpretation by Lembi et al. was that the unlabeled NPA was displacing the labeled NPA from binding sites in the pellet. The difference in counts (300 cpm, with data mentioned here) was considered to be the "specific binding." That the counts were binding to remnants of plasma membrane (rather than to other types of membrane present) was deduced from the observation that as the percentage of plasma membrane in the total membrane increased from 2 to 36% in various preparations, the amount of ^{14}C bound also increased. (The correlation coefficient of 0.949 was highly significant statistically.) Roland's (1969) stain, which was believed to be quite specific for plasma membrane, was used to estimate the percentage of plasma membrane in the fractions. IAA did not displace labeled NPA from the pellet – a disappointment, because they had hoped to show that NPA's effectiveness in reducing IAA transport through organ sections was causally related to mutual interference on the plasma membrane.

A year later the same general technique uncovered an "auxin-binding site" (Hertel et al. 1972), although the specific binding was much less than that for NPA (about 0.14% for IAA vs. about 25% for

NPA). The auxin binding was considered to be at a different site because NPA did not displace from the pellet ^{14}C that had been added as IAA, just as IAA had not displaced NPA from the earlier preparations. Both IAA and naphthaleneacetic acid (NAA) bound to this auxin site, judging by the ability of each to displace the other. Benzoic acid as a nonauxin did not bind, and 2,4-D bound weakly (only 0.04%). Displacement of ^{14}C-NAA by other compounds revealed a nice series with decreasing auxin activity being paralleled by decreasing efficacy of displacement (using 2,4-D, D-dichlorophenoxy-isopropionic acid, and L-dichlorophenoxy-isopropionic acid). Tri-iodo-benzoic acid (TIBA), well known as an antagonist of IAA movement through tissue sections, displaced ^{14}C-NAA when added at high concentrations. No displacement was found when GA-3, abscisic acid, or benzoic acid was tested.

Ray refined the techniques and found evidence that ^{14}C-NAA bound specifically to membranes of the endoplasmic reticulum (Ray et al. 1977; Ray 1977), the membranes being different from those to which NPA bound. Batt and co-workers had already reported two classes of auxin binding sites (Batt et al. 1976; Batt & Venis 1976).

The crucial question for these continuing investigations is, "Can it be demonstrated that this so-called specific binding is causally connected with the biochemical basis of auxin action?" (The binding might be specific but used only for auxin destruction or inactivation. For instance, Poovaiah and Leopold [1976] reported that Ca^{2+} increased NAA binding at concentrations that decreased coleoptile growth.)

All the preceding studies used *Zea*. Switching to *Cucurbita* hypocotyls, M. Jacobs and Hertel (1978), provided evidence for stronger binding of IAA on subcellular fractions that contained plasma membrane, judging by biochemical markers. The efficacy of displacement of ^{14}C-IAA from the pellets by nine auxin analogs showed general agreement with their efficacy in inhibiting IAA transport, 1-NAA and 2-NAA being exceptions. The authors of course suggest that this binding site may be involved in the polar transport of IAA.

Auxin and hydrogen ion secretion

Routine application of PESIGS rules would require investigators to determine if growth produced by IAA is specific for the IAA molecule or if it can be produced by the acid side chain alone. Although a few people in the 1930s showed that acid pH caused some elongation even if no auxin was added to excised sections, it has only been in recent years that detailed comparisons of acid and auxin effects have been made. These new investigations were encouraged by the studies of rapid actions of auxin, because acid can also give a fast increase in elongation. According to Evans's review (1974), the most obvious difference in the responses to acid and auxin is that the response to acid starts to fade away after an hour or so, whereas the auxin response continues. Supporting the interpretation that the acid response constitutes at least part of the response to auxin is the evidence that treatment with IAA quickly increases the acidity in a small volume of solution around *Avena* coleoptile sections (Cleland 1976) or even in the free space in *Zea* or *Pisum* sections as measured by inserted microelectrodes (M. Jacobs & Ray 1976). Other literature, much of it attempting to relate auxin and acid effects to cell wall metabolism, is described by Evans (1974).

Summary

To summarize this already brief account of the main areas investigated in the search for the biochemical basis of auxin action, it seems clear that half a century of investigation leaves our thirst for elucidation unsatisfied. Several of the areas mentioned are still being investigated and answers may still come from them. Part of the rationale for others has weakened during the intervening decades. For instance, the presumed clear distinction in chemical structure between IAA and animal hormones has disappeared. Serotonin (5-hydroxy-tryptamine), a close relative of IAA, has been found to be a neurohormone in animals, involved in transmission of nerve impulses across some synapses. Such synaptic transmission is one of the few polar phenomena known in higher animals that is roughly equivalent to the polar movement of IAA across cells in higher plants. It is intriguing that such similar molecules are used for polar transmission in these two groups of organisms.

6

Leaf and bud development and cytokinins

Although auxin could control the growth of stems, coleoptiles, and petioles, and to some extent the growth of fruits, it had no detectable effect on the growth of leaf blades, although auxin was produced in the blades of young leaves (Avery 1935). A search started, therefore, for the internal factors that do control leaf growth. This chapter describes the tortuous course of that search, which led 20 years later to the discovery of a new class of growth regulators, the cytokinins.

Early research on factors controlling leaf growth

Went (1938) observed that excision of the cotyledons of pea seedlings resulted in a 50% decrease in the growth of leaves above. He concluded that some factor or factors from the cotyledons controlled leaf growth. A major attack on the problem was then mounted by David Bonner as part of his Ph.D. thesis at California Institute of Technology.

His first paper in 1939 (Bonner et al.) used discs cut from leaves of radish seedlings. (The discs were circles cut with a cork borer from the same portion of many leaf blades. It was easier to measure the regular area of the circular discs than the irregular area of an entire leaf blade. In addition, various zones of a blade normally grow at different rates, so the use of discs from a specified zone provided more uniform experimental material.) A bioassay was worked out in which the discs could be photographed before and after treatment, with area measurements then made at the investigators' convenience from the photographs.

With this assay the workers found that in 30 hours control leaf discs gave a 30% maximal increase in area and wet weight. If they added water in which seeds of pea, radish, or corn had been soaked, they obtained even greater growth. Diffusate from intact, sterile leaves at low temperature also gave greater growth of the leaf discs than did the controls. Such stimulation was found from leaves of three genera, although adding a homogenate of the leaves did not work. Because it was so easy to soak pea seeds, the workers concentrated on finding the active material in the soak water from pea seeds. Sucrose represented about 40% of the dry weight of the material. When they substituted, accordingly, 2% sucrose or 2% glucose for the soak water from pea seeds, they obtained about a 50% increase in disc growth from these additives also.

However, there was something in the soak water in addition to the sucrose, because the addition of soak water to the optimal level of sucrose gave 20% more growth of the radish leaf discs than did the sucrose alone. During the preceding few years yeast extract had been successfully used as an additive for obtaining continuous aseptic culture of excised root tips, and it was therefore natural to try adding yeast extract to these leaf discs to see whether any of the many substances in yeast extract could replace the nonsucrose entity from the pea seeds. They found that yeast extract did have considerable activity in the leaf disc bioassay, and the many compounds known to be in yeast extract were then added off the shelf, one by one, to see if they were active. Thiamine, biotin, ascorbic acid, and riboflavin were not active by themselves, but Bonner and Haagen-Smit (1939) found that adenine did give greater growth of leaf discs, and so did its analogues, hypoxanthine and xanthine (Figure 6-1). At a concentration of 10 mg/liter adenine gave a 10% increase in growth of the leaf discs. As a final step in the chain of evidence, David Bonner in his 1940 thesis reported that hypoxanthine could be isolated from the diffusate of the pea seeds and therefore was a likely candidate for the nonsucrose factor in the pea seed soak water causing leaf growth. Bonner and Haagen-Smit (1939) also stated that intact plants, as well as isolated leaf discs, could respond to such factors: adenine added to mineral nutrients with which intact plants were watered caused greater growth of *Cosmos* plants. (No data were provided for this, only the photograph of the plants.)

During the succeeding years David Bonner switched to research on fungal biochemical genetics, and effects of adenine on the growth of shoots, as would be expected from Bonner's results, could not be found by several other investigators. For instance, Kruyt and Veldstra (1947) tried to repeat the earlier results and could not get nearly as strong an effect of adenine on leaf growth as had been reported, particularly with regard to the dry weight increases. They also felt that the adenine effect was probably not specific because of similar effects they obtained with naphthalene compounds. De-

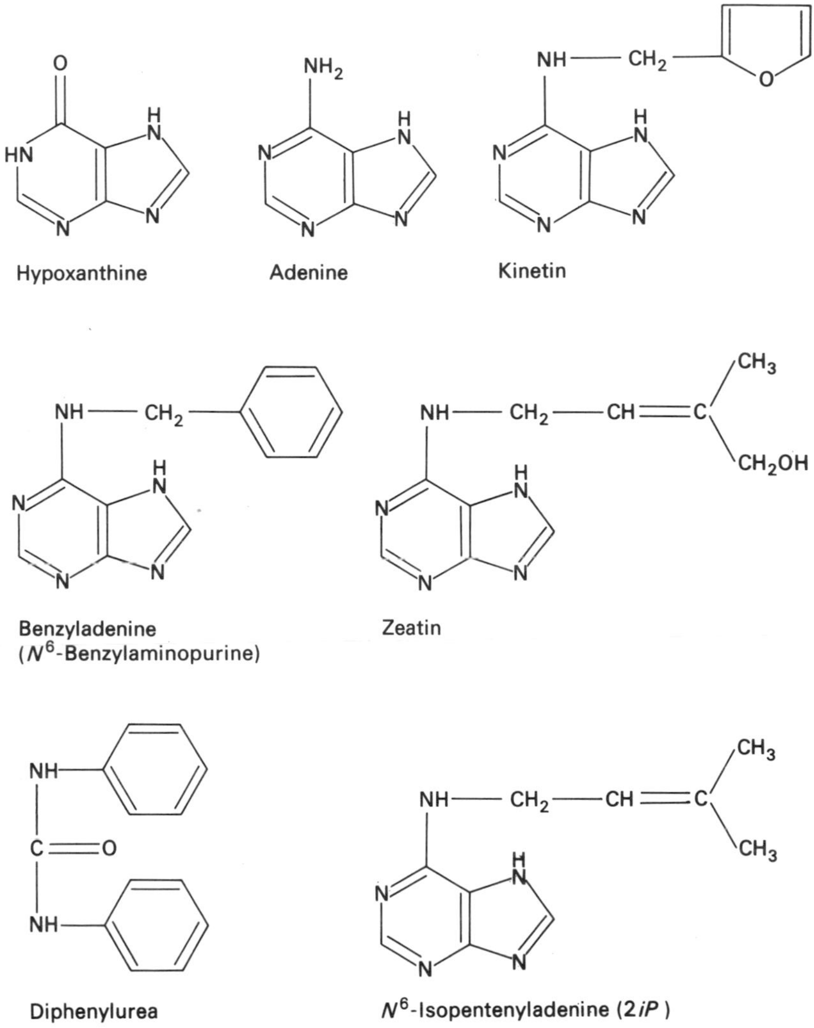

Hypoxanthine

Adenine

Kinetin

Benzyladenine
(N^6-Benzylaminopurine)

Zeatin

Diphenylurea

N^6-Isopentenyladenine (2*iP*)

Figure 6-1. The chemical structures of various compounds reported to have cytokinin activity.

Ropp (1945) also found that he could not get any effect of adenine on the growth of cultured shoot tips of rye. It was not until Skoog started on his postwar program of research that the string of negative results was broken.

Skoog's culture work on buds and cell divisions

Skoog's basic interest seemed to derive from his thesis work with Thimann and was concerned with what controls promotion and inhibition, an interest expressed first in his superb work on apical dominance (Thimann & Skoog 1934). In his research on the control of leaf and shoot growth, he has used almost exclusively the technique of tissue culture; that is, he used a cube of tissue cut from tobacco pith and grown in sterile nutrient medium. This has the advantage cited by Skoog of giving isolation of the tissue from secondary influences, but it does have the disadvantage that excision can artificially make substances limiting. Hence, one cannot tell from work done solely with tissue culture or organ culture whether the factor found limiting under that artificial condition is also limiting in the intact organism. In essence it is applying the isolation rule of the PESIGS rules without applying the others. Another disadvantage, of course, is the extra labor involved in using and maintaining sterile tissue cultures, but this disadvantage is undoubtedly more than outweighed by the fact that sterile conditions obviate the possibility that the added chemicals really affect the growth of microorganisms, which in turn produce metabolites that directly cause the developmental effects observed. (Such artefacts have confused researchers in various areas of development. See, for instance, Lonberg-Holm's [1967] criticism of Burdett and Wareing's paper.) Instead of using discs cut from young leaf blades, Skoog concentrated on the development of buds (very young leaves, the associated unelongated stem, and the apical meristem that produces both stem and leaves).

Skoog and Tsui (1948) confirmed that auxins inhibit bud development on cultured callus or stem pieces but found that adenine or adenosine, in the mg/liter range, stimulates it. Checking the chemical specificity, they reported (1951) that of the chemicals tested, adenine sulfate gave the largest number of buds per cultured piece of tobacco stem, with guanine second and adenosine monophosphate third. It was not the mere presence of adenine in the culture medium that was effective, but the *balance* between adenine and auxin (Skoog & Tsui 1951). If 40 mg/liter of adenine were added with a very low concentration of indole-3-acetic acid (IAA) (0.001 mg/liter), the cultures first formed buds at the apical end and only later formed roots. If the IAA concentration was increased to 0.02 mg/liter, rapid growth ensued but neither buds nor roots developed. IAA alone gave roots but no buds; adenine alone gave only buds as early responses. This concept of the balance between two substances controlling development became one of Skoog's favorite and most influential ideas.

That this response to adenine was not limited to tobacco variety Wisconsin No. 38 (the main research material of Skoog's laboratory) was indicated by data on the culture of horseradish roots – a species whose fleshy root was already known for its ability to regenerate buds (Lindner 1940). Adenine increased the number of buds regenerated as well as their rate of formation over the already considerable number in the control segments of horseradish root (Fig. 6 of Skoog & Tsui 1951), and IAA decreased the number. Nicely paralleling the ability of untreated horseradish roots to regenerate more buds and fewer roots than tobacco stems was their much lower content of auxin (as judged by auxin activity in ether extracts; see Tables 5, 6 of Skoog & Tsui 1951). The only noticeable difference between the roots and stems was that, for the roots, adding IAA along with the adenine had no effect on the bud formation resulting from the adenine (their Fig. 6). This is difficult for us to interpret, partly because the researchers did not give the concentrations of adenine or IAA being added in the experiment of their Fig. 6, but one might guess that the added IAA was not in high enough concentration when added to the low *endogenous* auxin level in horseradish roots to reach a threshold level sufficient to interact with the adenine.

Reverting to cultured pith of tobacco, Jablonski and Skoog (1954) studied changes at the cellular level. Controls on White's nutrient medium gave relatively slow growth (an 83% increase in dry weight in 16 days), but the addition of the optimal 2 ppm of IAA produced a tenfold increase in cell size and marked enlargement of nuclei, although this growth was unaccompanied by any detectable cell divisions. However, the addition to the cultures of tissue containing vascular strands produced many cell divisions in the cultured pith. (Haberlandt had reported years before that a slice of tissue, if it contained the phloem portion of the vascular tissue, could induce cell divisions in a slice lacking phloem [1913].) Various chemicals, including adenine, could not substitute for the vascular tissue in the cell dividing, or cytokinetic, effect, although extracts of malt were active.

The isolation of kinetin

Miller et al. (1955b) fractionated yeast extract, which was also active, assaying for cytokinetic activity with the tobacco cultures to which IAA had been added. They found that activity was always

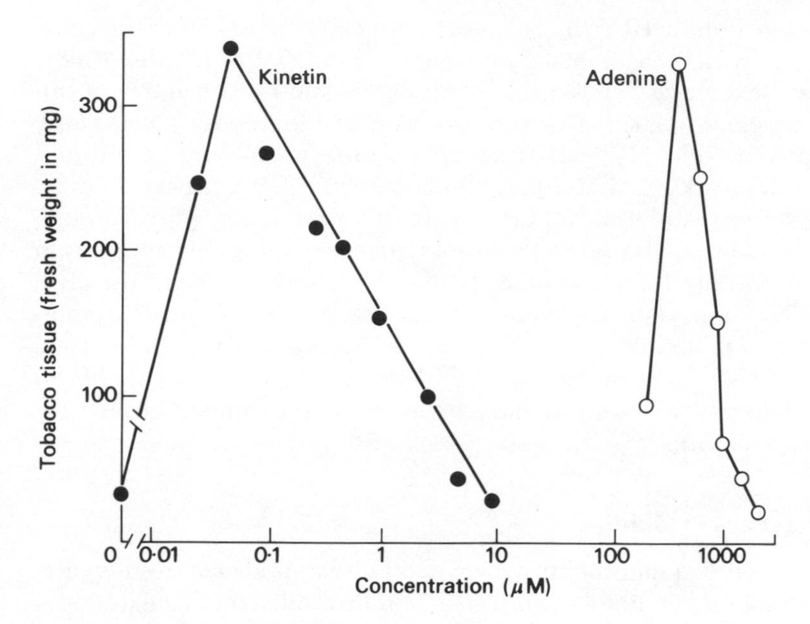

Figure 6-2. Dose-response curve of tobacco tissue culture to kinetin and adenine (after Miller 1968).

found at a spot absorbing at 268 nm. in ethanol. This, plus precipitation with silver nitrate, suggested that the active chemical might be a purine. As one of the easily available purine derivatives, a commercially prepared sample of herring sperm DNA off the shelf was tested and found active. Unfortunately, when that was used up, freshly purchased samples showed no activity. Tracking down the history of the first bottle revealed that it was 4 years old, and with that clue they discovered that activity could be produced in freshly purchased DNA by autoclaving it at pH 4.3. Presumably, the DNA in the original bottle had decomposed during its long storage period to give the active material that autoclaving could produce more quickly. The active factor was isolated (Miller et al. 1955b) and then synthesized (Miller et al. 1955a). It was called kinetin because it induced cytokinesis when added with IAA to tobacco cultures. Kinetin (6-furfurylaminopurine) is a purine derivative and can be considered to be an adenine with a specific side group (Figure 6-1). The speed of isolation and synthesis, in sharp contrast to the situation for auxins, was undoubtedly due to the fact that Skoog had the perspicacity to enlist the aid of top-notch chemists.

The strikingly increased effectiveness of kinetin over adenine in the tobacco callus bioassay is shown in Figure 6-2. Kinetin, with its optimum below 10^{-7} M, is more than 1000 times as effective as adenine, which has a sharp optimum at 4×10^{-4} M (Miller 1968).

The class name "cytokinins" is used for substances active at minute concentrations in bioassays such as the tobacco pith test. Many compounds related to kinetin and adenine were synthesized and tested for cytokinin activity, in the hope of finding a substance even more active than kinetin. Many synthetic compounds were found to be active, although the overwhelming majority were less active than kinetin (Strong 1958; Skoog & Leonard 1968). Benzyladenine (also called N^6-benzylaminopurine) is one of the few synthetic compounds with activity as high as kinetin's (Figure 6-1). Because of their high activity in bioassays and their commercial availability, benzyladenine and kinetin have been the most commonly used synthetic cytokinins.

Endogenous cytokinins

With kinetin so manifestly an artefact, a search also started for the presumed endogenous cytokinins. The first substance isolated and identified from a previously known natural source of cytokinin was 1,3-diphenylurea (Figure 6-1). The natural source was coconut milk, which is the coconut fruit's endosperm while it is still in a liquid stage. Coconut milk has been used as a valuable supplement to culture media for plant tissues ever since 1941, when Van Overbeek, Conklin, and Blakeslee had the brilliant idea of using coconut milk when they were faced with the problem of culturing very small embryos from *Datura* crosses. Reasoning that the endosperm would not be so firmly ensconced as a nutritive tissue for embryos among the angiosperms unless it was unusually valuable for embryo growth, they selected coconut milk as the endosperm to use because of the relatively large volume of endosperm per fruit, the existence of the endosperm in an early liquid stage, and the ease of removing the milk in sterile condition by piercing one of the eyes of the coconut with a sterile hypodermic needle. Embryos too small to develop in the usual culture media could be induced to develop when coconut milk was added (van Overbeek et al. 1941, 1942).

Van Overbeek et al. (1944) achieved a 170-fold enrichment of the fraction from coconut milk that promoted embryo development, but no active chemicals were actually isolated until after Steward and his co-workers mounted a major attack on the problem, switching to an assay using growth in culture of phloem parenchyma cut from carrot roots (Caplin & Steward 1948). They solved the recurring problem of those seeking to isolate hormones – namely, how to get the huge volumes of necessary starting material – by capitalizing on a Florida hurricane that knocked down hundreds of coconut trees near Miami. The 700 gallons of coconut milk collected was too big a volume to be processed in the university laboratories, but DuPont allowed one of their pilot plants to be used. It was from this windfall that the 1,3-diphenylurea was isolated (Shantz & Steward 1952, 1955).

An innocent reader in the cytokinin literature is apt to be surprised by the paucity of references to diphenylurea as the first endogenous cytokinin to be isolated. It is, in fact, not even mentioned as an *endogenous* cytokinin by some reviewers (e.g., Fox 1969; Kende 1971). And even the original isolators phrased their description of its existence with rare caution. Why is this? First, the early reports said that diphenylurea gave erratic results, but Bruce and Zwar (1966) demonstrated that impurities were the cause of the variable responses. When diphenylurea was more thoroughly purified, its effect on the tobacco pith assay became nicely reproducible with an activity somewhat less than kinetin's. Second, and more important, is the possibility that diphenylurea was an artefact. Shantz accidentally discovered that the pilot plant used at DuPont to process the hurricane's gift of coconut milk had also been used for synthesizing urea herbicides. Was the diphenylurea "isolated from coconut milk" merely a contaminant from a preceding batch of urea herbicide? We do not know. A smaller volume of coconut milk from a different country seemed to have only 20% of the diphenylurea in the big Florida batch, but of course the volume was too little for an actual isolation and coconut milk from different sources does vary in composition. There is no direct evidence for accidental contamination of the Florida batch; nevertheless, another isolation of diphenylurea is very much needed to answer more definitely the question whether 1,3-diphenylurea is the first *endogenous* cytokinin to have been isolated.

Various analogues of diphenylurea have been tested for activity, and it is clear that substituted ureas act as cytokinins in a variety of bioassays (e.g., cell division assays with tobacco pith, senescence-retardation, and stimulation of elongation of axillary buds [Bruce et al. 1965; Bruce & Zwar 1966]).

Aside from diphenylurea, it took 8 years from the elucidation of kinetin to the first isolation of an endogenous cytokinin in crystalline form. Part of this time was spent working out bioassays. One that involved measuring fresh weight increments of aseptic cultures of callus cubes derived from soybean cotyledons was devised by Miller (1963) and has been used by several others since. Miller followed the sensible procedure, suggested by the earlier work on endogenous auxin, of extracting seeds of a major commercial crop, *Zea mays* (thereby getting at relatively low cost a large volume of tissue that presumably would contain much stored "food" for the growth of the embryo). With this combined approach, he obtained an active fraction from the *Zea* seeds and concentrated it substantially (Miller 1961). Letham, meanwhile, was using young plum fruits as a readily available source of endogenous cytokinins and developed cultures cut from carrot roots as his bioassay (1963a). Much time is saved with the carrot assay, compared to the soybean callus assay, because carrots can be used fresh from the market, whereas soybean callus needs to be maintained in sterile nutrient medium as a source of callus cubes for the bioassay. However, with carrot cultures one loses the wide range of kinetin concentrations over which the soybean assay gives increasing response (Letham 1967, Table I). Using this different source material, different bioassay, and a different fractionation scheme, Letham found an active cytokinin fraction in extracts of the fruits (1963a). He then applied his techniques to Miller's source material and achieved the isolation of zeatin (Figure 6-1) in crystalline form from *Zea* seeds (Letham 1963b) and the elucidation of its chemical structure (Letham et al. 1964). Shaw and Wilson synthesized zeatin that same year. The cytokinin from *Zea* that Miller's procedure had concentrated was quickly shown to be zeatin also (Letham & Miller 1965).

This endogenous cytokinin is, like kinetin, an adenine derivative, but zeatin is 10–100 times more active than kinetin in various growth assays (Letham 1967a; Jacobs 1976). That is, zeatin is more active than kinetin as kinetin is much more active than adenine.

But there was cytokinin activity in the *Zea* seed preparations in addition to that which zeatin provided. Miller (1965) found activity in several zones of his chromatograms. Letham (1966, 1968) then isolated zeatin riboside and its 5′-monophosphate as active cytokinins from the *Zea* extracts, and Miller (1967) confirmed his work.

From seeds of a different genus (*Lupinus*) in a different angiosperm family, Koshimizu et al. (1967) extracted dihydrozeatin – zeatin with a saturated side chain; it is roughly one-tenth as active as zeatin in callus growth assays (see Schmitz et al. 1972).

Letham turned from extracts of whole seeds to another obvious potential source of cytokinins, the coconut milk that had been a classically rich source of "growth factors" ever since van Overbeek et al. (1941) first thought of using it for culturing excised embryos. Jablonski and Skoog (1954) had found it, in the form of coconut meat, to be cytokinetically active. Zeatin riboside was present in extracts of coconut milk (Letham 1968).

Following the creative precedent set by the earlier workers searching for endogenous gibberellins (see Chapter 7), several workers looked for effects in higher plants that resembled the effects from the addition of exogenous kinetin to the plant. For instance, the action of *Corynebacterium* on higher plants suggested that it might be forming a cytokinin (Samuels 1961), and it was found that added kinetin mimics the effect of *Corynebacterium* (Thimann & Sachs 1966). Klämbt et al. (1966) isolated another cytokinin from cultures of *Corynebacterium*. This was only slightly different in its side chain from the chemical structure of the zeatin and is known as both isopentenyl adenine and as 6-($\gamma\gamma$-dimethylallyl-amino) purine (see Figure 6-1). (According to later workers, such as Rathbone and Hall [1972], much of the cytokinin isolated from the *Corynebacterium* was an artefact resulting from acid hydrolysis of the transfer RNA.) The next year Klämbt extracted from *Agrobacterium*, the microorganism famous for inducing tumors on higher plants, an active kinin that was apparently this same isopentenyl adenine. (It was not obtained in crystalline form.) Various transfer RNAs provide isopentenyl adenine upon acid hydrolysis (Hall et al. 1967), and this repeatedly confirmed observation became the basis for one of the major hypotheses as to the biochemical basis of action of the cytokinin, namely, that cytokinins act by being built into transfer RNA (see later discussion).

Rhizopogon, a fungus that grows on roots and gives enlarged cortical cells on the host, was investigated by Miller (1968) on the grounds that it acted as though it were producing a kinetinlike factor. He found evidence that zeatin riboside was the agent causing this effect.

Most of the compounds used as cytokinins are derivatives of adenine. Kinetin and zeatin are prime examples of this group. Diphenylurea and its derivatives have already been discussed in terms of both their isolation from coconut milk and their activity in various bioassays. Wood and Braun are active proponents of a different line of thought. They believe that the usual cytokinins, such as kinetin and zeatin, are not the prime endogenous cell-division-promoting factors but that kinetin and its analogues act by causing the formation in the plant of substances that they have called *cytokinesins* (1967). (This is reminiscent of the belief in the 1940s of von Guttenberg's group that IAA acted by stimulating some other substance, which was the true endogenous auxin.) Cytokinesin is believed by Wood and Braun to be a different compound from the kinetin-zeatin group, and although their assertions about its probable structure have changed quite sizeably over the years, in recent papers they believe it to be a substituted hypoxanthine (Wood et al. 1974). As one would expect, the Skoog laboratory group, which discovered kinetin, are less than avid supporters of this hypothesized new compound. Miller (1975) has reported finding ribosyl-*trans*-zeatin in extracts of Wood's line of *Vinca* tumor tissue and he concludes that the evidence for cytokinesin as a separate class of compound is insufficient to warrant use of the term. As an outside observer, I do not find it reassuring that it has taken such a painfully long time – since 1963 at least – to get true identification of Wood's cytokinesin. It has still not been isolated and adequately chemically identified. However, it is an intriguing point, not mentioned in the literature, that the compound actually isolated by David Bonner from the soak water from pea seeds was hypoxanthine, another 6-purinone, and one that was active in his bioassay on radish leaf growth.

It is obvious that until there is actual isolation and identification of the hypothesized cytokinesin of Wood and Braun, their proposal that the cytokinins act by producing cytokinesins inside the cell is going to be very difficult to confirm.

Effects of cytokinins added to plants

Because kinetin was available commercially for many years before zeatin and other endogenous cytokinins were discovered, most of the currently known effects of exogenous cytokinins are based on the addition of kinetin or its synthetic counterpart, benzyladenine (Fig. 6-1). Kinetin is necessary in cultures of tobacco pith of the Wisconsin No. 38 cultivar, along with IAA, to produce a sizeable number of cell divisions. One ppm kinetin was close to optimal. Some mitoses occur in these cultures if only IAA is added (Das et al. 1956). (Note also that if kinetin is added and IAA is not, then the great burst of mitoses does not occur either. Hence, it is more or less an accident of the order of discovery of the auxins and cytokinins that kinetin has been called a cell division factor for that tissue.) The promotion of cell division by the addition of kinetin in the presence of IAA has been reported for various other plant systems such as pea root callus (Torrey 1958b), cocklebur stem callus (Fox & Miller 1959), and carrot root (Letham 1967a). Kinetin was quickly discovered, however, to be able to cause cell growth without cell division in some systems (Miller 1956). Harking back to the original bioassay of D. Bonner et al. in the 1930s, Kuraishi (1959) showed that kinetin could cause leaf growth even in old leaves of many genera. The ability of adenine and some of its derivatives to stimulate shoot bud initiation in callus cultures, which Skoog and Tsui had described, was shown to be a property of kinetin at much lower hormonal concentrations (Skoog & Miller 1957). Such effects were also quickly reported for root cultures of *Isatis* by Danckwardt-Lillieström (1957) and *Convolvulus* by Torrey (1958a), and also in excised leaves of *Saintpaulia* by Plummer and Leopold (1957) and in *Begonia* by Schraudolf and Reinert (1959). Most of these plants were already famous for their ability to regenerate buds without cytokinin treatment; hence, some question was raised as to whether kinetin could be considered a bud-*forming* substance. However, Doershug and Miller (1967) found that under their conditions lettuce cotyledons never formed buds in culture unless kinetin or high concentrations of adenine were added. As is so often the case, IAA had to be added also to get buds to form.

A creative adaptation of the results so far was made by Wickson and Thimann (1958), who reasoned that since kinetin caused many buds to be initiated and then develop without one bud inhibiting the others, perhaps kinetin would affect bud elongation of already differentiated buds (as contrasted solely to bud initiation) in situations where there might be apical dominance. By floating sections of pea stems on various solutions, they showed that IAA inhibited bud growth, as would be expected for legumes from most earlier work, and kinetin released the axillary buds from inhibition by IAA if both were present in a roughly 1 : 1 molar ratio. When kinetin was added at the base of an intact plant, it also gave elongation of the axillary bud to a level of 71% of that found in the decapitated control. Their conclusion was that apical dominance may be due to an interaction between auxin and a kinetinlike endogenous factor in the stem. Several other workers have reported stimulation of bud elongation by added kinetin (e.g., Lona & Bocchi 1957; Sachs & Thimann 1967).

Retardation of senescence by cytokinins

One of the most striking and unusual effects of added cytokinin is its antisenescent action on detached leaves. When a leaf is cut from the plant it typically shows a decrease in protein and chlorophyll content. Richmond and Lang (1957) showed that a detached *Xanthium* leaf when treated with kinetin maintained its protein level and lived longer. This important finding has since been confirmed by other workers in *Xanthium* and other genera (e.g., Mothes & Engelbrecht 1961; Kulaeva 1962). A well-known fact from the older literature was that if roots regenerate on a detached leaf, the leaf lives longer and does not senesce so quickly (e.g., Chibnall 1939). Hence, kinetin can partially substitute for the roots in maintaining excised leaves. The antisenescent effect of added kinetin is so clear-cut that, in fact, one of the bioassays for cytokinins has been based on the prevention of loss of chlorophyll by the addition of cytokinins to discs cut from *Xanthium* leaves (Osborne & McCalla 1961). In at least one case (Kursanov et al. 1964) kinetin has been reported to be able even to restore senescent leaves that have already lost their color.

Although Letham's (1967b) statement that cytokinins are the most effective known senescence retardants seems justified on the basis of the evidence available so far, we should not lose track of the fact that other classes of hormones can retard senescence in some genera. For instance, 2,4-D prevents senescence of parts of *Prunus* leaves on which the synthetic auxin is spotted (Osborne 1959); GA-3 prevents senescence in *Taraxacum* (Fletcher & Osborne 1965, 1966), *Tropaeolum* (Beevers 1966), and *Rumex* (Whyte & Luckwill 1966; Goldthwaite & Laetsch 1968); and benzimidazole prevents senescence in wheat leaves according to Person et al. (1957). We should be properly cautious when interpreting the brief papers that often report these effects for the first time. A substance reported as having no effect in a system may not have been tested in a wide enough range of concentrations to allow the activity to show up. An example of this sort seems to be a report by Whyte and Luckwill (1966) that kinetin (among a number of other substances) was not active in preventing senescence of *Rumex*. However, Goldthwaite and Laetsch found that kinetin was effective when added at around 40 mg/liter concentration and benzyladenine was nearly as effective as GA-3 when one considered the rate of chlorophyll loss after short periods of incubation. An additional caution, of course, is that many of these substances were tested before zeatin or isopentenyl adenine were known, and a system in which kinetin has no activity might very well show activity with the 100 times more active zeatin.

Auxin–cytokinin balance in leaf regeneration

Skoog's hypothesis that a balance between cytokinin- and auxin-controlled shoot and root development was based on results from cultures of tobacco pith. Convincing though the results were in this isolated system, it was important to see whether the results applied to the intact organism. An interesting series of papers by Heide made a big step in this direction. Heide investigated the regeneration from leaves of *Begonia*, a genus that is typically propagated by regeneration from leaf cuttings. Heide showed that higher temperatures caused more roots and fewer buds to regenerate and that continuous light (as contrasted to short days) gave fewer buds from the leaf cuttings. Adapting the Skoog interpretation of the tissue culture work, Heide hypothesized that the continuous light and high temperature were producing less endogenous cytokinin and more endogenous auxin and that this shift in the balance of endogenous hormones was what resulted in the observed changes in regeneration pattern. When auxin or cytokinin was added to the regenerating *Begonia* leaves, the results fitted this hypothesis (Heide 1965). The endogenous levels of auxin were estimated by extracting leaves with cold ether and running the acid fraction on paper chromatograms with isopropanol, ammonia, and water. Heide found one zone of auxin activity that ran to the Rf of IAA, as well as a zone of gibberellin activity. Short-day treatment had the expected effect on the presumed IAA. There was a steady decrease in the activity at the Rf of IAA as extra short days were given and no new zone of growth promotion occurred in chromatograms of such extracts. The effects of temperature were not quite so clear-cut: although the lowest temperature tested (15°C) did give less presumed IAA, there was not less at 20 than at 25°C. Nevertheless, the general trend was in the direction expected from the Skoog–Heide hypothesis (Heide 1967, 1968).

The evidence for the hypothesized changes in endogenous cytokinins is weaker than that for auxins – a regrettable fact, because this would have been the main new information from the series of experiments. Heide and Skoog (1967) extracted the *Begonia* leaves and tested the material for cytokinins on the tobacco callus assay. From leaves grown in natural day length they obtained 30–300 μg "kinetin equivalent" per kilogram fresh weight, and when paper chromatograms of the extracts were run, they found high cytokinin activity at Rfs corresponding to zeatin in six different solvent systems. According to Heide (1972), they have not yet checked the effect of temperatures on endogenous cytokinin, but the effect of an increasing number of short-day photoperiods has been followed. They conclude that an increasing number of short days (up to 16) "appeared to increase the extractable cytokinin content" but state that these "must be considered as preliminary results." The reason for this cautious statement is that although three of the four experiments gave an increase in cytokinin with increasing number of short days, the fourth experiment showed "no significant effect of day length." The data in their Table 1 are given as averages of all four experiments and those averages show no steady trend toward an increase; in fact, no statistical evidence is presented as to whether any of the means shown are significantly different from any of the others. One can hope that the data on extractable cytokinin will be worked on further, because the rest of the system seems to have been so clearly and nicely explored.

Another genus famous for the regeneration of its leaves was also explored by Heide (1972). Results from the leaves of *Streptocarpus* do not seem to fit the Skoog–Miller hypothesis: added cytokinins have in essence no effect on the regeneration, low temperature increases *both* roots and buds (as contrasted to increasing only buds in *Begonia*), and auxins increase both buds and roots (as contrasted to increasing only roots in *Begonia*), although it is true that the maximal concentration of auxin for increasing roots is one to two orders of magnitude higher than that for buds. Heide surmised that *Streptocarpus* leaves have large amounts of endogenous cytokinin and too little auxin, so that only the auxin is limiting regeneration of the two types of organs. This view is supported by extraction studies that showed very little auxin extractable from the *Streptocarpus* leaves, the amount being one-tenth to one-hundredth that found in *Begonia* leaves.

Sites of production of cytokinin in the plant

Seeds and fruits are obvious sources of cytokinins, as implied in the foregoing material on the isolation of zeatin and its derivatives from *Zea* kernels. Various other fruits and fruit parts also showed cytokinin activity (Bottomley et al. 1963, in unpurified extracts). Miller and Witham (1964) found substantial activity in extracts of *Zea* shoot tissue, young roots, and kernels, although kernels gave the most activity per unit dry weight. Extracts of young sunflower fruits also gave cytokinin activity at the same Rf as the *Zea* extracts.

The presence of cytokinins in roots would be expected from the observation that the antisenescent effect of roots developing on excised leaves could be mimicked by adding kinetin to excised leaves that did not have roots. If roots were present on an isolated leaf, the antisenescent effect of kinetin was more difficult to demonstrate (Mothes 1960; Kulaeva 1962), suggesting that roots supplied nearly optimal levels of cytokinin. Several works, starting with Kulaeva

(1962), showed that there was cytokinin activity in the root exudate of several species (Loeffler & van Overbeek 1964; Kende 1964; Nitsch & Nitsch 1965b; Klämbt 1968). (Root exudate, also called xylem sap or bleeding sap, is operationally less specific than it sounds; it is the liquid collected from the stump of a stem that has been cut a short distance above the soil line. The collectors typically hope that this xylem sap represents the liquid normally moving up from the root system in the transpiration stream of the intact plant.) Encouraging the interest of physiologists on the presumed contents of the transpiration stream was an experiment of Kulaeva's in which she steam-girdled a petiole and showed that kinetin added to that leaf had as little antisenescence effect as if the leaf were still functionally attached to the root system. The steam would be expected to kill all the cells in that region; hence, the only functional connection with the root would be through the transpiration stream.

More direct evidence for cytokinins' existence in roots had already been presented by Vardjan and Nitsch (1961), who bioassayed extracts from various regions of the fleshy storage root of endive (*Cichorium intybus* L.). This root was investigated because its well-known ability to regenerate buds suggested that it had a high concentration of endogenous cytokinin. Vardjan and Nitsch found a gradient of cytokinin activity in the root with the maximum at the root base near the root–stem transition region. Zeatin riboside was later isolated from the roots (Bui & Nitsch 1970). Evidence for nonfleshy roots was slower to appear. Zwar and Skoog (1963) found activity in water extracts of dried barley roots, and Miller and Witham (1964) found it in "purified" extracts of young roots of *Zea*. Both pairs of authors cautioned about the masking of cytokinin activity in crude preparations. Localization within nonfleshy roots was first shown by Weiss and Vaadia (1965). Following up the earlier reports that root exudate of sunflower contained cytokinins, they extracted root tips of young sunflower seedlings, ran the extracts on paper chromatograms, and assayed the zones with the soybean callus test. Extracts of the most distal 1 mm of 1200 roots showed two zones of activity – equivalent, they said, to 130 and 170 ng kinetin in the bioassay. Extracts of sections only 1–3 mm from the root tip showed "little if any" cytokinin activity (equivalent to only 1 ng kinetin from 1200 root sections). (The low activity in the bioassay of the extract of 1- to 3-mm sections could easily have been due to masking substances in their relatively crude extracts.)

Four cytokinins were identified by gas–liquid chromatography in extracts of pea roots: zeatin, isopentenyl adenine, and their ribosides (Babcock & Morris 1970).

A nice paper by Radin and Loomis (1971) revealed that there were three zones of cytokinin activity on thin-layer chromatograms of extracts of radish "roots" (really the root plus the fleshy hypocotyl), with the activity at two of these zones increasing 17 days after germination, a time when cambial activity began in the developing fleshy storage organ. Further evidence that the cytokinins in these two zones were functioning to initiate cambial activity in the intact root was that eluates of these zones caused a statistically significant increase in cambial activity when added to radish roots growing in sterile culture. Judging by location of the zones on the chromatograms and other characteristics, the cytokinin in one of these two zones was zeatin ribonucleotide whereas the other was zeatin, zeatin riboside, or both. The third zone of cytokinin activity differed in several ways: it showed no increase in activity until about 32 days after germination – long after cambial activity had started; it did not stimulate cambium when added back to the sterile root cultures; it was apparently localized in the xylem tissues of the fleshy "root" (unlike the presumed zeatin derivatives, which were present in roughly the same amounts in xylem tissues and the tissues external to the xylem); and there were no clues as to its chemical nature (except that its Rf was not that expected for either zeatin or isopentenyl adenine).

Zeatin and its riboside and ribotide were presumably the endogenous cytokinins of rice roots (Yoshida & Oritani 1971), whereas a fourth factor, which showed little activity in a carrot callus assay but much more in a chlorophyll retention test, was tentatively identified as zeatin glucoside (Yoshida & Oritani 1972).

Similar results were obtained by Short and Torrey (1972), who found four zones of activity, three of them presumably from zeatin, its riboside and ribotide, in extracts of the distal 1 mm of pea roots. (Extensive purification was necessary to reveal these cytokinins, and the negative results with pea roots reported earlier by Shibaoka and Thimann 1970 were guessed to result from their less thorough purification treatments.) There was 44 times more total cytokinin activity per unit fresh weight in the distal millimeter of the pea roots than in the next 4 mm.

In addition to the data on cambium in radishes of Radin and Loomis already discussed, there are several indications that cytokinins are present in relatively high amounts in the vascular tissues (phloem, xylem, and procambium–cambium). Severed vascular strands placed against tobacco pith tissue induced cell divisions (Jablonski & Skoog 1954), and the presence of internal phloem on a tobacco pith cube was sufficient to substitute for exogenous cytokinin (Bottomley et al. 1963). Extracts of scrapings of the cambial zone (which we would expect to include young cells of phloem and xylem in addition to cambium) showed cytokinin activity, whether taken from tobacco, pine, or *Eucalyptus* (Bottomley et al. 1963). We can also ferret out from Table 2 of Letham (1967a) data supporting the view that the "cambial zone" of carrot roots contains more cytokinin than zones specifically out in the phloem or specifically in the xylem: explants from the cambial zone grew significantly more without added cytokinin and showed a smaller increment from the addition of kinetin. Nitsch and Nitsch (1965a) extracted with methanol the actively growing cambial zone of a poplar tree and found, after removing water-soluble inhibitor(s) from the extract, that a single zone of cytokinin activity was apparent after paper chromatography. In the tobacco pith assay this zone showed activity equivalent to 1 mg kinetin/kg dry weight of cambial zone tissue. (They did not report the cytokinin content of other tissues, unfortunately, so we do not have direct evidence that the content of the cambial zone was either higher or lower than the average. However, their working hypothesis, nicely confirmed, was that cambial-zone tissue, famous in the tissue culture literature of France [Gautheret 1959] for its ability to grow without added cytokinin, would be synthesizing ample amounts of its own cytokinin.)

Biochemical basis of action of cytokinins

Like that of all the plant hormones currently known, the biochemical basis of action of cytokinins is still a mystery. Kinetin added to various systems, such as excised leaves, caused a fast increase in the amount of RNA and in some cases caused an increase in DNA and in total protein. It was therefore natural to think that kinetin might be fixed into RNA as the prime basis of its action – the more so as kinetin was discovered during the heyday of DNA–RNA research. This would also perhaps explain the localization of some kinetin effects (i.e., the observation that, by indirect evidence, kinetin does not seem to migrate much when it is added in spots to excised leaves). Skoog and his co-workers were strong advocates of the view that kinetin acts through being built into a transfer RNA. This possibility seemed much more probable after it was discovered that some transfer RNAs, upon acid hydrolysis, released a sizeable amount of cytokinin (Hall et al. 1967). It was found, in fact, that isopentenyl adenine was an integral part of transfer RNA and was in a position adjacent to the anticodon. (See Fig. 12-1 of Watson 1970 for a clear diagram of the "clover-leaf interpretation" of the structure of transfer RNA.) It was apparently this isopentenyl adenine that was released from transfer RNA upon acid hydrolysis. But were cytokinins active through being built into transfer RNA in this position and thereby influencing protein synthesis? An early paper (McCalla et al. 1962) reported that radioactive benzyladenine was not incorporated into RNA to any significant extent when the benzyladenine was added to leaf discs of bean and *Xanthium*. For *Xanthium* leaves benzyladenosine was the major soluble metabolite. In *Phaseolus* leaves, in contrast, they obtained a single metabolite only and did not obtain evidence as to its chemical structure. Fox (1966) felt that metabolites of the cytokinins should be searched for in tissues that were known to require cytokinins for growth, and he accordingly added labeled benzyladenine to such cultures of soybean or tobacco. After 35 days' culture the tissues were extracted and compounds containing ^{14}C were searched for. A total of 40% of the ^{14}C added was recovered. Of that 40%, 15% was in "total RNA" if ring-labeled benzyladenine had been used. Soluble RNA fractions contained only 0.5–2.5% of the ^{14}C. Although the rationale of using cytokinin-requiring tissues is intelligent, 35 days of treatment is a painfully long time to wait before trying to find where the ^{14}C has gone. The radioactive label would be expected to have been transferred, in that long period, to many compounds other than the one or ones on which the primary action of cytokinin was based. And, of course, very little ^{14}C was found in the soluble RNA if its presence there is the basis for cytokinin activity. Kende has presented evidence against the hypothesis that kinins act by being built into transfer RNA, most notably in a paper on moss protonema (Brandes & Kende 1968). Benzyladenine causes moss protonema to start differentiating "buds" (Gorton & Eakin 1957) and Brandes and Kende found that the benzyladenine effect can be easily washed out of the protonema – certainly not expected if the cytokinin has been built into transfer RNA. On the other hand, mosses are so far removed evolutionarily from angiosperms that one should be skeptical about assuming exact parallels.

There has been controversy about whether cytokinins are directly incorporated as intact molecules into the transfer RNAs from which they can be obtained by acid hydrolysis and about whether they are active as cytokinins while incorporated in the transfer RNA. One of the recent papers from Skoog's lab (Walker et al. 1974) refers to earlier literature on this point and uses double labeling of benzyladenine to obtain evidence on the mode of incorporation. They added the benzyladenine to a tobacco callus that requires cytokinin for growth and harvested the tissues after 24 to 28 days (again a rather long period before checking). They found that a very small amount of labeled benzyladenine was incorporated intact into benzyladenosine that they recovered from transfer RNA hydrolysate but that the level of "benzyladenosine recoverable . . . is far lower than that of endogenous cytokinin-active ribonucleosides. Hence, any role that the incorporated [benzyladenosine] in tRNA might play in growth regulation would appear to be other than as a replacement for endogenous cytokinins in their known functions as constitutents of tRNA."

Techniques of cytokinin bioassay

As with auxins and gibberellins (see Chapter 7) a variety of bioassays for cytokinins have been proposed and used. The advantages and disadvantages are discussed in the reviews of Miller (1963), Letham (1967a, b), and Skoog and Armstrong (1970), The bioassays most often used measure the growth of aseptic tissue cultures or leaf growth and senescence effects. The tissue culture assays, whether based on callus from tobacco pith (e.g., Linsmaier & Skoog 1965) or from soybean (Miller 1963) or on little cylinders of secondary phloem from carrot (Letham 1967), have the major disadvantage of requiring 21–35 days' growth. In addition the first two require the callus stocks to be maintained in aseptic culture – a prime way of gobbling up time and space. However, the culture assays have the sizeable assets of responding to few or no substances other than cytokinins and of responding to very low concentrations of cytokinin. And their aseptic requirements, although obnoxiously time-devouring, do leave you with more assurance that your results are not due to bacterial interactions with the bioassay. I have used the soybean callus bioassay and have found it reliable and essentially as described; typical dose response curves for zeatin and kinetin are shown in Figure 6-3 (Jacobs 1976).

The chlorophyll retention effects of cytokinins are measured in Osborne and McCalla's (1961) bioassay. This has the great advantage of speed; only two days are required for the assay. Originally proposed for *Xanthium,* it has been used by Letham (1967) with several other genera. Osborne and McCalla noted that *Xanthium* responded to several compounds other than standard cytokinins (such as sugars, benzimidazole, and – to a slight extent – various purines). We have found this *Xanthium* chlorophyll retention bioassay to be easy and reliable.

The tissue culture assays as contrasted to antisenescence assays for cytokinins illustrate in an extreme form the principle that holds

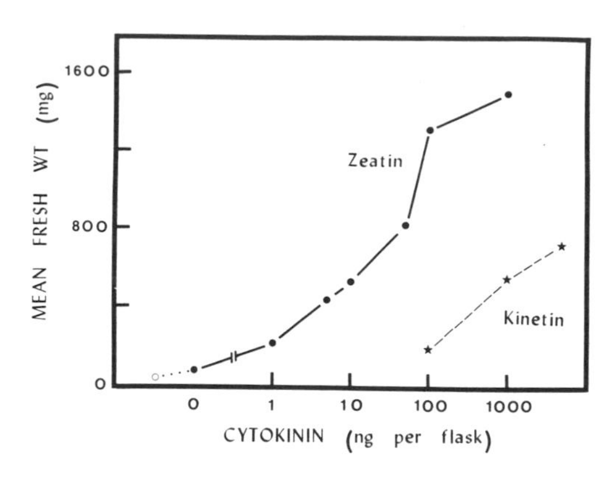

Figure 6-3. Dose-response curve of soybean callus culture to the cytokinins zeatin and kinetin (Jacobs 1976).

true for auxin and gibberellins – namely, that relative potency of hormones can vary drastically with the bioassay used. Zeatin is much more active than kinetin in callus-culture assays (Figure 6-3), but kinetin is often more active than zeatin in "leaf-senescence" assays (Letham 1967).

7

"Flowering hormones" and gibberellins

From the viewpoint of evolutionary history the angiosperms have apparently taken over most of the land surface of the globe from various less highly evolved groups of vascular plants and have even started to reinvade fresh water and the edges of the salt seas. The success of the 12,500 genera that constitute the angiosperms would be explained, according to typical evolutionary thinking, as being the result of clusters of mutations that in several different, unrelated ways gave the angiosperms selective advantages over the gymnosperms that preceded them in evolutionary history. The flower was presumably one of these evolutionary advantages, for it is so ubiquitous among the angiosperms that the "flowering plants" is frequently used as an alternate name for the group.

Flowers are efficient devices for ensuring sexual reproduction. Developmentally they result from the changed activity of the apical meristem of shoots. Instead of continuing to develop leaves, the meristem forms sex organs. But in the development of the "typical flower"[1] there is a gradual transition from making leaves to making sex organs (Figure 7-1). The first organs developed in this "typical flower" are the clearly leaflike sepals, which are green and flat. The next organs to develop are the petals: leaflike in their flatness but often with colors other than green (apt to be red if birds transfer the male sex cells, other colors if insects are the pollinators). A full transition to nonleaflike organs is made with the stamens (the male sex organs). It is only from the leaflike stamens in flowers of what are presumed to be the most primitive angiosperms that the evolution of the stamen from a leaf can be surmised (Bailey & Smith 1942). Finally, with the development of the carpel (the female sex organ in the center of the "typical" flower) the apical meristem completes

[1] The word "typical" means less than usual, of course, when applied to the angiosperm flower, because the flower has evolved into such diversified forms among the many angiosperm genera.

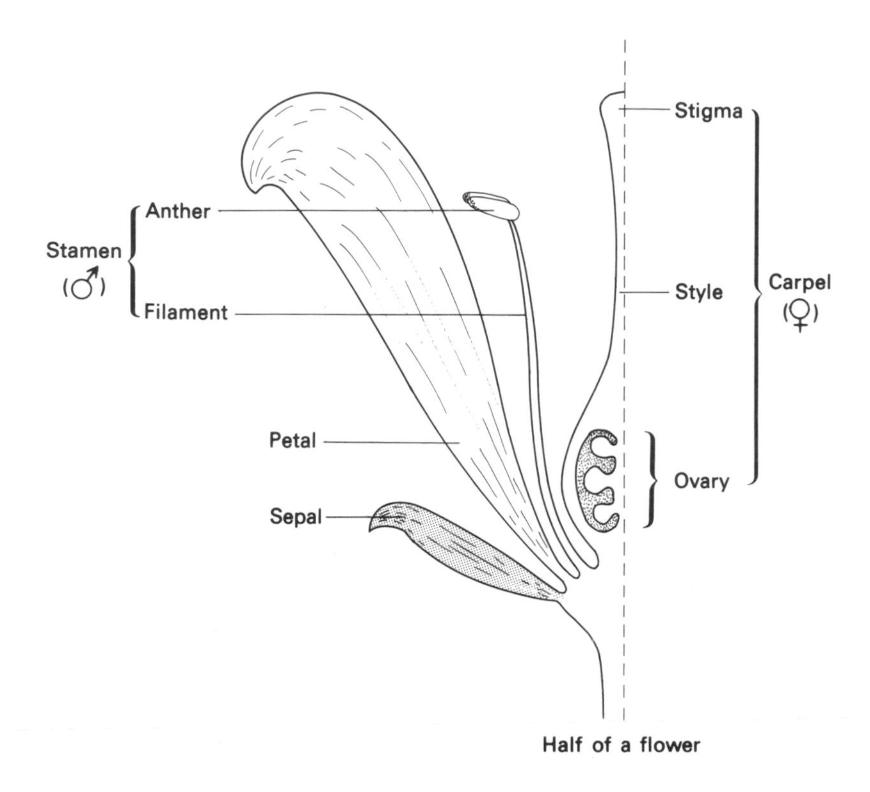

Figure 7-1. Diagram of half of a "typical" flower, showing the successive organs differentiated as the apical meristem develops sepals, petals, stamens, and carpels. (Only one organ of each type is shown, although there are usually several to many of the first three.)

both the typical flower and its own meristematic activity. Evidence that the carpel is also an evolutionarily modified leaf (that has folded over its edges so as to enclose and protect the spore cases) also comes from the flowers of some primitive angiosperm trees (Bailey & Swamy 1951).

Incredible floral diversity has evolved among the angiosperms. Flowers can be formed from the apical meristem of the main shoot (thus stopping its growth) or solely from the meristems of lateral buds (so the main shoot can continue its growth) or even from buds that grow out from the bark of the tree trunk (typically these are bat-pollinated flowers). Many genera have separate flowers developing in a clearly demarcated cluster along an axis, making what is called an inflorescence. The inflorescence may be so highly evolved that it looks like a single flower (as in the "flower" of a daisy, a dandelion, or a dogwood). Some of the angiosperms have, in the course of evolution, not only dropped the tree habit of their primitive angiosperm ancestors but eschewed the pollinating dependence on flying animals by reverting to wind pollination, with associated loss of the bright petal color that attracts us animals, winged or not. The grasses (such as *Avena* and *Zea*, whose coleoptiles have been so important in auxin investigations) are prime examples of such extreme evolution. Their flowers and inflorescences are so extremely specialized that a separate vocabulary has been invented to describe them.

Early research on factors controlling flowering

What controls the onset of flowering, this transition of an apical meristem from making leaves to making floral parts? This is a question of practical as well as theoretical interest. If the grower can cause all the plants in his crop to flower simultaneously, then the timing of fruit development will be so close that a single harvesting will usually be feasible. For many years, an increase in the "C/N ratio" (*C*arbohydrate/*N*itrogen) was pushed as the factor that edged plants over into flowering (Hicks 1928). In retrospect, this hypothesis seems to have been held more from desperation than from inspiration. "Ripeness to flower" was another hypothesis favored by some; flowering was believed to occur after a certain number of leaves had developed. In 1914 Tournois reported that decreasing the hours of light to which *Humulus* or *Cannabis* plants were exposed in 24-hour periods speeded flowering. But before he could exploit this discovery, he became one of the sacrifices to the much larger experiment in which French generals were testing whether *élan vital* was sufficient armament against German machine guns. Hence, the first sizeable break in understanding the control of flowering physiology was the discovery and elucidation of photoperiodism by Garner and Allard (1920 and many later papers).

The discovery of photoperiodism by Garner and Allard

Garner and Allard had been working on varieties of tobacco for the U.S. Department of Agriculture in its Maryland station (tobacco being a major crop in states of the Old South) and had come across a tobacco that did not flower even after a whole summer's growth. Because of its giant growth form, they named it Maryland Mammoth. After bringing it into the greenhouse, in the hopes of keeping it alive over the winter, they found to their surprise that Maryland Mammoth flowered. Step-by-step checking of the discernible differences between a summer field and a winter greenhouse (e.g., lowered light intensity, different temperatures) finally left them with *day length* (the number of hours of light in each 24-hour period) as the obvious remaining possibility. To their delight they found that day length did control flowering in Maryland Mammoth tobacco: plants grown in less than 12 hours of light/24 made the transition from vegetative to sexual development. If day lengths of more than 18 hours of light were given in each 24-hour period, the plants continued their vegetative development. Garner and Allard recognized the importance of this observation and, by pursuing the subject, discovered that developmental changes following changes in day length occurred quite generally. They gave the name "photoperiodism" to the response of an organism to the relative length of day and night. As their definition implies and as their own research to some extent demonstrated, photoperiodic treatments can affect a variety of developmental phenomena (e.g., leaf abscission, tuber formation, shoot growth). However, the overwhelming majority of subsequent plant research has been on photoperiodic control of flowering. Four years later, photoperiodic effects were reported in animals too (in aphids by Marcovitch [1924] and in birds by Rowan [1926]).

Types of photoperiodic responses

Garner and Allard discovered that there were short-day plants (SDP) which flowered when the days were "short," and long-day plants (LDP), which flowered when the days were "long," as well as many day-neutral plants (DNP), which were relatively unaffected by day length. (DNPs may show slightly faster flowering under long-day conditions as a result of more photosynthesis allowing faster vegetative growth.) Among both SDPs and LDPs, the preferred photoperiod may be an absolute requirement, with no flowering resulting in the unfavorable photoperiod even after several years, or the preferred photoperiod may merely cause somewhat faster flowering. The variety of *Xanthium* (cocklebur) most often used is a famous example of a SDP with an absolute requirement for SD. *Perilla* is one of the most thoroughly investigated SDPs with a quantitative response to photoperiod. That the photoperiod controls flowering in nature as well as in the greenhouse was indicated by the observation that SDPs typically are plants that flower in the autumn, whereas LDPs typically flower in the sum-

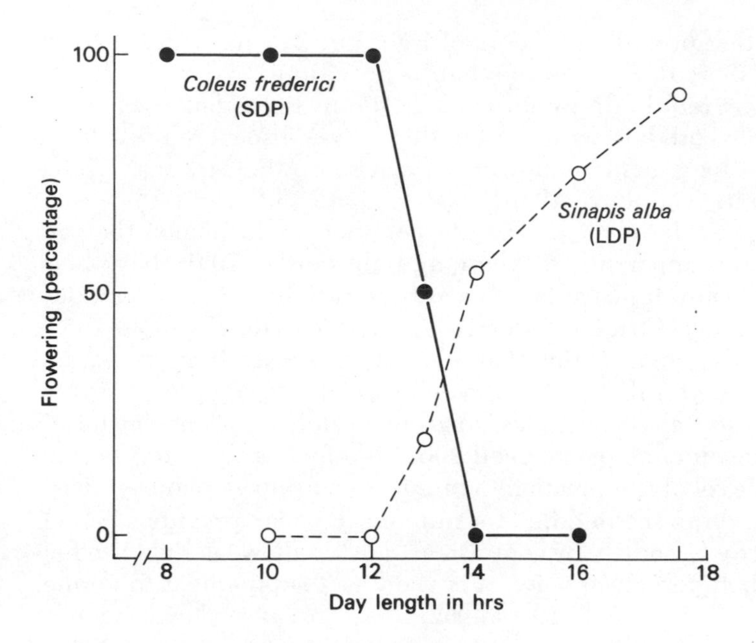

Figure 7-2. Demonstration of the critical day length for *Coleus frederici*, a SDP (after Halaban 1968), and for *Sinapis alba*, a LDP (after Bernier 1969).

mer, when the days are long (Garner & Allard 1923). More recently, plants have been found with a double photoperiodic requirement: there are LSDPs that require LDs followed by SDs, as with *Bryophyllum* (Dostál 1950; Resende 1952) or *Cestrum* (night-blooming jasmine, investigated by Sachs 1956), and SLDPs, the latter including some winter varieties of grasses that show faster floral development if SDs precede LDs (e.g., McKinney & Sando 1935).

The term "short-day plant" turned out to be somewhat of a misnomer on two counts. First, for strict SDPs flowering does not occur under photoperiods longer than the critical daylength, for LDPs it will not occur under photoperiods shorter (Figure 7-2). Some SDPs, in fact, have a longer critical day length than some LDPs. If one selects one's examples carefully, one can have a specific SDP being induced to flower in the same photoperiod that induces a specific LDP. For example, the most commonly used variety of *Xanthium* (an SDP) has a critical day length of 15.0–15.5 hours of light per 24 (Hamner & Bonner 1938), whereas the LDP *Hyoscyamus* has a critical day length of 10–11 hours (Lang & Melchers 1943). The second reason for considering SDP a misnomer is that it was realized that providing 8 hours of light per 24 (the typical experimental "short day") was inevitably providing 16 hours of dark. Which was required? By growing SDPs in light–dark cycles that combined short days with short nights, it was discovered that SDPs need the "long night" to flower. In that sense they are really long-night plants (Garner & Allard 1931).

In contrast, however, LDPs are not short-night plants; their requirement is apparently for the long light period. LDPs have been reported to flower just as fast in continuous light as in any long-day, short-night cycle tried (Naylor 1941, Snyder 1948, Takimoto 1955).

Obviously, angiosperms that are not in the seedling stage, and thus growing on the food reserves in the seed, need to photosynthesize to stay alive, much less to grow. Merely to balance materials used in respiration, plants need 200–1700 foot-candles and require a higher level of illumination to provide a surplus of photosynthate. Full sunlight in the middle of a summer day may provide as much as 10,000 foot-candles. By contrast, astonishingly weak light is effective in photoperiodism when it is used as a supplement to normal daylight. As little as 5–10 foot-candles can cause flowering in a LDP, when it is given as an 8-hour supplement to 8 hours of daylight (for a 16:*8* period of light–*d*ark).

But, as Katunskij (1936) first discovered for a LDP and as Hamner and Bonner (1938) reported, apparently independently, for a SDP, a short light break at or near the middle of a long night can be just as effective as 8 hours of supplemental light for these photoperiodic phenomena.

Commercial flower growers use these short light breaks in the middle of a long night as a cheaper method of controlling the onset of flowering than burning light bulbs for 8 hours to give "true long days." It is unfortunate that more researchers on the physiology of flowering do not use light breaks. Many papers in the literature purport to study photoperiodic effects but actually compare the effects of SD (8:*16*) cycles with LD (16:*8*) cycles. Such an experiment ineluctably confounds true photoperiodic effects with the effect of all the photosynthesis that can occur during that extra 8 hours of light in the 16:*8* cycles. This is not merely a hypothetical difficulty. Several researchers have reported differences in chemical composition of plants grown under SD as contrasted to those under LD only to discover later that the effects disappeared when flowering was controlled by light-break treatment instead (e.g., Taylor 1965, Jacobs 1978a). Hence, routine use of a short, low-intensity light break in the middle of a long dark period should be used instead of the conventional true long day, and the true short day should have added to one end of its high-intensity period a low-intensity light equivalent in duration and intensity to the light break. Such an arrangement provides exactly the same total light per 24-hour cycle and precludes confusing photosynthetic effects with photoperiodic ones.

The hypothesis that photoinduced leaves produce a flowering hormone

By masking various aboveground organs of plants and exposing other organs to photoperiods known to induce flowering when the whole shoot was exposed, researchers found that leaves were apparently the perceptive organs (Knott 1934, and many later authors), a rather unstartling finding in view of the many and obvious adaptations of leaves for receiving light. Hamner and Bonner (1938) found that SD treatment of a single leaf could induce flowering in *Xanthium*, even though all the other leaves and the stem were being exposed to LD. In the climate of opinion of the 1930s it was natural to interpret these results as the action of a hormone produced in the SD-treated leaf and thence transported to the apical meristem where the hormone caused the meristem to stop making leaves and to start making flowers. Chailakhyan formulated this hypothesis and coined the term "florigen" for the presumed flowering hormone (1937).

The photoinduced leaf is permanently changed, judging by grafting experiments on the most thoroughly investigated genera. If a single leaf of SDP *Xanthium* was photoinduced while on the plant (i.e., exposed to short days until the plant flowered), then that leaf was cut off and grafted to a *Xanthium* plant permanently kept in the noninductive LD conditions, the grafted leaf would induce flowering in the host plant (Lona 1946, Thurlow 1948). The standard florigen interpretation was that the induced leaf continued to produce florigen even after it was grafted and had therefore been placed in LD conditions. There are many such cases in the literature, including LDPs as well as other SDPs, since transmission by grafts was first shown in 1936 (Chailakhyan; Kuyper & Wiersum). Such a photoinduced leaf of *Perilla* not only can induce the first *Perilla* plant it is grafted onto but can induce successive *Perilla* hosts if it is cut off each host and regrafted (Zeevaart 1958). Its inductive ability apparently remains with it until it senesces and dies. An induced *Xanthium* leaf also shows such evidence of a permanent change. Induced *Xanthium* leaves, however, have the further unusual property of being able to "infect with flowering" the noninduced leaves of the hosts. After *Xanthium* is flowering as the result of such graft induction, any other leaf from the nonflowering host can be excised and transmit flowering to a new host plant even though neither this second leaf nor the second host has been exposed to anything but LDs. Such indirect inductions have been carried out for seven successive grafts with no reported sign of diminishing inductive power (Lona 1946; Thurlow 1948). *Silene* (LDP) also shows indirect induction like *Xanthium*'s, although the literature does not give the impression that widespread searches have been pushed (Wellensiek 1969).

Such grafting experiments indicate that, once it is photoinduced, the leaf is a self-contained inductive unit. But can isolation be carried further, that is, does the leaf need to interact with the rest of the plant to become photoinduced? For the SDP *Perilla* isolation of photoinduction can occur to the extent that a rooted leaf cutting can be induced; the roots can then be cut off and the leaf grafted onto a *Perilla* plant under LD with subsequent flowering on the host (Lona 1949a; Zeevaart 1958). Hence, in this case the only possible interaction would be with the roots on the petiole base. *Xanthium* leaves could not be photoinduced as excised leaves unless the lateral bud was also present (Lona 1949a; Carr 1953; Zeevaart 1958).

In the 40 years or so since the florigen hypothesis was advanced, hundreds of extracts of flowering plants have been tested for flower-forming activity on plants in noninductive photoperiods, and more hundreds of chemicals have been added from the chemical shelf in the hope of stumbling on an active compound. There has been painfully little success. Let us review the development of ideas about florigen, the critical experimental results, and the modified hypotheses that resulted. Most experiments have been done with plants whose flowering can be controlled by photoperiod, although some researchers have concentrated on plants that require a cold treatment instead. ("Vernalization" is a term to describe these cold treatments, often weeks long, that some plants require for flower development. In nature, such plants would be vernalized by living through the winter, proceeding to flower only during their second year of life.)

Evidence favoring the florigen hypothesis

Several lines of evidence led researchers to think that the flowering hormone of short-day, day-neutral, and long-day plants was the same. Grafting experiments were particularly convincing. Several cases have been reported where day-neutral donors grafted to SDPs held under LD induced flowering in the host. This not only happened between varieties of the same species (e.g., *Xanthium*, soybean, and tobacco) but even between different species of a genus (*Nicotiana tabacum* and *N. sylvestris* [LDP]). And several different laboratories have reported that induced donors of one photoperiodically sensitive genus can induce flowering in plants of a different genus with the opposite day-length requirement. For instance, LDP *Hyoscyamus* can induce SDP Maryland Mammoth tobacco when the tobacco is kept in LD conditions (Melchers & Lang 1941). The LDP *Sedum* can similarly induce SDP *Kalanchoe* (Carr & Melchers 1954; Zeevaart 1958), and any one of three other LDP Compositae genera can induce SDP *Xanthium* (Okuda 1954). Such graft transmissions were interpreted as showing that DNPs, LDPs, and SDPs all made the same flowering hormone and that the florigen was not species-specific or even genus-specific.

In addition to the evidence from grafting, other facts supported the view that SDPs and LDPs must be very similar. Nullification of the effect of a long night shows the same action spectrum in LDPs and SDPs (Borthwick et al. 1950). Opposite reactions to photoperiod have often been found between varieties of the same species, indicating that photoperiodic reaction might be controlled by a small number of genes. In fact, the first SDP discovered, Maryland Mammoth variety of tobacco, is an SDP due to a single recessive gene (Lang 1942). Several other cases have also been found where a single gene controls the photoperiodic reaction (Bremer 1931, for lettuce; Little & Kantor 1941, for sweet pea; Laibach 1947, for *Coleus frederici*). And even though other plants are known in which three or more genes control flowering (e.g., Cooper 1954, for *Lolium*), the fact that, at least in some cases, one gene can change the reaction of photoperiod supports the view that SDPs and LDPs can be very similar.

Can florigen from a photoinduced leaf cross a diffusion gap? Apparently giving way too easily to an expectation that florigen would act like auxin, the only previously discovered plant hormone, Hamner and Bonner (1938) concluded that an induced *Xanthium* plant, grafted to a noninduced plant, could cause flowering even before the cells grew together through an interposed piece of lens paper. Several sharp-eyed readers noticed, however, that Hamner and Bonner had checked tissue contact only up to 14 days, whereas they left the grafted plants in contact for as much as 33 days before checking floral induction. When tissue contact was checked during the entire graft-induction period, it was found that floral induction was transmitted only after the grafted cells had grown together with the cells of the host (Withrow & Withrow 1943). No transmission across a diffusion or water gap has ever been found (e.g., Galston 1949a).

There are a few cases where people have reported plant extracts that would cause flowering in short-day plants. Those by Roberts (e.g., 1951) can be ignored; from critical discussions it is clear that he did not even understand what adequate controls should be, much less use them. Bonner and D. Bonner (1948) in a sad little note reported that the extract of one particular inflorescence of a *Washingtonia* palm tree caused intiation of floral primordia in *Xanthium* kept in a 20:4 long day, but they admitted that the result was not reproducible with any other tree.

Is florigen an "antiauxin"?

With this depressing lack of success in finding an active extract or a floral-inducing chemical off the shelf, it was natural to turn to hypotheses other than florigen for an explanation of flowering. In most, if not all, cases where physiological experiments can be interpreted as showing the action of a stimulator, one can just as rationally interpret the findings as being the result of "release from inhibition." Instead of a florigen acting to stimulate flowering directly, the plant may be ready to flower whenever some inhibitor of flowering, present in the vegetative phase, is counteracted. This switch in viewpoint became popular in the early 1950s (e.g., Bonner & Thurlow 1949; Harder & van Senden 1949; von Denffer 1950). They proposed as the vegetative inhibitor the only hormone well known at that time, namely, auxin. Dostál and Hosek (1937) had already demonstrated that added auxin could inhibit flowering of the LDP *Circaea*.

A staggering variety of tests was run, adding auxins of different chemical formulations or adding chemicals believed to affect the levels of endogenous auxin (and hoped to be specific). Surprisingly, there were relatively few examinations of changes in endogenous levels of auxin that accompanied photoinduction. Partly this was because it is so much easier to splash on a solution of synthetic auxin than to extract and bioassay endogenous auxin, but it was also partly the result of special difficulties from inhibitors that researchers encountered when they tried to bioassay auxin extracts of such famous SDPs as *Xanthium* (e.g., Bonner & Thurlow 1949; Bonde 1953). Of course, the investigations were not helped by the fact that results due to photoperiod were often confounded (in the statistical sense) with those due to 8 or so hours of extra illumination. The results of this work were summarized by Lang (1961). The general conclusion was that release from auxin inhibition was not the prime controlling mechanism for flowering, although auxin could, of course, modify the flowering process. In a few cases, soon shown to be unusual, auxin *could* initiate flowering (pineapple and litchi); auxin sprays are now a standard method for inducing flowering (and thereby fruiting) in field cultivation of pineapple.

The discovery of gibberellins and their effect on stem elongation

With physiologists at an impasse in investigating photoperiodic control of flowering, help came from an unexpected quarter. Phinney had been investigating the physiological basis of dwarfing in *Zea mays* L. Among the hundreds of mutants known in *Zea* are more than 20 that cause dwarf growth. The stem of the mutants is only about half as long as normal at maturity; the normal number of leaves develop, but average cell length is half that of the normal plants. Nine of these dwarf mutants have been shown to be from single-gene mutations, all recessive and nonallelic. Following the Beadle–Tatum hypothesis that one gene controlled one enzyme, Phinney was trying to discover if one growth-limiting chemical had its production curtailed in these single-gene dwarf mutants.

In the 1930s it had been asked whether auxin – that being the only growth hormone then known in plants – was depleted in the dwarf forms, thereby causing the dwarfing. Van Overbeek (1935, 1938) demonstrated that less auxin was indeed obtained by diffusion from the coleoptiles of several dwarf corns than from normal coleoptiles. However, many years later Phinney and his co-workers tried adding IAA to dwarf corn to see if growth could be brought up to the normal level by substituting the hormone hypothesized to be in short supply due to the mutation. No amount of added IAA could do so (Harris 1953). Apparently, therefore, the plants were producing less auxin because they were not growing, rather than vice versa.

At about this time the Western world "discovered" gibberellins. Since the turn of the century, a disease of rice plants had been studied in Japan. It was the result of infection by the fungus *Gibberella*. The active chemical, which caused excessive elongation of the rice plant, was isolated in the late 1930s. It was very active; 1 mg/liter gave great stimulation of rice growth. It is called gibberellic acid (GA-3). Its structure is shown in Figure 7-3. "Gibberellin" (GA) is a more general term for the class of chemicals.

Although many papers on gibberellins were published over the years by Japanese scientists, no one in Europe or America paid any

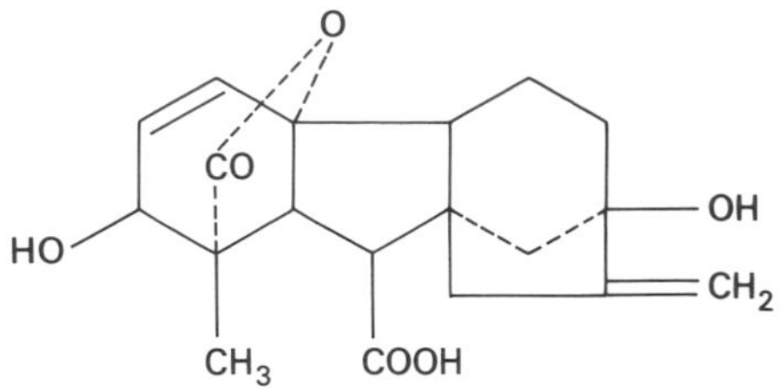

Gibberellic acid (GA–3)

Figure 7-3. The chemical structure of gibberellic acid (GA-3).

attention to them – partly because publication was usually in Japanese, partly because World War II stopped communication with Japan. However, in the early 1950s the British chemical company I.C.I. and a U.S. government laboratory started programs to isolate the active compounds.

The first report from the Western world was that by Brian and Hemming (1955), who were working for I.C.I.: a few micrograms of GA-3 added to dwarf pea plants gave greater elongation than GA-3 added to normal tall forms of pea. Phinney quickly followed suit and applied GA-3 to his single-gene dwarf mutants of corn. In five of the mutants he could fully restore normal growth by adding a few micrograms of GA-3 to the shoot tips every few days (1956). If he stopped adding the GA-3, the growth rate declined once more. A particularly nice feature of his results was that the minimum amount of GA-3 that would restore normal growth of the dwarf plants had no effect on normal corn. (However, if he boosted the GA-3 level one-hundred fold, then even the normal plants would show increased growth from the added GA-3.) The obvious conclusion was that something like GA-3 (which had been isolated from the fungus) must be needed for growth of these angiosperms. And it was likely that these five single-gene mutations caused dwarfing in corn by somehow reducing the level of a native gibberellin. This was further supported by study of the gibberellin endogenous in corn. Extracts of both corn seeds (Phinney et al. 1957) and normal corn shoots (Phinney 1961) contained gibberellin, judging by activity in a bioassay. The five mutants that responded to added GA-3 were said to contain either less than half as much total gibberellin as the normal (d_1 and d_2 mutants) or no detectable gibberellin (d_3, d_5 and an_1 mutants) (Phinney 1961). Two zones of gibberellin activity were found in chromatograms of extracts of normal corn shoots, one being close to the Rf of GA-3.

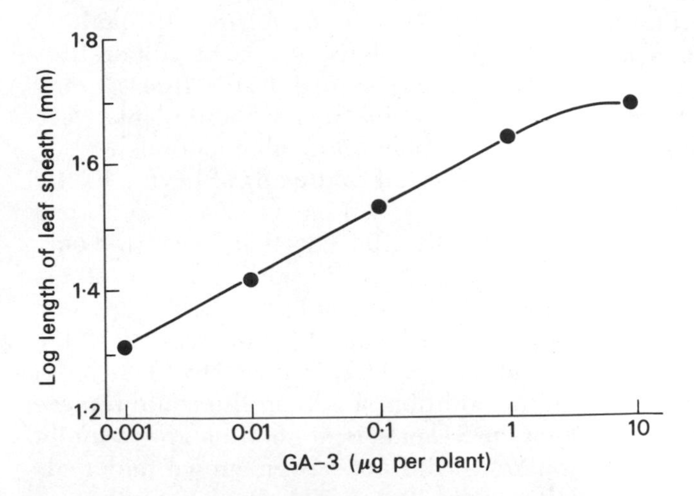

Figure 7-4. The dose-response curve of leaf length of dwarf *Zea* d_1 to gibberellic acid (after Phinney 1961).

The response of the corn mutants to added GA-3 is so regular that it has been made into one of the most useful bioassays for gibberellins. The average length of the first leaf sheath of dwarf d_1 shows a linear response to added GA-3 over the range of 0.001 to 1.0 μg per plant when the logarithm of length is plotted against the logarithm of dose (Figure 7-4).

Gibberellin's effect on plants sensitive to photoperiod

The stimulation of stem elongation by GA-3 in the dwarf forms of pea and corn inevitably pushed botanists into looking for similar situations. Lang, a colleague of Phinney's at the University of California at Los Angeles, was one of the first to do so. As a long-time investigator of flowering, he was, of course, aware of the fact that most LDPs show striking elongation of the stem after photoinduction (Lang 1957) – so striking, in fact, that it is called *bolting*. Under SD such plants develop such extremely short internodes that the growth form is called a *rosette*.

For those interested in the physiology of flowering, the striking stem elongation caused by gibberellin in dwarf pea and *Zea* suggested that gibberellin might be involved in the bolting of the rosette LDPs. Three laboratories leapt to this conclusion. Lang, who had the advantage of having as colleagues both Phinney and West (the latter a major investigator of the chemistry of gibberellins), rushed out several fast notes in *Naturwissenschaften* (1956a, b, c) and *Proc. Nat. Acad. Sci.* (1957): an impure mix of gibberellins caused bolting and flowering in the annual LDPs *Hyoscyamus niger, Crepis, Samolus,* and *Silene* even though these plants were kept in clearly noninductive conditions [9:*15* photoperiods].

This startling success could have led to dreams of having finally found the flowering hormone. However, Lona (1956a, b) and Lona and Bocchi (1956) reported that of the 10 LDPs that they tested only three flowered from the added gibberellin, although five others showed increased elongation of the stem. Particularly important was the work of Bünsow and Harder (1956). Working with an LSDP *Bryophyllum* (i.e., a plant that required LD followed by SD in order to flower), they found that the addition of gibberellin could replace the LD requirement but not the SD one. *Bryophyllum* grown under LD and treated with gibberellin did not flower; grown under SD and treated with gibberellin it did flower. This has been fully confirmed (Penner 1960; Zeevaart 1969).

In *Coreopsis grandiflora,* a plant with dual requirement for day length but with the sequence reversed from that required in Harder's *Bryophyllum,* GA-3 again could replace the LD but not the SD requirement for flowering (Ketellapper & Barbaro 1966). In another SLDP Wellensiek (1960) found elongation but no effect on flowering from added GA.

One can always argue – and frequently does, if a favorite hypothesis is being undermined – that the case that does not fit is an exception. However, for those who wish to believe that gibberellin is the flowering hormone, aside from *Bryophyllum* and *Coreopsis,* it is difficult to explain away the fact that SDPs cannot be made to flower in LD by the addition of gibberellin (*Xanthium* by Lona 1956a; Lang 1957; Biloxi soybean by Lang 1957 and Chailakhyan 1957; Maryland Mammoth tobacco by Chailakhyan 1957; *Perilla* by Lona and Chailakhyan; *Kalanchoe* by Harder & Bünsow 1958; and many others). The sole exceptions uncovered so far are the SDP *Impatiens balsamina* (Nanda et al. 1967) and the SLDP *Scabiosa* reported by Chouard (1957) to flower on LD when GA is added. If there is a single flowering hormone, non-species-specific, as many grafting experiments suggest, then gibberellin cannot be it. Although gibberellin causes flowering in more species than any other chemical known (Table 7-1), its general inability to substitute for a SD requirement, as well as its ineffectiveness with LDPs that have a nonrosette shoot, is strong evidence against gibberellin being florigen. In quite a few cases, particularly in woody perennials such as

Table 7-1. List of genera of which at least one species has had its flowering initiated by the application of gibberellin(s)

Long-day plants (in short days)		
Anethum	Lactuca	Polemonium
Arabidopsis	Lapsana	Raphanus
Blitum	Lolium	Rudbeckia
Brassica	Myosurus	Samolus
Chrysanthemum	Nicotiana	Silene
Cichorium	Papaver	Sinapis
Crepis	Petunia	Spinacia
Hyoscyamus	Phaseolus	
Long-Short-day plants (in short days)		
Bryophyllum		
Short-Long-day plants		
Coreopsis (in short days)		
Scabiosa (in long days)		
Short-day plants (in long days)		
Impatiens		

Source: Summary tables in Lang 1965, Vince-Prue 1975.

Citrus, GA-3 actually inhibits flowering (see Goldschmidt & Monselise, 1972, for a list of examples).

From the work on photoperiodically sensitive plants, the primary effect of added gibberellin seems to be to cause stem elongation. In some LDPs that elongation can, in turn, trigger flowering. (Even in the first spate of fast publications, Lang [1957] pointed out that LDPs treated with exogenous gibberellin showed stem elongation before floral development, whereas photoperiodically induced plants seemed to have the two phenomena occurring together.)

Endogenous gibberellins and their activity

With this gibberellin from a fungus having such impressive action on so many angiosperms, various laboratories searched for gibberellins or gibberellinlike substances native to angiosperms. Repeating the earlier history of IAA, which was isolated from a fungus (Thimann 1935a) before it was found in angiosperms, GA was found in extracts of higher plants by bioassays (Radley 1956, 1958, 1959;

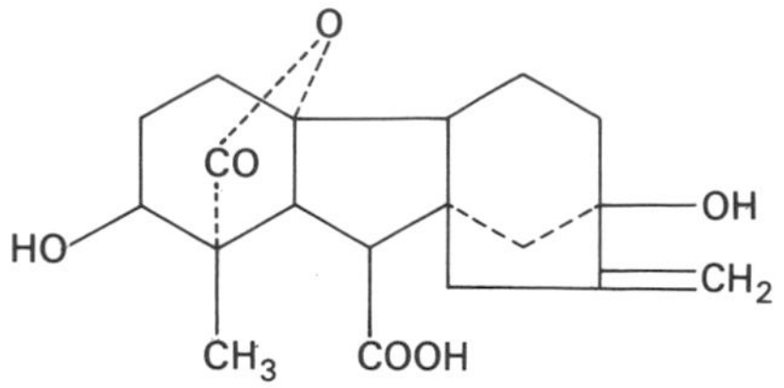

Figure 7-5. The chemical structure of a few of the more than 50 gibberellins isolated from plants.

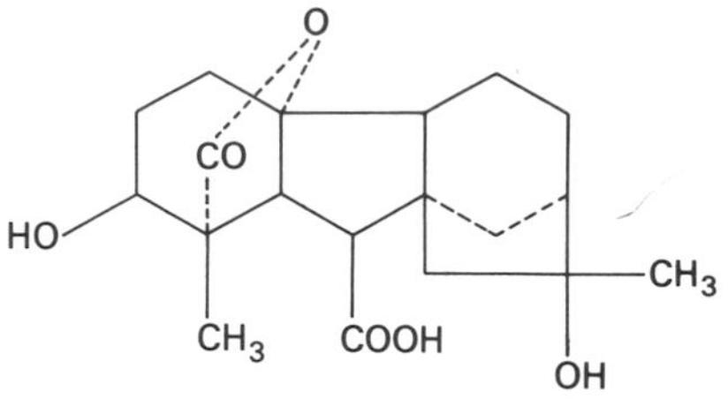

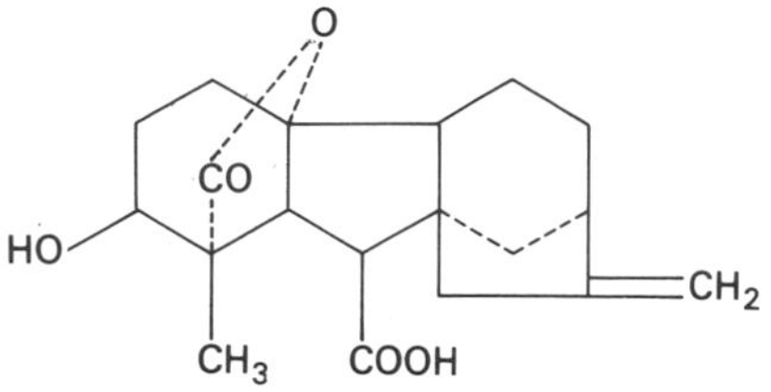

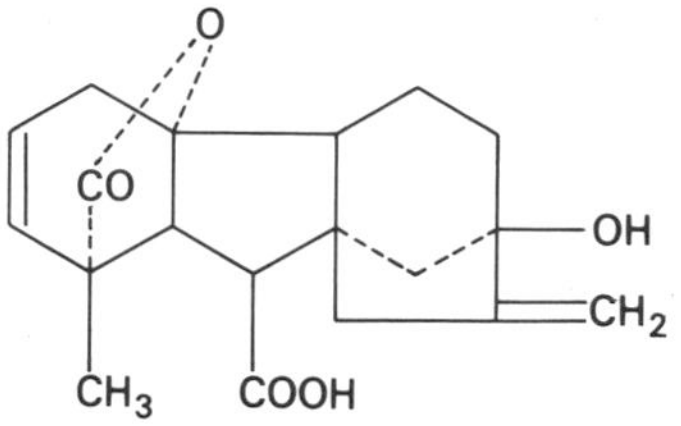

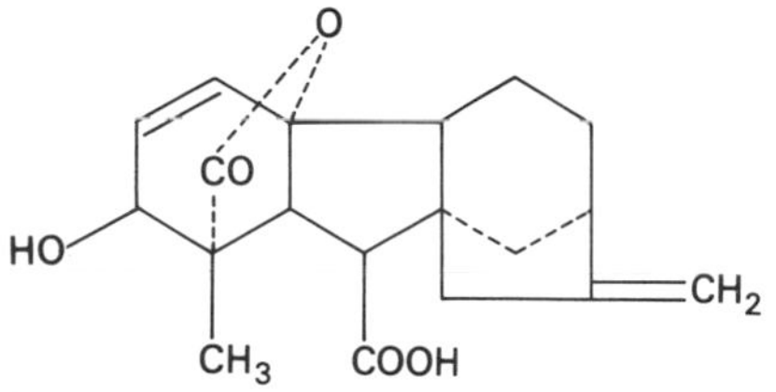

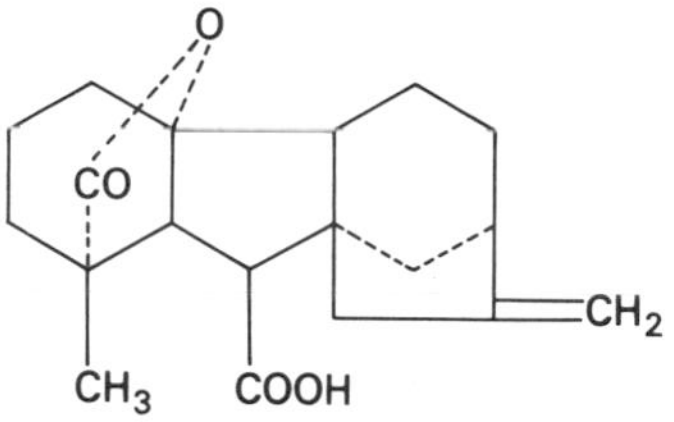

Radley & Dear 1958; Phinney et al. 1957). The early work on the endogenous gibberellins of angiosperms paralleled the earlier work with auxins: extracts were typically run on paper chromatograms and various zones were then tested for bioactivity. However, quite quickly an important difference became apparent between the two classes of hormone. In contrast to the ubiquity of IAA as the major endogenous auxin of angiosperms, an astonishing array of gibberellins was found, many in the same plant (Figure 7-5). More than 50 had been identified by 1977. Gibberellic acid (GA-3) is the main gibberellin found in the fungal medium and is the only gibberellin widely available commercially. All gibberellins are based on the gibbane skeleton, and differences can be so slight as to involve shifts in the position of hydroxyl groups or saturation of double bonds in the ring (e.g., GA-1 compared to GA-3). Even aside from the vexing and important question as to what the plant is doing with all these gibberellins, the multiplicity of gibberellins poses difficult technical problems. First, a chromatogram will be of little help in identifying unknown gibberellins. There is not much point in running one-dimensional chromatograms of a plant extract, then saying that there is bioactivity at the same Rf as GA-3, when more than 40 other gibberellins might be in the extract too. Gas chromatography combined with mass spectrometry has been advocated as a more secure method of identification (MacMillan 1972), but the cost of GC-MS is so great that only a few laboratories in the world can afford it (e.g., MacMillan at the University of Bristol). Clearly, MacMillan is going to be coauthor of many papers.

Problems of interpretation are also posed by the multiplicity of endogenous gibberellins. Since one can usually buy nothing but GA-3, that is the gibberellin typically added to plants. But are negative results with added GA-3 merely due to the fact that GA-3 is not one of the endogenous gibberellins of that particular plant? Evidence that this was not solely a theoretical difficulty was found with even the first four gibberellins available (e.g., Bukovac & Wittwer 1961; Phinney 1961; Halevy & Cathey 1960; Hashimoto & Yamaki 1960; Lockhart & Deal 1960). More detailed evidence was presented by Wittwer and Bukovac (1962) and Michniewicz and Lang (1962), who used gifts of nine gibberellins to test the reactions of several plants to all of them. Some of Wittwer and Bukovac's data, with statistical significance determined by Duncan's multiple-range test, are given in Table 7-2. The much greater effect of GA-4, GA-7, and GA-9 than of GA-1 and GA-5 on the elongation of cucumber hypocotyls is striking. This was confirmed for other varieties of cucumber and for other species of the Cucurbitaceae family.

Gibberellins 1–9 were tested for bolting and flowering effects on more plants by Michniewicz and Lang (1962), although the amounts of the various gibberellins given to them were so small that only preliminary data on elongation could be provided. Lang (1957) had already made the important discovery that gibberellin (applied as a mixture of GA-3 and either GA-1 or an unknown contaminant) could replace the cold requirement for flowering in several biennial species growing in LD. Michniewicz and Lang now found that some of the added gibberellins could substitute for the cold requirement in two more such species (*Myosotis* and *Centaurium*), or for the LD requirement in the LDPs *Crepis* and *Samolus*, or the LSDP *Bryophyllum*. Again, however, the nine GAs differed con-

Table 7-2. The relative effectiveness of nine different gibberellins in causing stem elongation in pea, beans, and cucumber[a]

	No Gibb.	Gibberellin added								
Dwarf pea		A-1	A-4	A-3	A-7	A-5	A-2	A-6	A-8	A-9
	3.1	15	14	13	13	12	11	9.6	8.4	6.9
Phaseolus		A-4	A-3	A-5	A-6	A-1	A-7	A-9	A-2	A-8
	1.7	21	19	18	16	15	15	12	7.9	1.7
Cucumber		A-9	A-4	A-7	A-3	A-2	A-1	A-5	A-6	A-8
	0.3	4.5	4.4	4.2	1.5	1.3	0.9	0.7	0.6	0.5

Source: Wittwer & Bukovac 1962.
[a] Stem elongation is given in centimeters. Values connected by the same underlining are not significantly different statistically.

siderably in their flower-forming effectiveness and the order of effectiveness was not the same in all five plants. They found GA-7 to be the most effective floral inducer in two of the five species. Elongation of the stem typically occurred when one of the gibberellins was effective in causing flowering, but in most cases GA resulted in more elongation than did induction by photoperiod or cold treatment – fitting the view that elongation is the primary effect of gibberellins.

GA-1–27 were tested in nine bioassays (Crozier et al. 1970) with similar variation in the effectiveness of the different gibberellins.

With 50 or more gibberellins identified by 1977 in extracts of various plants, how many of these gibberellins is the investigator likely to find in any one plant? Seeds have been favored material for chemists specializing in the isolation and identification of gibberellins, as was true for those trying to identify the native auxins. (Seeds can provide large amounts of tissue without the expense of actually growing plants. The disadvantage, seldom mentioned, is that a seed is not apt to be a hotbed of hormonal activity.) Eight gibberellins have been identified in extracts of seeds of *Phaseolus coccineus* (GA-1, 4, 5, 6, 8, 17, 19, and 20) (MacMillan & Pryce 1968). How typical in this respect seeds are of plant organs in general is, of course, unknown – even aside from the question of how typical *P. coccineus* is of the 170,000 or so other species in the angiosperms. We can take a bit of solace perhaps from this slight evidence that not all the gibberellins are apt to be in one organ. As hormone physiologists we can take more solace from the fact that there is no evidence that the overwhelming majority of the known gibberellins are actually hormones in the sense that they move around the plant.

This problem of identifying the native gibberellins of a particular plant might explain why exogenously added GA-3 has not caused flowering in SDPs held in LD (with *Impatiens* the only known exception) or even in most of the nonrosette LDPs that have been tested.

Parallel variation between endogenous GA and flowering

Aside from that problem, is there evidence that endogenous gibberellins show parallel variation with a cold- or photoinduced flowering condition? Several laboratories have reported that they do.

Harada and Nitsch (1959a) extracted the "shoot apices with the young leaves enclosing them" of plants that had been photo- or cold-induced for successive weeks. In the rosette LDP *Rudbeckia speciosa*, treatment with continuous 18:6 photoperiods resulted in the appearance on the chromatograms of a new zone ("*E*") with gibberellin activity in several bioassays (Harada 1960). This activity first appeared in the third week's collection, with bolting not becoming striking until the next week (Figure 7-6). The activity declined for the next 3 weeks. After 8 weeks of continuous LD a second period of fast elongation occurred as the pedicel of the inflorescence grew: preceding that by 1 week was another increase in the bioassay activity of that same chromatogram zone. The cold-requiring *Chrysanthemum morifolium* Ram. var. Shuokan gave extracts with sizeable bioassay activity at similar locations on the chromatograms. Substance E, the new zone of gibberellin activity found in photoinduced *Rudbeckia,* was present in low amounts even in Shuokan plants not given the cold treatment but increased markedly after 3 weeks of cold treatment – the minimum period required to cold-induce Shuokan, according to the authors. The fact

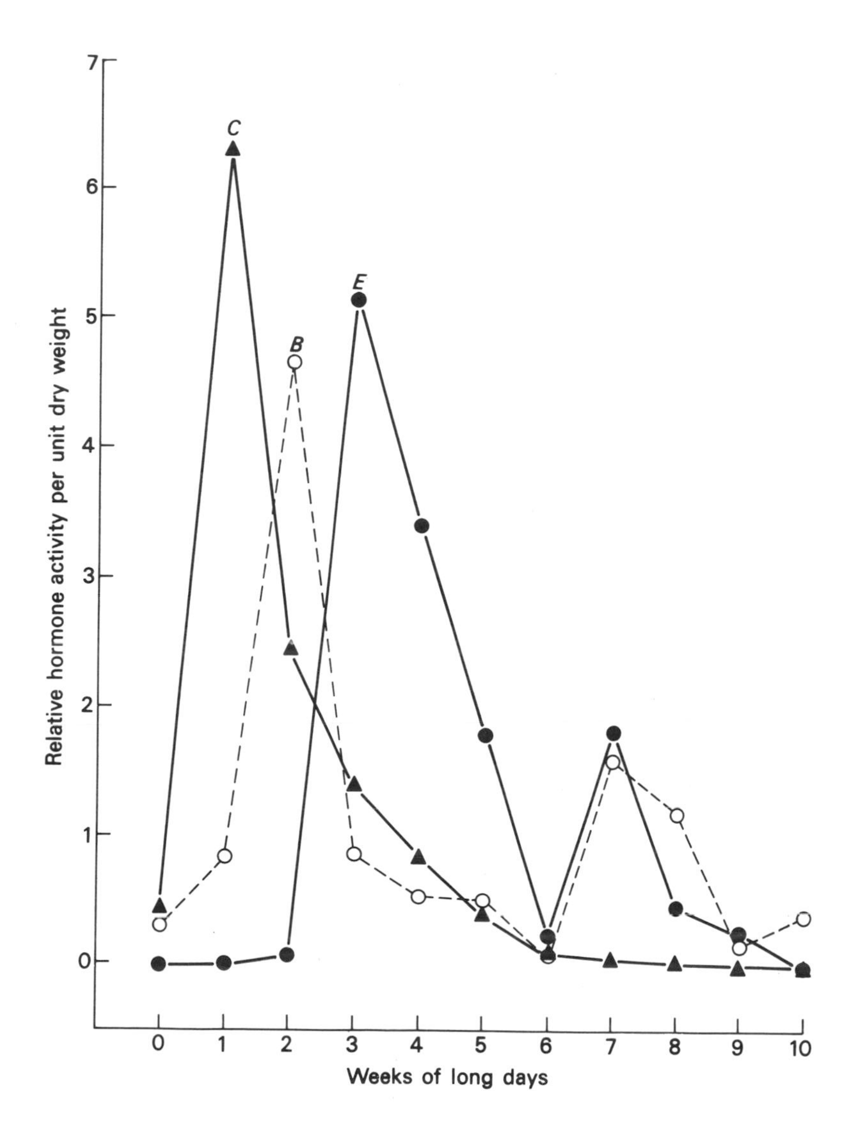

Figure 7-6. The time course of appearance of three different zones (*C*, *B*, and *E*) of gibberellin-auxin activity in chromatographed extracts of LDP *Rudbeckia* treated with successive weeks of long-day photoperiods (after Harada & Nitsch 1959a). The coordinates of the *Y*-axis differ for each of the three zones. Because the activity at the E zone appeared only one week before bolting occurred, Harada and Nitsch considered E likely to be the substance causing bolting.

that some Substance E was present even in the plants not treated with cold seemed to them to fit the fact that "after a very long time" Shuokan plants would bolt and flower even if kept always above 16°C. If there is one florigen and Substance E is it, then photoinduction of a SDP should cause more of this substance too. With this in mind Harada and Nitsch tested the SDP Shasta variety of *Chrysanthemum*, but unfortunately they found so much inhibitory activity in the chromatogram zones where Substance E was expected that no sign of Substance E could be seen in either SD- or LD-treated plants.

The Substance E zone of chromatographed extracts of *Rudbeckia* was eluted and added to *Rudbeckia* plants kept vegetative with 12:*12* cycles. Seven of the eight treated plants showed stem elongation and six of the elongated seven were in flower by the end of the experiment. None of the controls bolted or flowered (Harada & Nitsch 1959b).

To identify Substance E Harada and Nitsch (1961) switched to a cold-requiring biennial variety of *Althaea rosea* (hollyhock), chromatographed extracts of which revealed the presence of Substance E (operationally this means that bioassay activity occurred at the same general zone of chromatograms at which such activity had appeared in chromatographed extracts of *Rudbeckia* or Shuokan *Chrysanthemum*). Activity in this zone was low in extracts of *Althaea* not given cold treatment but increased substantially after a winter of cold. About 30,000 "apices" (probably apical buds) were harvested in the spring, after the winter cold but before bolting had occurred (Harada & Nitsch 1967). From the purified extracts of these shoot tips GA-1, GA-3, and GA-9 were identified by a variety of criteria including mass spectrometry. Unfortunately, the relation of these chemicals to the zone of gibberellin activity earlier called Substance E is not known, although there seems to be a silent implication that these three gibberellins are what gave the gibberellin activity of Substance E. The glucoside of GA-8 was later identified in such extracts also (Harada & Yokota 1970).

Radley (1963) reported that on the first day after she switched her culture of spinach (a rosette LDP, according to the earlier literature) to LD [24:*0*] from SD [8:*16*], there was a big increase in total GA activity in the shoot – 10 to 30 times as much, according to her bioassay data. By the second LD the level had declined again. However, in a later note (1970) she admitted that the results were not consistently reproducible. Zeevaart (1971a), in one of his typically thorough papers, reinvestigated spinach and found no such big spurt of total GA activity after a single LD cycle (actually 24:*0*),

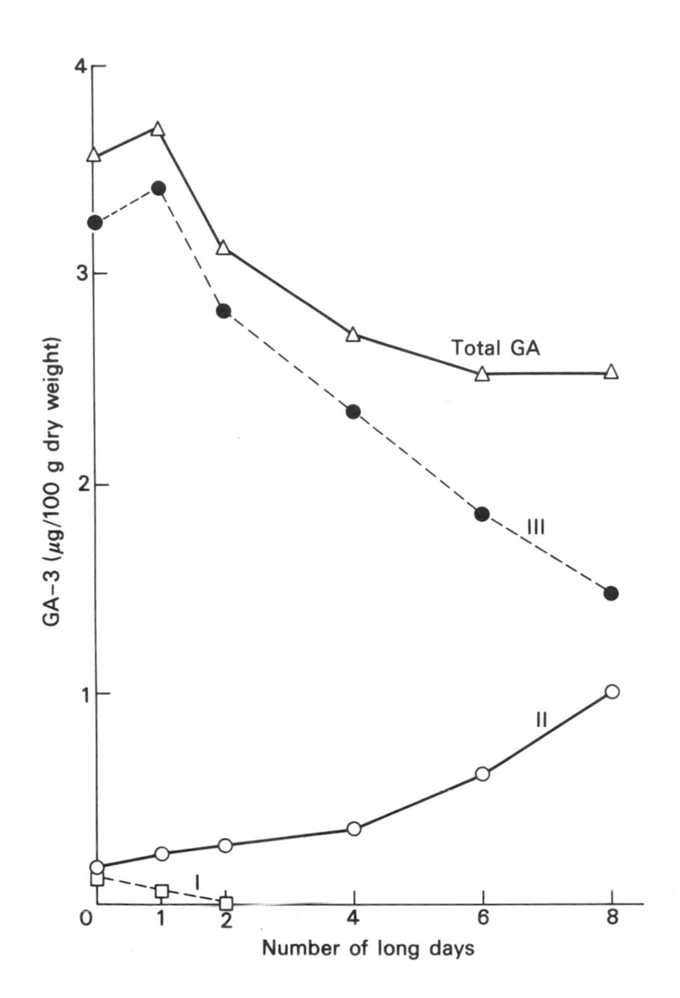

Figure 7-7. The time course of gibberellin activity at three different zones of chromatograms of *Spinacia* (LDP) treated with increasing numbers of long days (after Zeevaart 1971a).

although his Fig. 5 does show a small increase. (Comparison of Radley's and Zeevaart's results is made more difficult by the facts that they used different bioassays and, for the most part, different cultivars.) He kept track of three different TLC zones showing GA activity (Figure 7-7). His results emphasize the logical weakness of following nothing but "total GA activity," as Radley did, when there are at least three zones changing activity in different directions with increasing inductive cycles. Although the "total GA" generally decreased, one zone ("GA-II") showed a steady increase up to eight LD cycles – enough LDs to induce reproduction and to start stem elongation. Unfortunately, for our interest in floral control, his cultivar of spinach would flower eventually even if kept in SD, and GA-3 added to his plants in SD provided only a slight (and probably nonsignificant) stimulation of flowering. His interpretation was that the elongation following 8 LD cycles resulted from the steadily increasing level of "GA-II" (Figure 7-7).

Trifolium pratense L. (red clover), an LDP, showed somewhat similar changes with increasing LDs in the gibberellin activity found with the barley endosperm assay in three different zones of the chromatographed extracts (Stoddart 1962; Stoddart & Lang 1968). The ^{Rf}GA-3^{Rf} showed a big increase per leaf after three LDs (the earliest it was checked after the transfer from SD), with ^{Rf}GA-7^{Rf} and ^{Rf}GA-9^{Rf} increasing to their maxima after 12 LDs. All three zones declined in activity by the next collection (after 28 LDs).

Silene was one of the first LDP species shown to flower in SD after GA application (although it took many weeks of high daily doses of Lang's 1957 GA mix to achieve this and he noted that stem elongation prior to flowering was considerable in the GA-treated plants). Cleland and Zeevaart (1970) thoroughly investigated *Silene*. Unlike Michniewicz and Lang (1962), they found that GA-3 was not less effective than GA-7 in causing stem elongation and some flowering, although they never obtained 100% flowering from added GA under their conditions. Treatment with true LDs revealed that four or more LDs were sufficient to induce 100% flowering in their strain but that six or more LDs were needed to give maximal stem elongation. Treatment with lower doses of GA would, by contrast, produce stem elongation but no flowering – the reverse in timing of normal photoperiodic induction. Endogenous gibberellin activity was followed, using the dwarf corn bioassay either directly or on eluted TLC zones of shoot extracts. Two to five LDs resulted in more "total GA activity" in three of the four experiments. When the extracts were chromatographed, they found the same two or three zones of activity (depending on which solvent system was used in the TLC) from both SD- and LD-treated shoots, but the level at each zone was higher after the LD treatment. Two of their zones of activity co-chromatographed with GA-3 and GA-5, whereas the third zone always ran lower than GA-7 in their solvents. Cleland and Zeevaart felt their results indicated that in *Silene* gibberellins act primarily on stem elongation whereas the "flowering process is largely, if not entirely, independent of GA control." In terms of the evidence this seems a bit strong. They do not know what GAs are really represented in those three active zones, so the fact that GA-3, which co-chromatographs with one zone, and GA-7, which co-chromatographs with none, do not cause flowering in all plants when added from outside is not strong evidence for the lack of floral effectiveness of the endogenous gibberellins.

The preceding papers are among the best on the subject of endogenous gibberellins and their relation to photo or cold induction of flowering. Yet every one of them has inexplicably used long days to photoinduce, thus confounding the effects of photoperiodic induction (which could be tested using short light breaks near the middle of a long night) with effects of the many hours of extra light used to create a LD.

In no case can we feel assured that all the endogenous gibberellins have been measured, so imprecise still are the extraction methods and so insensitive to some of the known gibberellins are most of the bioassays used. Yet this brief survey of the more thorough reports in the literature leaves little doubt that, in some cases at least, some endogenous gibberellins do increase in plants that have been pushed into the reproductive state by LDs or cold treatment.

Chailakhyan's modification of his "florigen" hypothesis

Even with all the problems of interpretation of data on endogenous gibberellins, the current evidence does seem to require a modification of the original florigen hypothesis and Chailakhyan (1958) has suggested such a change. He hypothesized that two types of substances are needed for flowering, one, the gibberellins and the other, "anthesins" – the latter being substances, still hypothetical 20 years later, that would be limiting flowering in all those plants not limited by gibberellins. In his view LDPs would contain anthesin under SDs and LDs but would produce enough gibberellin for stem elongation and flowering only under LDs. Conversely, SDPs would make gibberellin under either LD or SD (and hence would not be in a rosette form in LD) but would produce the necessary anthesins only under SD. Day-neutral plants would produce sufficient anthesin and gibberellin under either LD or SD. Chailakhyan and Lozhnikova (1960) concluded that gibberellin production occurred in leaves of SDPs as well as LDPs and that LDs brought about increased gibberellin levels irrespective of the photoperiodic reaction of the plants. However, their experimental data in support of their conclusion was relatively weak: they tested the raw acetone extract of squeezed leaves, with no chromatographic separation of components, and with *Rudbeckia* elongation and date of flowering as a semiquantitative measure of gibberellin activity in their small sample. Although extracts of Mammoth tobacco did show gibberellin activity even from leaves in SD, the only other SDP tested (*Perilla*) showed no activity at all from extracts of SD leaves and only a bit of stem elongation from extracts of LD leaves.

In their 1966 paper Chailakhyan and Lozhnikova improved the techniques and added studies on the effect of light breaks as compared to LDs and SDs. It was particularly appropriate that they should be the ones to make the useful light-break determinations, because their countryman, Katunskij, was the discoverer of a light break's effectiveness 30 years before. Light breaks gave more gibberellin activity in leaf extracts than was obtained with SD treatment and consistently gave less gibberellin activity than LDs provided. For the SDP *Xanthium* the data showed no sign of gibberellin activity from leaves that had been exposed to SD. Hence, although the light-break data were particularly useful, the effect of SDs on the SDPs *Xanthium* and *Perilla* did not support the hypothesis that SDPs produced ample GA even under SDs.

By contrast, Harada's and Zeevaart and Lang's clever experiments with grafting have shed new light on the role of gibberellins in the induction of flowering. If a cold-requiring, day-neutral cultivar of *Chrysanthemum* was induced to flower at normal temperatures by treatment with GA-3, grafts from that flowering plant induced flowering in plants of a SDP cultivar of *Chrysanthemum* kept in LD (17:7) conditions (Harada 1962). (Direct treatment of the SDP cultivar with GA-3 did not induce flowering – the typical lack of response of SDPs to added GA-3.) Harada's interpretation was that GA-3 brought about the formation in the cold-requiring cultivar of a flowering hormone that could then be transmitted by grafting to the SDP cultivar.

Parallel experiments were done in the LSDP *Bryophyllum* (Zeevaart & Lang 1962). Plants kept in SD and induced to flower by GA-3 ("GA substituting for the LD requirement") could in turn induce flowering in *Bryophyllum* graft partners kept in LD. The latter plants would not flower if treated directly with GA-3 or if grafted to a nonflowering *Bryophyllum*. Again, the interpretation was that GA-3 causes the formation of a flowering hormone that can be transmitted.

To summarize the role of gibberellins in flowering: although GA causes flowering in more cases than any other known chemical, it typically has no florigenic effect on SDPs or nonrosette LDPs. And even in those LDPs, LSDPs, or cold-requiring plants where it does cause flowering, its effect is likely to be indirect. The striking correlation between GA and bolting in the rosette LDPs seems one more manifestation of the great power of GA to cause stem elongation in a broader range of plants. (See the general discussion of GA later in this chapter.)

Effect of the noninductive photoperiod on the inhibition of flowering

By this time other investigators were beginning to feel frustrated with the florigen hypothesis, modified or not – a state that might best be expressed by saying, "We seem to have cut our own throats with Occam's Razor!"

Some returned to the hypothesis of release from a flowering inhibitor as an important element in the photoperiodic control of flowering. Coupled with this was an idea that the noninductive photoperiod might be not merely neutral, as was often the tacit assumption of earlier researchers, but actively inhibiting flowering.

Indications that the noninductive photoperiod was actively inhibiting came early. If one branch of a Y-shaped plant of SDP *Xanthium* or Biloxi soybean were photoinduced, the other branch would flower also, although kept in LD. However, it would not flower if mature leaves still remained on it (e.g., Hamner & Bonner 1938; Borthwick & Parker 1938). The plant did not have to be a SDP, nor did the leaves on the "wrong" photoperiod have to be on a separate shoot to inhibit photoperiodic induction: Withrow et al. (1943) showed that the LDP spinach formed floral primordia more slowly when one or three leaves were given 24:*0* and the remaining leaves were all in SD than when the remaining leaves were excised. In *Perilla* (SDP) a single leaf in true LD delayed flowering most if it was on the same stem but above a single leaf in SD; it delayed flowering least if it was below the induced leaf; flowering was delayed to a middling degree if the LD leaf was at the same node (Chailakhyan 1947; Chailakhyan & Butenko 1957).

The early investigators interpreted this inhibiting action of LD leaves on flowering of a SDP as the result of a stream of photosynthate coming from the LD leaf and opposing the upward movement of florigen from the SD leaf. This was pure guesswork for many years. However, in 1947 Chailakhyan published an often cited paper presenting evidence in *Perilla* that a 3% sucrose solution substituted for a LD leaf could largely replace the inhibiting action of the LD leaf. Perusal of the original paper reveals that the evidence is less secure than one would like. Not only have the data been presented without statistical confirmation of the reality of the

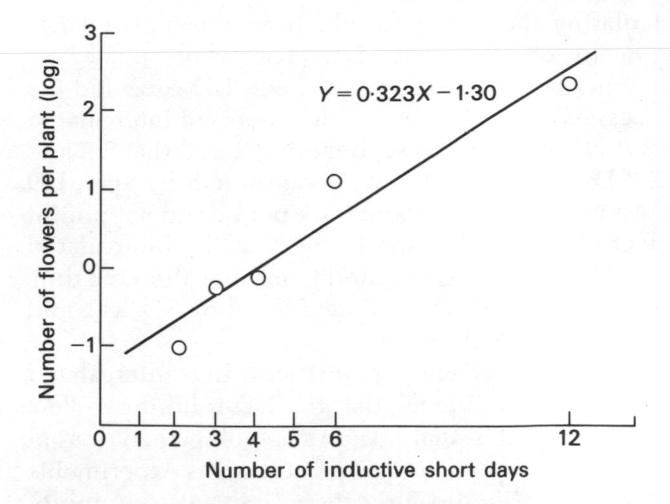

Figure 7-8. The effect of increasing numbers of short days on the number of flowers developed in the SDP *Kalanchoe* (after Schwabe 1956).

differences (and in this the paper is typical of Chailakhyan's work as well as that of many workers in floral physiology), but the experiment was apparently run once only, and with small samples. Considering the importance of the conclusion, it is regrettable that this experiment has not been repeated by others. However, if we are to lean on this frail reed, we should note that Chailakhyan's similar experiment with LDP *Rudbeckia* (1947) showed no replacement whatsoever of the inhibiting action of SD leaves by sucrose substituted for the leaves. Hence, by this evidence if inhibition of SDPs by LD leaves is due to their photosynthate, something else from leaves in SD is inhibiting LDPs.

Long (1939) intercalated LDs in a series of photoinductive SDs for the SDPs *Xanthium* and Biloxi soybean and reported that one LD could prevent two SDs immediately preceding it from having any cumulative effect. As he interpreted the results, "the effect of two long dark periods does not persist for more than 24 hours." This technique of intercalating noninductive LD photoperiods was taken up by several others, notably Schwabe, who developed an elegant quantitative treatment and interpreted the results in terms of LD inhibition rather than decay of the SD stimulation. Schwabe (1956) found that the SDP *Kalanchoe* shows a linear increase in the logarithm of flower number as the number of consecutive inductive cycles is increased (Figure 7-8). Each SD increased the number of flowers 2.1 times. The linear regression equation was then used as the basis for calculating the extent to which an interpolated LD inhibited induction by the series of SDs. (The whole plant was placed in the photoperiods.) On the average, one LD canceled the effect of 1.4 SD. Less extensive studies of the effects of intercalated LDs on the SDPs *Perilla* and Biloxi soybean indicated that 2.2 and approximately 2 SDs, respectively, were annulled by one LD (Schwabe 1959). An intercalated 24-hour dark period did not inhibit the inductive effect of SDs, indicating to me that the intercalated LDs were actively inhibiting, as contrasted to merely allowing time for the decay of the flowering stimulus engendered by SD, as could be expected in the 24-hour dark period.

Of course, such treatments of whole plants with LDs intercalated in series of SDs gives us no evidence that the LD-inhibitory effect moves around the plant – that is, that it shows one of the major properties that define a hormone. Evans (1962) felt that his experiments with the strict SDP *Rotboellia* did indicate a "transmissible inhibitory substance," because LD given to leaves *below* those given inductive SDs had an inhibitory effect and Evans thought it unlikely that the LD leaves in that position would be "acting as sinks for the stimulus" from SD leaves. The fact that Evans had no actual data as to the effect of his experimental arrangement on the movement of photosynthates in *Rotboellia* weakens his argument that the LD inhibition was from a "transmissible substance." (The descriptions of Withrow et al. 1943, in LDP spinach make clear that they also had obtained inhibition from leaves below the induced ones.)

If leaves in a noninductive photoperiod can inhibit the floral stimulating action of leaves in inductive photoperiods, the thought quickly occurs to one that perhaps one does not need SD leaves on a SDP except to counter the floral inhibition from LD leaves. Perhaps the shoot apex will convert autonomously to the reproductive state, if it is not being inhibited by leaves under noninductive photoperiod. The simplest way to test this is to see if plants will flower in noninductive photoperiod with all leaves excised or if placed in continuous dark. In such cases one can expect to have to add basic food, such as sugar, to replace the photosynthate the leaves would be making in the light. Gentcheff and Gustaffson (1940) found that spinach (an LDP) would flower in total darkness if supplied with sucrose. If the leaves were all excised, *Hyoscyamus* (LDP) formed flowers equally rapidly in LD, SD, or continuous darkness (Lang & Melchers 1943). If the leaves were not excised, flowering occurred in LD only. This was a thoroughly convincing paper: sample sizes were large, differences were tested for statistical significance, and there was even a second SD treatment added in which the light intensity was roughly doubled so that the total light per 24 hours would be equivalent to that provided in LD. Note that *Hyoscyamus* plants only flowered after defoliation if prior to that they had grown a sizeable storage root. Leopold (1949) noted that both potato and "Alaska" pea plants, which both formed floral primordia when grown in total darkness, contained sizeable amounts of stored food.

Several short-day plants have been shown to react similarly. *Chenopodium* plants that had grown outdoors before treatment formed floral primordia in LD if the leaves were excised and if sucrose was added to the mineral solution in which the roots were placed (Lona 1948). If plants were grown in the field and then placed in continuous darkness, the leaves could be left on and floral primordia could still be formed in seven of the eight plants in the sample (Lona 1949b). *Perilla* also formed floral primordia if transferred to continuous dark after it was old (i.e., after 14 or more weeks in LD) (Lona 1949b), but the significance of this is lessened by Lona's observation that the 10–14 SDs required when plants of his strain of *Perilla* were young decreased with age until only three SD cycles would cause flowering in plants 16 weeks old. The SDP *Fragaria* showed 100% flowering in continuous light if the leaves were excised, no flowering if they remained intact. When they were left intact but placed in the dark, 8 of the 15 plants flowered (Thompson & Guttridge 1960).

Evaluation of the foregoing evidence in light of the PESIGS rules suggested to Raghavan and Jacobs (1961) the need for more isolation of the reacting system (viz., the shoot tip). Their shoot tips of *Perilla* were grown in aseptic nutrient culture to test the effect of photoperiod on such isolated parts. If the portion cultured included not only the apical bud, with its average of three folded-up leaf pairs, but also the first two pairs of unfolded leaves below the bud, then the cultured tips reacted like intact plants in the greenhouse: in SD they formed flowers; in LD they continued to form leaves. However, if the older pair of unfolded leaves was excised before culturing, the shoot tips formed inflorescence and flower primordia even under LD (18:6). In SD they developed further, to the stage of open flowers – a stage reached sooner the more leaf pairs remained on the cultured shoot tip. The results were interpreted as evidence that the *Perilla* shoot tip will autonomously initiate flowering if it is not inhibited by unfavorable photoperiod acting on leaves past a certain stage of development – and, of course, if it is getting sufficient general nutrient to support the greatly increased activity involved in floral development (Jacobs & Raghavan 1962). To proceed from that first stage of floral initiation to the stage of differentiating sporogenous tissue and open flowers required SD in any of the treatments they tried. Their general conclusion was that flowering in *Perilla* apparently was regulated by a balance between a floral promoter produced in SD and a floral inhibitor produced in LD in leaves above a certain size.

An interesting side point arose from their attempt to culture the excised unfolded leaves in the same flask with the apical buds. Their hope was that under inductive photoperiod the excised leaves would release floral promoter into the medium, whence they could extract it. This hope proved futile. Instead of hastening the date of flowering of the buds, the excised SD leaves inhibited it. The lack of transmission of the promoter is in depressing agreement with all other attempts to transmit florigen across nonliving systems. The inhibition, however, was probably from IAA. Adding IAA to cultures gave progressively more inhibition of flowering the greater the IAA concentration, with as little as 1 mg/liter preventing flowering altogether even in apical buds cultured in SD (Raghavan 1961). No direct determinations of endogenous auxin were made, but the developmental stage of the second unfolded leaf of *Perilla* was the same as that of leaf 2 in the other Labiatae *Coleus*, which produces more diffusible auxin than any other leaf stage. It seems a reasonable guess that excised cultured leaf 2 of *Perilla* was diffusing auxin into the medium, thus inhibiting flowering, but could not transmit florigen from lack of tissue connection.

The isolation of floral inhibitors

With all these experiments, plus many others of a similar nature, suggesting that the "unfavorable" photoperiod is really actively inhibiting, what luck have people had isolating endogenous inhibitors hypothesized to be formed under LD in SPD or under SD in LDP? The first step in isolating floral inhibitor from SDP *Kalanchoe* leaves was taken by Schwabe (1972). He pressed juice from induced or noninduced leaves and then injected the juice into a photoinduced leaf of a plant otherwise in LD to see how much, if at all, the resulting flower number would be reduced from the injection. Sap from LD leaves reduced the number to a highly significant degree by statistical test; sap from SD leaves had no significant effect compared to controls injected with water. Blake (1972) developed a bioassay for floral inhibitors by partially inducing whole plants of LDP *Viscaria* and then culturing the excised shoot tips (said to consist of the true apical meristem and two leaf pairs). Extracts of leaves from SDP *Kalanchoe* were added through Millipore filters. After 3–4 weeks of culturing in SD the apices were dissected and scored for degree of flowering. Extracts of leaves from *Kalanchoe* in LD significantly increased the percentage of cultured *Viscaria* buds that were vegetative (12/16, as contrasted to 3/15 in untreated controls or 4/14 with extracts of SD *Kalanchoe* leaves added). Gallic acid was isolated and identified in extracts of *Kalanchoe* leaves (Pryce 1972) as the presumed inhibitor in the *Viscaria* bioassay. Treatment of *Kalanchoe* with continuous light was said to give at least 10 times more gallic acid from dialyzed aqueous extracts than if SD had been used. Fifty percent inhibition of flowering in *Viscaria* shoot tips from exogenously added gallic acid was claimed with 100 mg/liter, although 500 mg/liter was required for similar reduction of flowering in whole plants of *Kalanchoe* induced with SD. (No actual data on degree of inhibition were provided in Pryce's paper, only a table showing + for "significant inhibition" and – for "none.") Extraction with ether gave inhibition whether SD or continuous light leaves were used, a difference from results with aqueous extracts that Pryce ingeniously explained as perhaps meaning that in SD the gallic acid was fastened to a large molecule (from which it might be separated by ether extraction), whereas in LD it was not. The hypothesized large molecule might be florigen, which would stimulate flowering by inactivating the floral inhibitor!

Does gallic acid function in the plant as a natural inhibitor of flowering when *Kalanchoe* is placed under LD? Pryce quite rightly expressed concern about the high levels of gallic acid (500 mg/liter) that had to be added to inhibit flowering in *Kalanchoe*. Still more worrisome is the fact that his *Kalanchoe* was prevented from flowering with continuous light rather than with the shortest possible interrupted night. (Taylor has already shown that some phenylpropane derivatives are formed in *Xanthium* is true LDs but not in interrupted nights [1965].) Unfortunately, a recent report from the laboratory where Pryce did his work states that his inhibition results with gallic acid cannot be confirmed: the injection of even as much as 2000 mg/liter of gallic acid "gave no consistent inhibitory effects" (Schwabe & Wimble 1976).

Extraction of floral promoters other than gibberellin

There are two reports of success in extracting floral stimulators from SDPs. The work by Lincoln and co-workers (1961; Mayfield 1964; Hodson & Hamner 1970) is unusual in having been confirmed by other laboratories. A methanol extract of flowering *Xanthium* plants, after further manipulation, was added to leaves of vegetative *Xan-*

thium kept under LD (16:8), resulting in an early stage of development of the terminal male inflorescence. Carr (1967) confirmed their finding, adding the caution that any change that he made in the procedures described by Lincoln led to negative results. The publication rate by Lincoln et al. has been slow. A major difficulty in trying to isolate active fractions from the crude extract has been the lack of a fast and accurate bioassay (Mayfield, personal communication). At an international colloquium on "The Physiology of Flowering," held in 1978 at Gif, France, Lincoln reported results of a recent three-year joint effort with Hamner and West to resolve these difficulties. Consistent florigenic activity could not be obtained.

As a last experiment to describe attempts to isolate florigens (out of the many we have not space to describe), the ingenious experiments of C. F. Cleland have been selected. Noting that many physiologists had concluded that florigen moved in sieve tubes, Cleland decided to use aphids – organisms more skillful than man at inserting tiny tubes into the sieve tubes of the still living plant. Some aphids feed exclusively on sieve-tube "sap," others on xylem "sap." The former insert their proboscis into sieve-tube elements. The rear of the aphid can then be snipped away, leaving the proboscis as the equivalent of a tiny hypodermic needle from which exudes material from the punctured sieve tube. Or the aphid can be left intact in feeding position and the material that it excretes (euphemistically called honeydew) can be collected. This technique had been developed by earlier workers and had been used in valuable studies of the movement of radioactively labeled hormones (see Chapter 9). Cleland borrowed this technique in the hope that aphids feeding on sieve-tube contents would be inadvertently removing florigen and excreting it in their honeydew (1974a). Chromatograms of extracts of honeydew collected from aphids that had been feeding on photoinduced *Xanthium* (SDP) contained one zone that caused flowering in *Lemna gibba* G3 held in SD – this *Lemna* being a LDP that Cleland and Briggs had previously worked with (1969). Unfortunately, however, the extracts of honeydew from vegetative *Xanthium* was just as effective, as far as one could tell from the *Lemna* bioassay. Further discouragement lay in the fact that *Xanthium* itself was not induced to flower by this extract, nor was the SDP *Pharbitis*, although only preliminary experiments were done with *Pharbitis* (Cleland 1972).

The stimulating material was not produced by the aphid (in contrast to material found in an inhibitory zone of the chromatograms), as evidenced by the fact that acid hydrolysis of an extract of *Xanthium* leaves and buds gave a similar stimulatory zone and that

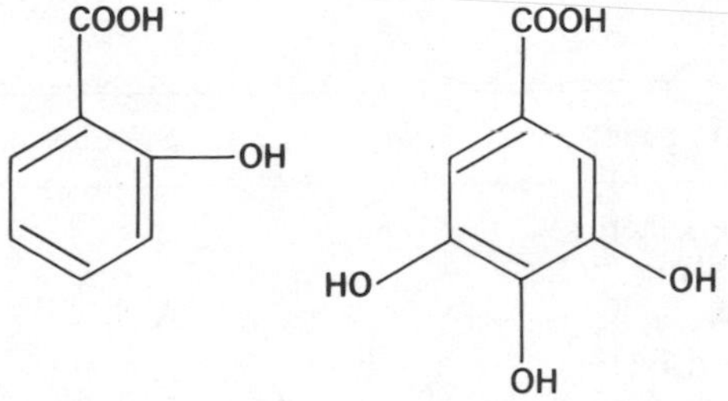

Salicylic acid (2-Hydroxybenzoic acid)

Gallic acid (3,4,5-Trihydroxybenzoic acid)

Figure 7-9. The chemical structure of gallic acid and salicylic acid.

aphids fed on a synthetic diet did not provide a stimulatory zone (Cleland 1974a). The active chemical was identified, by GLC-MS and other techniques, as salicylic acid (Cleland & Ajami 1974). Maximal stimulation of flowering occurred with about 6 μM. Figure 7-9 shows the structural formula of salicylic acid: it is the active ingredient of aspirin after the acetyl side chain of the latter has been removed; in fact, acetyl-salicylic acid stimulates *Lemna* flowering as well as salicylic acid does (Cleland 1974b). Gallic acid, the putative inhibitor of flowering isolated by Pryce from *Kalanchoe*, has the same basic benzoic acid structure but with two more hydroxy groups on the ring (Figure 7-9). Both Oota (1975) and Scharfetter et al. (1978) have confirmed the stimulatory effect of salicylic acid on the flowering of genera of the Lemnaceae.

The aphid–sieve-tube–florigen experiment was such a creative idea (and constituted so much work) that it deserved to succeed. It will be most disappointing if salicylic acid turns out to stimulate flowering on no species but *Lemna gibba* G3.

Current views on flowering hormones

There seems to be more and more evidence that does not fit the hypothesis that there is one flower-stimulating hormone among the angiosperms. Even the literature on grafting contains results of apparently satisfactory experiments that do not confirm the view that flowering can regularly be transmitted by grafting. Most physiologists currently working on flowering problems seem to favor a hypothesis involving both a stimulator and an inhibitor, even several of each – not, of course, that such a majority vote is necessarily a guide to Ultimate Truth! Bernier has given a particularly clear

Table 7-3. Flowering factors

	Flowering factors	Species no. 1	Species no. 2
Chailakhyan's first theory (florigen)	Required factor	A	A
	Nonlimiting factor	–	–
	Limiting factor	A	A
Chailakhyan's modified theory (florigen and anthesin)	Required factors	A + B	A + B
	Nonlimiting factor	A	B
	Limiting factor	B	A
Bernier's theory (1976)	Required factors	A + B + C	B + C + D + E
	Nonlimiting factors	A	B + C + D
	Limiting factors	B + C	E

Source: Bernier 1976.
Note: Table provides a summary of hypotheses explaining the control of flowering by one (florigen), two (gibberellin and anthesin), or three or more substances.

explanation of this sort of hypothesis (1976). Table 7-3, based on his Fig. 7, diagrams the hypothesis. The factors limiting flowering are assumed to be different for different plants; hence, as the diagram shows, there will be species or cultivars that can cause flowering in another species but, that cannot be induced to flower themselves by that other species – a situation that has been found in grafting experiments (van de Pol 1972).

A viewpoint that I have not heard expressed among the physiologists, but that seems to support the hypothesis that many factors are involved in the control of flowering, comes from consideration of the presumed evolution of the angiosperms. If the most widely held views of evolution are correct, what I have been referring to as "flowering" involves the condensation in angiosperms of a much more extended developmental phase in the ancestors of the angiosperms. Even in some other living vascular plants the gametophyte is an indpendent, free-living plant. If the evolutionists are correct, this independent gametophyte became reduced in size, longevity, and independence so that in angiosperms it represents that brief phase in floral development between the formation of pollen grains or embryo sac and the fertilization of the egg. Under these conditions, it would not be expected that all 170,000 species of angiosperms would, in evolving that drastically condensed gametophyte, end up with the limiting factor being the same individual chemical.

Other effects of gibberellins

The role of GA in germination of barley seeds

The action of gibberellin in barley seeds is probably the most thoroughly investigated of all GA effects, other than the increase in stem elongation already discussed. The seeds of barley (*Hordeum vulgare*) have been the focus of investigations by both academic physiologists interested in general principles and scientists trying to make better beer, cheaper beer, or both. The results of their combined labors provide a consistent picture of the hormonal physiology of seed germination as well as the basis for one of the most widely used bioassays for gibberellins – the barley endosperm (or barley half-seed) test.

Most of the volume of a barley seed is composed of the starchy endosperm, a mass of cells packed with starch. Surrounding this starchy endosperm is a thin layer of endosperm cells only one to three cells thick, called the *aleurone layer*. In one end of the barley seed is the embryo. Soaking the seed in water starts germination; the embryo grows at the expense of the stored food in the endosperm. The starch in the endosperm breaks down into maltose, which is used by the growing embryo. What controls the breakdown of starch? For many years it was thought that the embryo produced an enzyme that moved to the endosperm and there broke down the starch (e.g., Brown & Morris 1890). However, that has been shown to be too simple to fit the facts.

The involvement of gibberellin was first indicated by Hayashi's report that GA-3 added to barley seeds stimulated germination and increased amylase activity. Published in 1940, during World War II, this paper was neglected, as were the other early gibberellin papers. It was not until 1958 that Yomo provided evidence that something from the embryo controls amylase activity in the endosperm: the isolated embryo gave no amylase activity unless the endosperm was incubated with it. Extracts of embryos contained gibberellin, judging by bioassay, and a mixture of GA-1 and GA-3 could replace the embryo in stimulating amylase activity in the endosperm (Yomo 1960). Paleg (1960a), apparently independently, confirmed this action of added GA-3 on barley endosperm, and because heating the seeds completed nullified the GA effect, he concluded that an enzyme is probably involved. Added GA did not increase amylase activity if added to the medium after incubation, suggesting that GA was not merely increasing the activity of amylase already present. Also fitting the hypothesis that GA caused the formation of more enzyme was the 68% increase in protein found in the medium after endosperm was incubated with GA-3 (Paleg 1960b).

Somewhat puzzling if the endosperm was actually synthesizing enzymes under the influence of GA was the low rate of respiration of the endosperm. MacLeod and Millar (1962) provided the first specific evidence for the crucial role of the thin layer of aleurone cells that constitute the outer portion of the endosperm: GA-3 added to aleurone layer separated from the starchy endosperm resulted in increased enzymatic activity. Paleg (1964) and D. E. Briggs (1964) quickly confirmed this, as have many others since. Paleg (1964) further demonstrated that the aleurone layer as such was respiring at a reasonable rate but that measurements of respiration in the whole endosperm – dominated by the much larger volume of minimally respiring starchy endosperm – obscured the high activity of the aleurone cells (confirmed by Varner 1964). Starchy endosperm, with the aleurone layer removed either by shaving off thin slices or by soaking the seed then stripping off the aleurone layer, showed essentially no sugar release in response to added GA-3 (Paleg 1964; Varner 1964).

It is obvious that aseptic aleurone layers provide a particularly satisfying system for studying gibberellin action, representing as they do an extreme case of isolation. The aleurone layer is histologically simple, composed of nondividing and nonphotosynthesizing cells.

Such isolated aleurone layers have been used to study enzyme activities resulting from GA treatment. Varner and others presented evidence from density labeling with $H_2{}^{18}O$ or D_2O concerning de novo synthesis of the following enzymes: α-amylase (Filner & Varner 1967), protease (Jacobsen & Varner 1967), ribonuclease (Chrispeels & Varner 1967; Bennett & Chrispeels 1972). Synthesis of the amylase and protease is markedly stimulated by GA-3, whereas the ribonuclease and glucanase have their secretion but not their synthesis substantially increased by GA-3. Despite the initial hopes, the synthesis of neither α-amylase nor protease is apt to represent the primary action of GA, because it is only after 10 hours that their levels start to rise (Jacobsen & Varner; Jones 1971). Ribonuclease appears in the medium even later (about 20 hours after GA-3 is added). GA-3 increases the release of glucanase 8 hours after treatment. The complexity of the reactions of even such a "simple" isolated system as the aleurone layer has been further emphasized by Pollard (1969), who followed the timing of many activities in barley "half-seeds." ("Half-seeds" is the standard term in the literature for barley seeds cut in half transversely with the half containing the

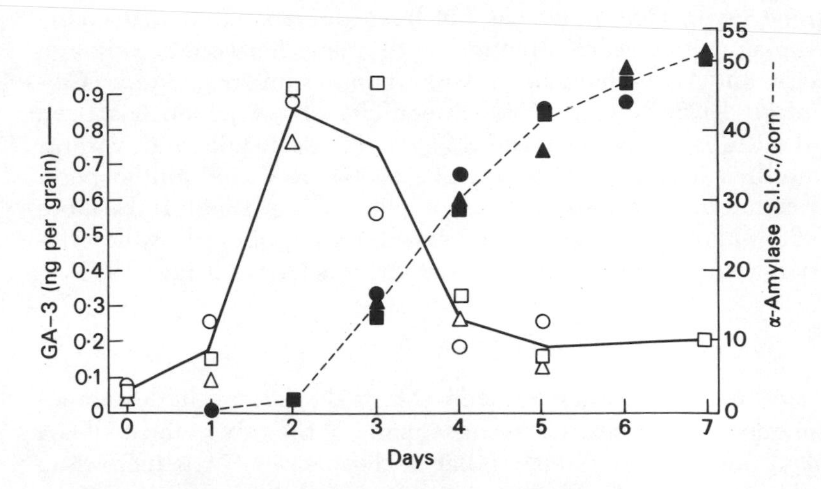

Figure 7-10. The time course in germinating barley seeds of gibberellin activity (calculated as "GA-3" by the lettuce hypocotyl assay) and α-amylase formation (after Groat & Briggs 1969).

embryo discarded: the half utilized, therefore, consists mostly of starchy endosperm with the aleurone layer around it, and the several thin layers of cells exterior to the endosperm that constitute the husk, pericarp, and testa [cf. Figs. 1 and 2 of Briggs 1973].)

The PESIGS rules applied to GA's germination effects. The evidence nicely satisfies the PESIGS rules for the conclusion that gibberellin from the embryo is the controlling factor for the release of starch-digesting enzyme from the aleurone layer. "Parallel variation" is indicated by the presence of GA-3 in the germinating barley seed, judging by the isotope dilution technique of Lazer et al. (1961) or in the "immature ears" by the more thorough evidence of Jones et al. (1963). I can find no paper reporting identification of GA-3 in the embryo per se – an unfortunate lack in the chain of evidence. However, Groat and Briggs (1969) found that the gibberellin content of germinating barley (estimated as GA-3 by a lettuce hypocotyl bioassay after paper chromatography) paralleled the time course of the rate of α-amylase formation with the gibberellin maximum preceding that of α-amylase by 14–18 hours (Figure 7-10). Excision of the embryo, as the presumed source of GA, results in no release of sugars from the endosperm. Substituting GA-3 for the embryo replaces the embryo's effect. Isolation of the GA-3 effect has been shown with barley half-seeds and even with shavings of the aleurone layer. Generality has not been demonstrated with many other plants, the usual situation with plant hormones, although *Triticum* and *Avena* have been worked on to some extent (e.g., Collins et al. 1972; Naylor 1966). Specificity of GA action has been tested with barley endosperm: if sugar release is followed, various gibberellins show activity (Paleg et al. 1964), as do helminthosporol and residue from various organic solvents (Briggs 1966). If the more specific α-amylase release is measured, only gibberellins, helminthosporol, and helminthosporic acid show activity (Briggs 1973).

GA effects related to auxin

The fact that IAA cannot replace GA in the latter's barley endosperm effects is interesting because many of the other effects of GA on plant development suggest that in those cases GA is increasing the effective level of endogenous auxin. In addition to indirect evidence from phenomena such as intensified or longer-lasting apical dominance when GA-3 is added along with IAA, direct extractions showed increased auxin from GA treatment (e.g., Nitsch 1957; Nitsch & Nitsch 1959) or an increase in diffusible auxin (Kuraishi & Muir 1964a). GA also causes more labeled IAA to be transported basipetally (see Chapter 9), with the timing in one case reasonably related to the concomitant increase in apical dominance (Jacobs & Case 1965). No one has as yet unraveled the relative contribution to these GA effects of stimulation of IAA synthesis, inhibition of IAA destruction or inactivation, and direct stimulation of transport.

The involvement of GA in apical dominance and in geotropism I shall not discuss, because of lack of space and because recent reviews exist on these subjects. The movement of GA and its interactions with the movement of other hormones is covered in Chapter 9. For GA effects on abscission see Chapter 8. GA in roots is discussed further in Chapter 10.

Growth retardants as anti-gibberellins

Various chemicals have been found to retard plant growth, suggesting they act by countering endogenous GA. Widely used in GA research, these chemicals include 2′-isopropyl-4′-trimethylammonium chloride)-5′-methylphenyl piperidine carboxylate (AMO-1618), β-chloroethyltrimethyl-ammonium chloride (CCC, Cycocel), tributyl-2,4-dichlorobenzylphosphonium chloride (Phosfon D). Although there is clear evidence from fungal cultures that some of these growth retardants can block GA synthesis, they are usually added to angiosperms in millimolar amounts, making it unlikely that their action will then be specific on hormones such as GA that are present in micromolar concentrations or less. The variety of effects reported when such growth retardants are added to angiosperms confirms the expectation that they do not function exclusively as anti-gibberellins (see the review by Lang 1970).

Sites of formation of gibberellins

The presence of gibberellins in extracts of various seeds and shoots has been described. More specifically, much more GA activity has been extracted from young leaves than from older leaves (e.g., Lang 1960; Cleland & Zeevaart 1970). But the presence of a substance in a given organ does not necessarily mean that the substance was synthesized there. (The well-known case of nicotine serves as a caution, the alkaloid being formed mainly in the roots of tobacco whence it is moved to the leaves for "storage.") Jones and Phillips (1964, 1966) applied to gibberellins the technique for determining where auxins are formed that was devised by van Overbeek (1941). Jones and Phillips used two sets of sunflower plants. One set had a given region extracted and chromatographed for GA bioassays; the other set was diffused for 20 hours onto agar and then both the agar and the diffused region were extracted separately to determine their GA content. Apical buds diffused sizeable amounts of GA in that 20 hours but still had as much extractable GA in them after diffusion as had been in the directly extracted buds. The conclusion was that the buds must be actively synthesizing gibberellin. In contrast, internode sections were not synthesizing GA (by the same criteria), although some GA activity was obtainable from them by either extraction or diffusion. Root tips were synthesizing gibberellins, but zones 4–8 mm proximal to the root tip were not.

Jones and Phillips bioassayed the GA produced by various leaves and internodes, using the diffusion technique. Results showed most GA diffusing from the youngest leaves and internodes tested, with a declining gradient down the shoot (Figure 7-11). The parallel variation between GA and internode elongation was clear-cut.

Kaufman and co-workers have carried the analysis of GA and elongation a major step further in their papers on *Avena*. Using GC-MS and other analytical techniques, they have identified GA-3 as the major GA in extracts of the inflorescence (although there are other zones on the chromatograms showing GA activity) (Kaufman et al. 1976). Excision of the inflorescence and the last leaf to develop caused a sharp drop in elongation of the internode below, and sub-

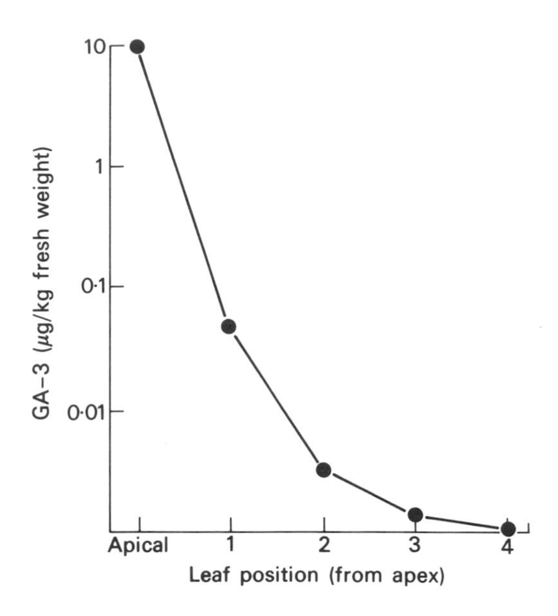

Figure 7-11. The amount of diffusible gibberellin (calculated as "GA-3," using a dwarf pea epicotyl bioassay) from successively older and lower leaves of sunflower (after Jones & Phillips 1966).

stitution of GA-3 at 30 μM for the excised organs fully restored the elongation of the internode (Koning et al. 1977).

Using bioassays of chromatogram zones, several groups have reported on gibberellins obtained by diffusion compared with those obtained by extracting tissue. Jones and Lang (1968) first called attention to the fact that two zones of GA activity were found with extracts of pea buds, but only one zone by diffusion. The typical result has been that one less zone of GA activity is in diffusates (Cleland & Zeevaart 1970 with *Silene* shoot tips; Wylie & Ryugo 1971 with peach shoot tips; Frydman & Wareing 1973 with *Hedera* root tips). It seems to have been unappreciated by the investigators of gibberellins that this greater chromatographic purity of diffusible GA is just what would have been expected from the reasoning and results of the earlier work on diffusible auxin (Scott & Jacobs 1964; see Chapter 3). On the assumption used in the auxin investigations that only the *hormonal* auxin is likely to be found in diffusates, it seems reasonable to hypothesize that only some of the gibberellins extracted from plants are hormones; namely, the smaller number typically found in diffusates.

Techniques of gibberellin bioassay

Bioassays are less useful with gibberellins than with the other plant hormones, because of the unusually large number of gibberellins already known to exist in plants. Gas chromatography, preferably combined with mass spectrometry (GC-MS), seems to be the method of choice at present. However, gibberellin bioassays are still routinely needed at various stages in gibberellin research. For example, even with the best possible GC-MS equipment one still needs to know if there is gibberellin activity in the original extracts that is not accounted for by the gibberellins identified by GC-MS.

Dozens of bioassays for gibberellins have been suggested in the literature, with individual tests being of widely varying use, levels of development, and dependability. More than 30 bioassays were listed in Bailiss and Hill's review (1971). As with the many auxin bioassays in the literature, most of the gibberellin bioassays are superfluous. One needs a bioassay with known specificity to chemicals that is easy to perform and widely known and used. (Those groans, as of the damned in hell, that passersby sometimes hear emanating from research laboratories, are apt to come from a hormone physiologist who has just read a paper with potentially important results based on a "brand-new bioassay" that the author has just created. The moral is: select a well-known bioassay. Your readers will know what the results mean, and a reasonable estimate of chemical specificity will be possible.)

Reeve and Crozier (1974) summarized the responses of five bioassays to approximately 40 gibberellins. (The book in which their chapter appears [Krishnamoorthy 1974], however, has such an unusual profusion of misprints that you should check any point or number of specific interest.) The lesson learned with auxin bioassays was borne out in testing the much larger number of gibberellins; namely, the relative activity of the hormones varies in different bioassays.

One of the most solidly established GA bioassays is the mutant dwarf-corn bioassay developed by Phinney and co-workers. The response of one of the *Zea* mutants was shown in Figure 7-4, and details of running the bioassay are given in Phinney and West (1961). The virtues of this assay are that solutions are conveniently added to the intact seedlings and that the single-gene mutation supports the hope that the response is quite directly allied to GA metabolism.

The barley endosperm assay has also been widely used. It exists in two forms: one developed by Coombe et al. (1967a, b), which

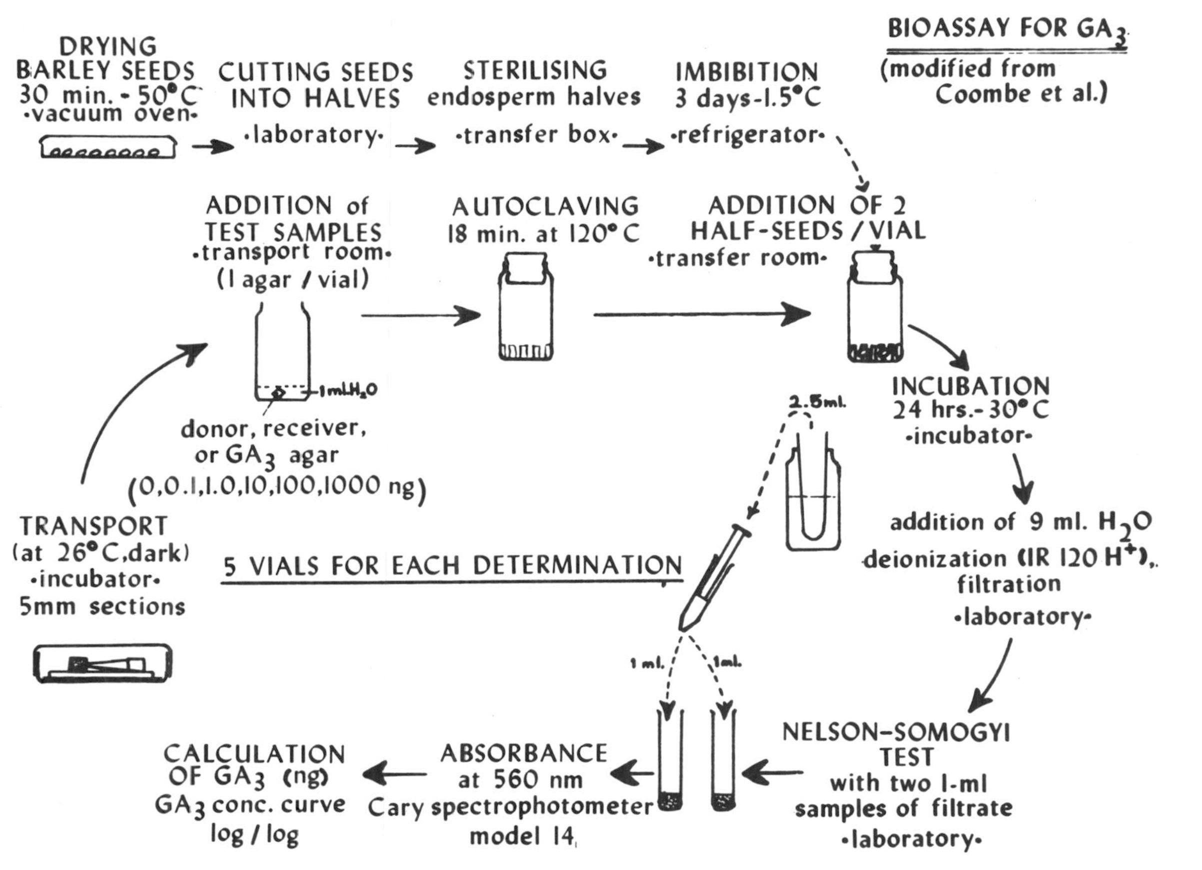

Figure 7-12. Flow diagram of procedures used by Jacobs and Pruett (1972) for measuring GA in transport receiver blocks using a modified barley endosperm bioassay.

measures increase in "reducing sugars" produced by barley half-seeds that have been treated with GA, and a later one by Jones and Varner (1967), which measures decrease in starch. The major assets of the barley endosperm test (or barley half-seed test, as it is also called) are the isolation and histological simplicity of the reacting system, the very high sensitivity to added GA-3, and the small number of gibberellins to which it responds. My own experience is limited to the Coombe *et al.* bioassay, but I have found it to be reliable. Our flow chart for the barley endosperm test is shown in Figure 7-12 (Jacobs & Pruett 1972).

A bioassay using a dwarf form of rice (the Tan-ginbozu cultivar) is popular because it responds to most of the known gibberellins (Murakami 1968a; Reeve & Crozier 1974). Hence, it is useful for screening for gibberellin activity in general.

Summary

In brief, the basis of GA action has been sought in the areas you might expect from our earlier discussions of auxins and cytokinins; namely, areas that represent a combination of what is reasonable and what is fashionable. Such areas were nucleic acids in the 1960s and early 1970s, and membrane effects in the later 1970s. Membrane effects are currently being investigated with particular enthusiasm (see Wood et al. 1972) – in obvious relation to GA effects on secretion of enzymes from the cells of the aleurone layer. Binding sites are being searched for, encouraged by the relative success of such efforts with the auxins (see Chapter 5). There is an excellent summary of hormone binding in Kende and Gardner's 1976 review, but the more recent report by Jelsema et al. (1977) seems to be the first major success with GA binding.

8

Senescence, abscission, and abscisic acid

Senescence is usually defined as the degenerative changes that occur in an ageing organism. Angiosperm shoots provide unusually favorable material for the study of senescence. From the tip to the base of a typical shoot are arranged a progression of leaves of identical genotype and of progressively increasing age. The oldest leaves typically change color (often turning yellow as the declining amount of green pigment unmasks the yellow-orange carotenoids) and finally fall off, as the result of abscission – an active process of being "cut off" and discarded (Jacobs 1962).

Although biologists are interested in studying the senescence of many types of organisms, there is a major practical problem in deciding what aspect of senescence to study. So many degenerative changes occur that it is neither feasible nor esthetically satisfying to study them all. Also, many of the changes occur gradually, with no clear-cut end point (Figure 8-1). Abscission, by contrast, is a very discrete and easily measurable event: the time when a leaf falls is incontestable. For these reasons, abscission has been the most thoroughly investigated aspect of senescence in angiosperms, with *days to abscission* being the usual unit of measure.

Also promoting and supporting studies to control leaf ageing and abscission have been the war departments of various countries, interested in either preventing leaf abscission (e.g., so that boughs cut from trees and used to camouflage tanks, trucks, and artillery would not lose their leaves) or causing it (e.g., so that soldiers in woods could be seen and attacked from the air).

Fruits develop from the carpel of the typical flower after the eggs in the carpel have been fertilized. Many plant morphologists and evolutionists think that the typical carpel of angiosperms evolved from a leaf that carried megasporangia (e.g., Bailey & Swamy 1951). If that is so, we could expect that fruit abscission would have many points in common with leaf abscission. In fact, it does. And the

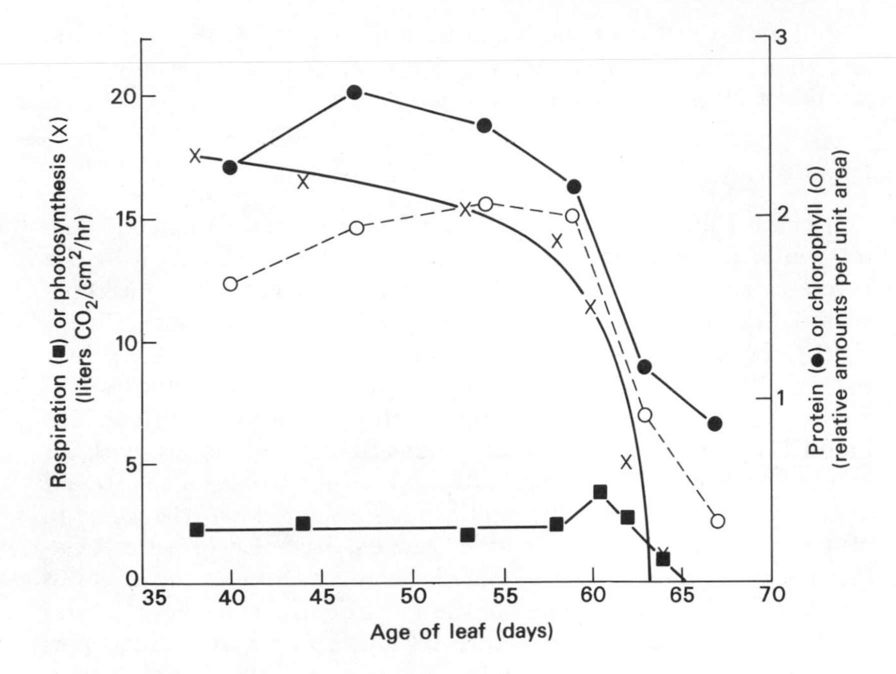

Figure 8-1. The time course of respiration (■), photosynthesis (x), and content of protein (●) and chlorophyll (○) in a *Perilla* leaf from maturation to abscission (after Woolhouse 1967).

economic importance of fruits has led to extensive experimentation to find ways of controlling fruit abscission, both to prevent fruit drop from occurring before the fruit is fully ripe and to initiate fruit drop at the optimal picking time. A particularly important fusion of these interests occurs in cotton growing. The switch from the human to the mechanical cotton picker meant the loss of any discrimination between leaves and cotton bolls. (The cotton of commerce comes from long hairs that grow from the surface of the cotton fruit.) The mechanical picker claws up anything remaining on the stem. If leaves are there along with bolls, the cotton is so dirty that expensive cleaning must be done. Hence, cotton farmers need to induce the abscission of all leaves with a cheap treatment that will not cause the abscission of the fruits. The control of abscission is so crucial to successful cotton farming that the research of one of the major investigators in abscission (F.T. Addicott) has been supported for many years by California cotton growers.

In this chapter we describe the gradual development of understanding of the way in which hormones control abscission, starting with the prevention of abscission by hormones from the leaf blade and then proceeding to the still unresolved questions about the internal factors that speed abscission.

Hormones preventing abscission

More than 100 years ago it was noticed that when the blade of a leaf was eaten away by an insect or when the blade was cut off by the investigator, fast abscission of the remaining petiole (leaf stalk) occurred (Wiesner 1871). This has been confirmed many times. Küster (1916) carried such experiments major steps further. He made a survey of plants to find optimal material for abscission studies and finally selected *Coleus*, particularly the red-leaved cultivar. He found this genus to be unusually sensitive to deblading, with all leaves abscising within 5 days after deblading. *Coleus* also showed a very smooth, basipetally declining gradient down the stem in normal preabscission time of intact leaves. Since Küster's time, *Coleus* has been investigated by more laboratories than any other genus and has also been the subject of the most quantitative work. Küster made another major contribution; he noted that a very small percentage of the total area of the leaf blade would prevent abscission of the remaining petiole. From this fact he deduced that "chemical correlations" rather than photosynthate must be controlling abscission. This introduced the idea that hormones were involved in controlling abscission.

Went's 1928 discovery of auxin reopened this area of investigation, as it did so many others. Laibach (1933) showed that the pollinia of various species of orchids, which he had shown were natural sources of auxin (1932), could increase the longevity of debladed leaves. Mai published the full report from Laibach's laboratory in 1934. He reported *Coleus* to be particularly responsive to pollinia substituted for the leaf blade, although pollinia were not nearly as effective in delaying abscission as was the intact leaf. Thirteen other genera were less responsive, pollinia doubling the preabscission time of their debladed leaves on the average, rather than increasing it fivefold, as in "Coleus-red." *Cannabis* and *Ampelopsis* were essentially unaffected by pollinia substituted for the blade. Mai, of course, attributed the antiabscission effect of the pollinia to the auxin they produced, although an auxin-containing extract of pollinia was even less effective than pollinia in fully replacing the excised leaf blade.

By this time indole-3-acetic acid (IAA) had been isolated from fungal medium as the first naturally occurring auxin and soon thereafter several notes reported antiabscission activity from IAA.

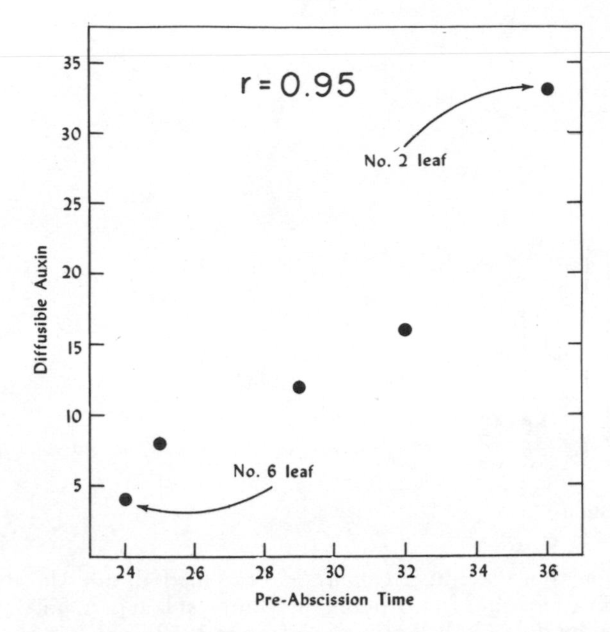

Figure 8-2. Correlation between preabscission time and production of diffusible auxin in *Coleus* leaves of various ages (no. 2 leaf being the youngest shown, no. 6 the oldest; data from Wetmore & Jacobs 1953, figure from Jacobs 1962).

Myers's fine paper (1940) considerably improved the experimental treatment of abscission. He considered both the anatomy and auxin physiology of abscission in *Coleus*, particularly noting changes with leaf age. For instance, he found that more diffusible auxin was produced by young than by old leaves, and young leaves normally showed greater preabscission times. He rather inexplicably concluded that, although auxins did influence the process, the "relation [of auxins] to abscission phenomena does not seem to be a simple one."

PESIGS rules applied to auxin and leaf abscission

Jacobs and co-workers started their series of abscission papers on *Coleus* by building directly on Myers's work, but with more quantitative methods (e.g., analyzing results statistically and matching more closely by developmental age). Parallel variation was demonstrated between the amount of diffusible auxin produced per leaf blade and the normal preabscission time (Figure 8-2; data from Wetmore & Jacobs 1953). Excision of the blades, as the sources of

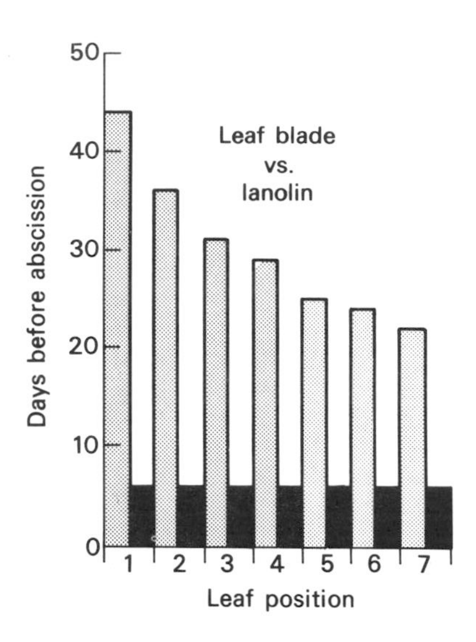

Figure 8-3. Preabscission time for intact *Coleus* leaves (shaded) and debladed leaves (black) of various ages (no. 1 being the youngest leaf pair, one member of each pair being debladed; Wetmore & Jacobs 1953).

endogenous auxin, caused fast and uniform abscission of the debladed leaves (Figure 8-3). Substitution of synthetic IAA (at 1% in lanolin) for the leaf blades restored normal longevity (Figure 8-4; and Kaushik 1965). Wetmore and Jacobs (1953) concluded that "diffusible auxin from the leaf blade is the factor which normally controls the pattern of leaf abscission" once the leaf has grown to the size of leaf pair 2 (the leaf producing most auxin). Scott and Jacobs (1964) presented evidence that the endogenous auxin of this *Coleus* is IAA. (See Chapter 3.) Isolation of the system, in the form of treating debladed leaves attached to an excised section of internode ("explants"), showed that IAA could prevent petiolar abscission in *Coleus* even when roots and shoot meristems were absent (Luckwill 1956; Wright 1956; Gorter 1957; Jacobs et al. 1965; Halliday & Wangermann 1972). Specificity of the chemical was tested extensively by Gardner & Cooper (1943): each of nine auxins inhibited abscission; 156 nonauxin organic chemicals did not. Myers (1939) had already found that three auxins inhibited abscission, with rough equivalence to their auxin activity, and that the acetic and butyric acid side chains of two of them were without abscission-inhibiting activity when added by themselves. The synthetic auxin naphthaleneacetic acid was a bit more effective than IAA (Wetmore & Jacobs 1953; Gaur & Leopold 1955).

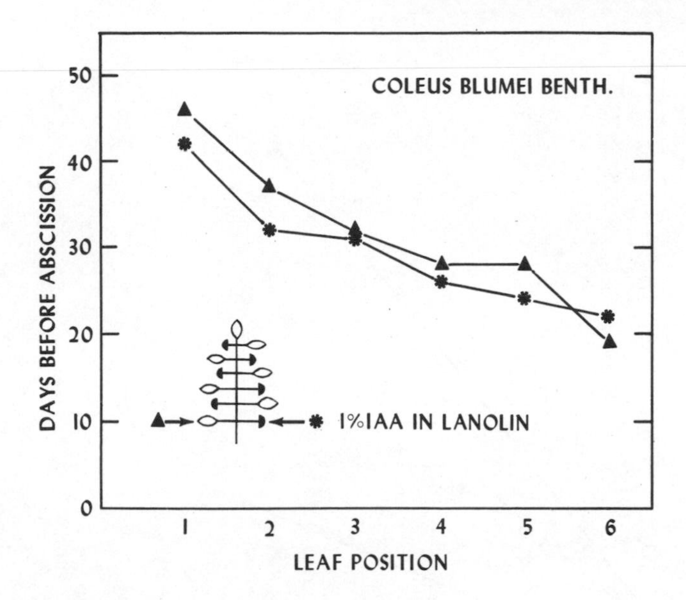

Figure 8-4. Effect of substituting IAA for the leaf blade in restoring preabscission time in *Coleus*. The insert diagram shows the pattern of treatment (data from Wetmore & Jacobs 1953, figure from Jacobs 1962).

The generality of IAA as an endogenous abscission-inhibiting factor has not been checked as thoroughly and quantitatively in genera other than *Coleus*. However, in additon to Mai's use of many genera to inhibit abscission with pollinia, synthetic auxins have been demonstrated to inhibit abscission of leaves and fruits of many genera. *Phaseolus* leaves have been the favorite abscission material of several laboratories (Addicott's, Leopold's, & Osborne's) – particularly in the form of isolated explants – and cotton has been studied for the economic reasons mentioned earlier (Addicott et al. 1964).

IAA transport and its relation to abscission

An initially puzzling feature of the results of Myers and of Wetmore and Jacobs was that 1% IAA in lanolin, when substituted for leaf blades of various ages, restored essentially normal longevity to all debladed leaves. If available IAA were the factor preventing abscission, why did not 1% IAA-lanolin, which provided essentially the same amount of auxin as young petiole 2 (as evidenced by Myers's direct bioassay), keep all the older debladed leaves on as long as 2? Jacobs (1962) guessed that the ability of petioles of various ages to

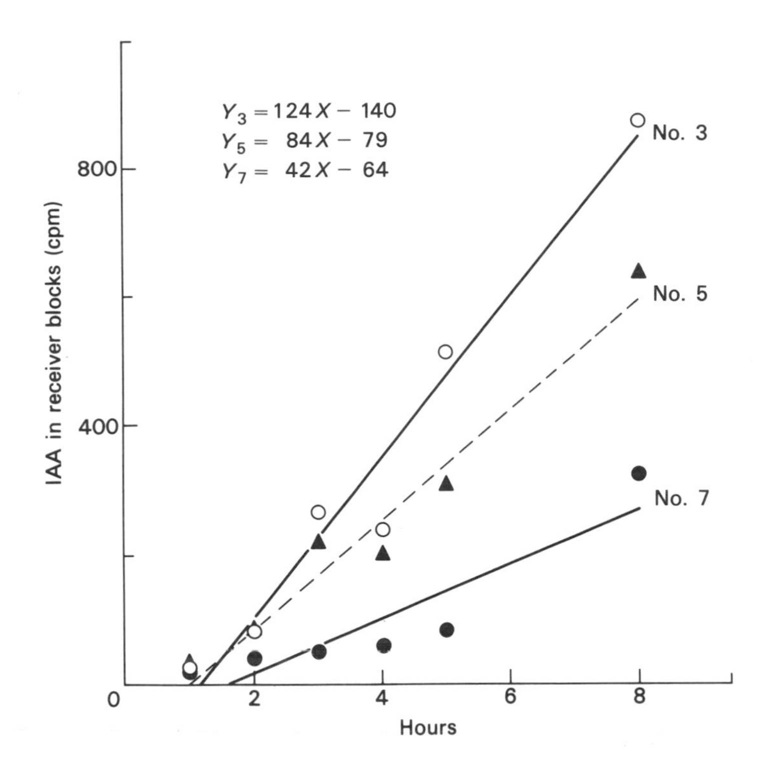

Figure 8-5. Time course of basipetal movement of ^{14}C-IAA through *Coleus* petioles of three different ages (no. 5 is approximately two weeks older than no. 3, no. 7 two weeks older than no. 5). The best-fitting straight lines and their regression equations are also given (raw data from Werblin 1966, senior thesis, regression results unpublished).

transport IAA normally matched the available supply from the leaf blade. Direct tests confirmed this hypothesis (Figure 8-5 and Werblin & Jacobs 1967). The results show that when the same amount of ^{14}C-IAA was added to the leaf-blade end of petiolar transport sections, the amount of IAA that moved through the section in the direction of the basal abscission layer was progressively less the older the leaf from which the section was cut. Veen and Jacobs (1969a) confirmed the decrease in basipetal IAA transport through petioles with increasing age and demonstrated that progressively older petioles immobilized progressively more of the available IAA by conjugating it with aspartic acid (Figure 8-6; Jacobs 1972) – a likely explanation for the decreased transport.

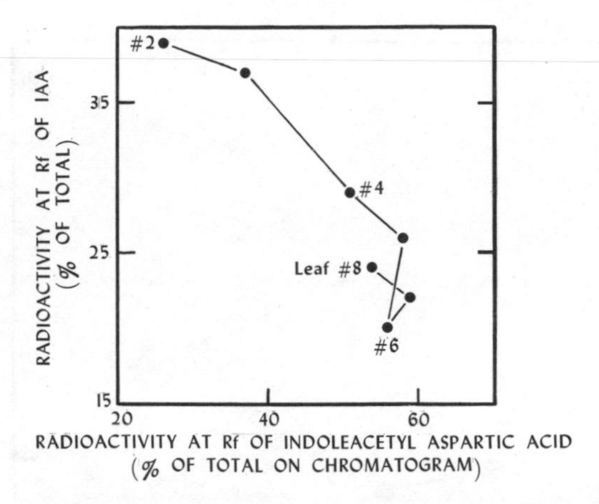

Figure 8-6. Decrease with leaf age (no. 8 being the oldest leaf) in free ^{14}C-IAA and increase in its conjugate, indoleacetyl aspartic acid, extracted from *Coleus* petioles that had been used for ^{14}C-IAA transport tests (data from Veen & Jacobs 1969a, figure from Jacobs 1972).

Laibach and Mai, in their early papers on the use of auxin-containing pollinia, noted that the added pollinia not only prevented abscission of the debladed leaves but also increased their elongation. They suggested that auxin is somehow preventing abscission through its classical effect of stimulating growth. Some later workers tended to ignore this hypothesis (despite its being preferable according to Occam's Razor) in favor of the hypothesis that auxin prevented abscission by directly affecting the abscission layer. Both Myers and Wetmore and Jacobs noted that petiolar elongation was in general correlated with preabscission time, whether the leaf blade or synthetic IAA was the source of auxin. A more quantitatively detailed examination of elongation in all ages of *Coleus* petioles showed that, although petioles of young intact leaves did elongate more than older petioles (as expected from their longer preabscission times) and although IAA did cause greater elongation of debladed leaves than occurred in debladed controls, IAA substituted for the blades did not completely restore normal elongation of younger leaves, even though it did, in the same experiments, completely restore normal longevity (Figure 8-7; Jacobs et al. 1964). And when older debladed leaves were treated with IAA so as to replace exactly the normal preabscission time, the IAA provided *more* elongation than occurred when the leaf blade was present. For the younger petioles, which grew more when intact than when IAA

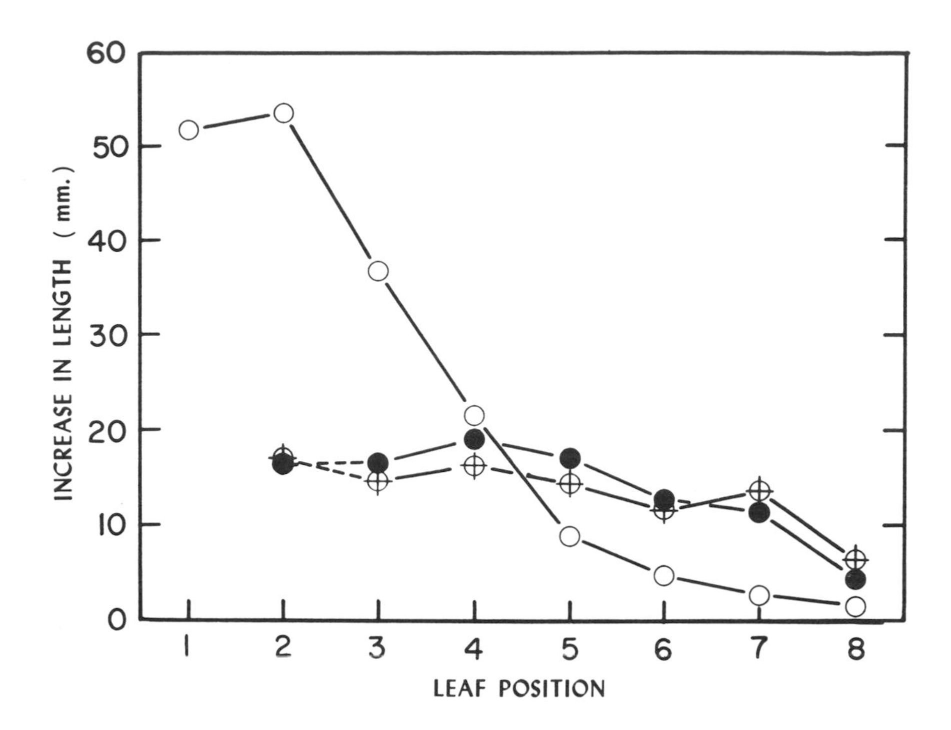

Figure 8-7. Increase in length of petioles of *Coleus* leaves of various ages (no. 8 being the oldest) in intact leaves (open circles) and in leaves that had IAA substituted for the leaf blade (Jacobs et al. 1964). (The solid circles and the open circles with a cross represent two different patterns of deblading the leaves.)

was substituted for the blade, a time-course study revealed that IAA fully replaced the blade during the first week after deblading. It was in the later weeks that elongation ceased in the IAA-treated young petioles but continued in the intact ones (Figure 8-8; Jacobs et al. 1964).

Hence, although longer preabscission times are generally associated with more petiole elongation, there is not a quantitatively exact correlation between the two. The occurrence of some petiolar elongation may merely be a reflection of the correlation of basipetal IAA transport with elongation (see Chapter 9): that is, in order to move the IAA down to the abscission layer, some elongation may be necessary.

Pertinent evidence came from studies of the effects of gibberellic acid (GA-3) on abscission and elongation in *Coleus* (Jacobs & Kirk 1966). GA-3 speeded the abscission of young debladed leaves,

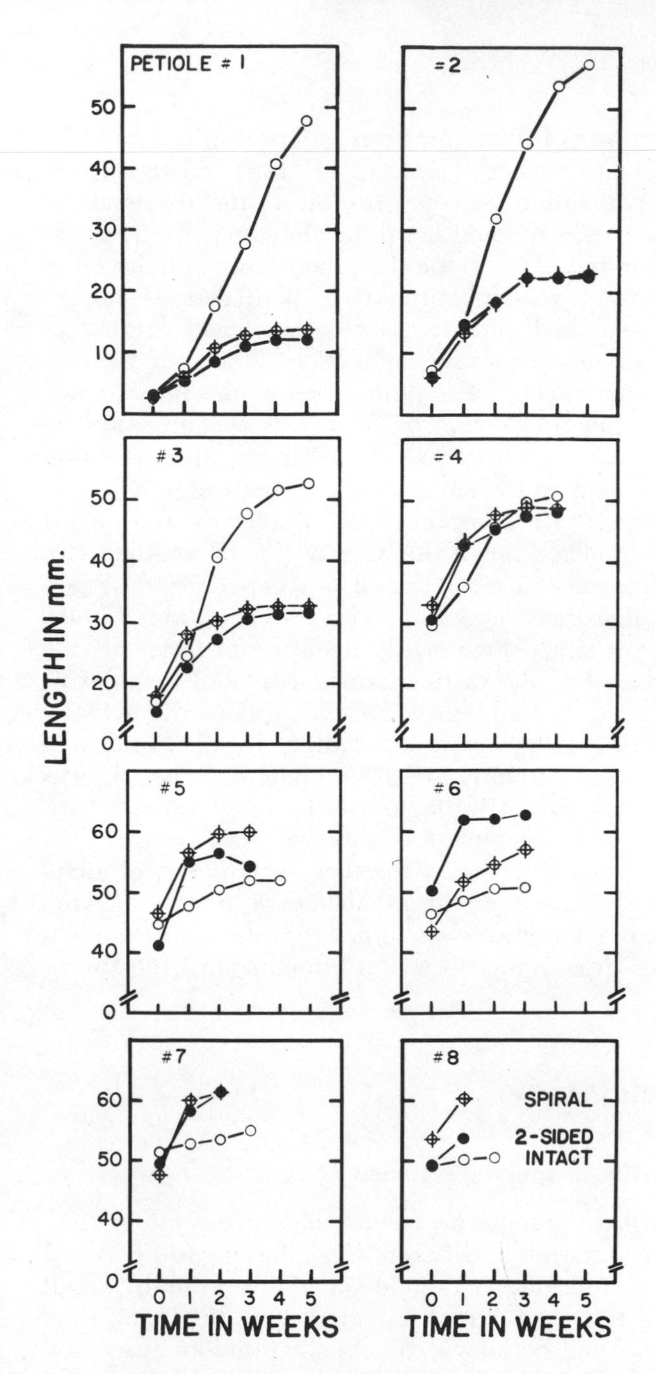

Figure 8-8. Time course of petiolar elongation in various ages of *Coleus* leaves with intact blade (open circles) or IAA substituted for the blade in a two-sided (solid circles) or spiral (open with cross) pattern of deblading (Jacobs et al. 1964).

abscission speeding by GA-3 having been reported in isolated cotton and bean explants earlier (Carns et al. 1961, Chatterjee & Leopold 1964). When added IAA was inhibiting the abscission of debladed *Coleus* leaves, GA-3 had a striking effect in partially countering that inhibition but at the same time increasing petiolar elongation. When only GA-3 was substituted for the blades, abscission was still speeded to a small but statistically significant degree, although the small increase in average elongation of the young leaves was not significant statistically. The point of particular interest was that the abscission-speeding action of GA-3 was accompanied by *stimulation* of elongation – the reverse of the relation apparent from IAA. By adding ^{14}C-IAA to the intact leaf and seeing the effect of GA-3 on the subsequent distribution of ^{14}C, Kaldewey and Jacobs (1975) felt they had tracked down the basis of this effect: the GA-3 caused more ^{14}C from IAA to be retained in the petiole (thus presumably causing the extra elongation) and less to be retained in the basal abscission layer (thus presumably resulting in faster abscission). (Earlier workers had shown that radioactive label from auxins added to the outer end of debladed leaves or explants did actually appear in the area of the abscission layer, although, like Kaldewey and Jacobs, they did not identify what compound(s) the ^{14}C was associated with in that zone [Rubinstein & Leopold 1963; Jacobs 1964; Jacobs et al. 1966; Rasmussen & Bukovac 1966].)

Obviously, to decide with more assurance if the locus of auxin action is on the abscission zone itself, we need experiments showing whether auxin can prevent abscission under conditions that allow it to stimulate petiolar growth but prevent it from reaching the abscission zone.

Hormones stimulating abscission

Auxin's ability to speed abscission

In the early 1950s experiments on factors that speed, rather than slow, abscission were started at different laboratories. Addicott and Lynch (1951) discovered that IAA could speed abscission by about 1.5 days if applied to the proximal, rather than to the distal (leaf blade) side of an isolated petiolar explant containing an abscission zone. Distal application of IAA inhibited, as would be expected from the literature, and proximal application was without effect if IAA was also available distally. In a less isolated system a debladed leaf on an otherwise intact greenhouse plant was likewise found to abscise faster if IAA was applied proximal to the abscission zone, although proximal IAA actually retarded abscission still more if the blade had been left on.

By the use of various deblading patterns in *Coleus,* Rossetter and Jacobs (1953) discovered that intact leaves speeded abscission of nearby debladed leaves, particularly if the intact leaves were young and directly above the debladed ones. The abscission-speeding effect of the distal young leaves could be quantitatively replaced by IAA substituted for them (Jacobs 1955, 1958) and was, therefore, apparently a natural case of the speeding from auxin that Addicott and Lynch had first reported for exogenous auxin in *Phaseolus.* One percent IAA in lanolin was shown by the *Avena* curvature bioassay to replace exactly the endogenous supply of IAA from the shoot tip (the apical bud plus the first unfolded leaf pair) (Figure 4-11): this same amount of IAA exactly substituted for those young organs in speeding the abscission of debladed leaves below (Jacobs et al. 1959). Leaves opposite or below debladed leaves also had abscission-speeding action, but this was due not to their producing their own auxin (judging by the ineffectiveness of IAA-substitution) but to their providing nonauxin materials for the faster growth and faster auxin production of the younger leaves above. This abscission speeding by IAA did not occur solely with debladed leaves – IAA substituted for the side branches in *Coleus* speeded the abscission of the fully intact leaves on the main stem (Jacobs 1958). Most of these experiments have been repeated and confirmed by others (e.g., Laudi 1956; Laudi & Gerola 1956; Terpstra 1956; Vendrig 1960; Valdovinos et al. 1967). These experiments led to the interpretation that in the intact plant IAA from the leaf blade normally inhibited abscission, but when that supply of IAA became low – whether from normal ageing or from artificial deblading – the abscission-stimulating effect of auxin from the young leaves became operative. It was not known whether the speeding effect of IAA on the abscission of debladed leaves actually resulted from IAA's activity at the abscission zone or whether the IAA acted locally and only indirectly affected the abscission of the debladed leaf.

Osborne's "senescence factor"

A quite different approach was that of Osborne (1955). She reported that something diffused out of petiole sections cut from old, "senescing" leaves of several genera that speeded the abscission of young petioles on an explant of *Phaseolus.* Petiole sections from young leaves of *Phaseolus* did not show this abscission-speeding

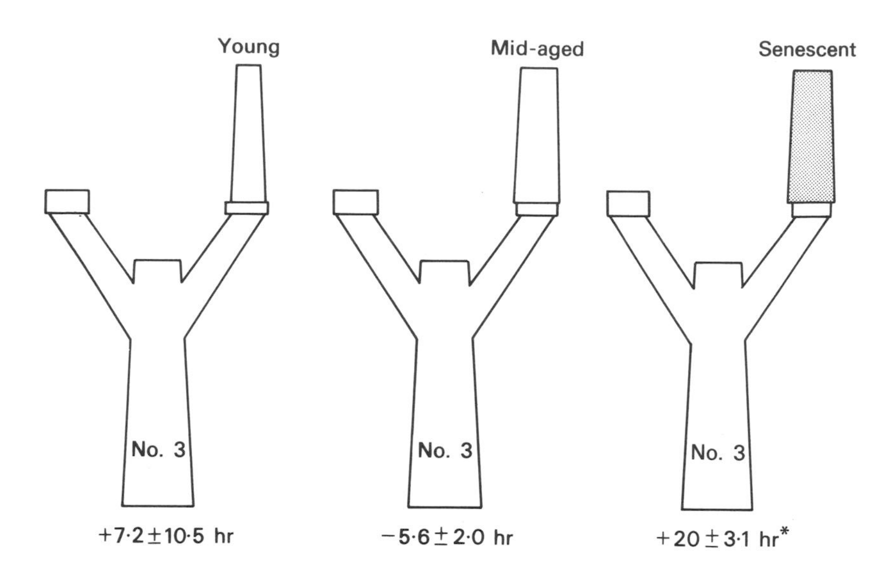

Figure 8-9. Average difference in hours to abscission of *Coleus* no. 3 petiolar explants treated with sections from young no. 2, mid-aged no. 5, or senescent no. 8 petioles. Only the senescent petiole sections caused significantly faster abscission (as indicated by the asterisk; Jacobs et al. 1962).

action. Osborne called the material(s) from old petioles *senescence factor*. Preliminary results indicated that senescence factor could counter the inhibitory effect on abscission of added IAA. Osborne proposed that in the intact plant, senescence factor caused abscission of old leaves by similarly countering the abscission-inhibiting effect of endogenous auxin.

Operationally, senescence factor is the substance (or substances) that comes from a section of senescent petiole by either diffusion or extraction. This "substance," when added to a debladed leaf on a young explant, speeds the abscission of that debladed petiole by about one day. Contrary to Osborne's original view, senescence factor does not diffuse from a petiole until after the oldest leaf has visibly faded in color (Figure 8-9) – at least this is the case in *Coleus blumei* where this point has been specifically checked (Jacobs et al. 1962) and in *Catharanthus* too (Prakash 1976). The diffusion of senescence factor from senescent leaves of *Coleus* has been confirmed in *C. rehneltianus* also (Böttger 1970b).

Although Osborne followed the sensible path of enlisting the aid of chemists, in 1977 senescence factor had still not been isolated and chemically identified.

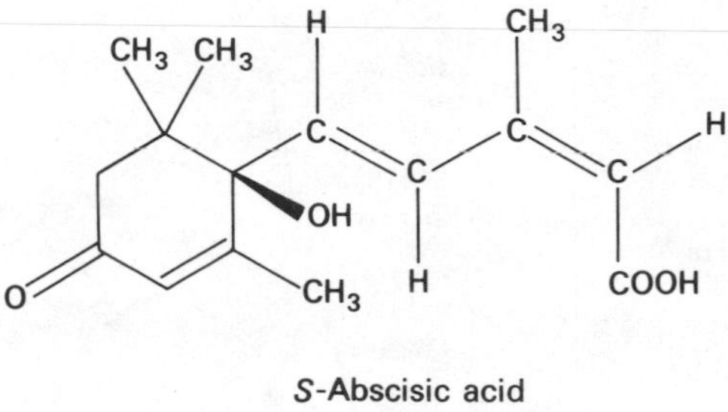

S-Abscisic acid

Figure 8-10. The chemical structure of abscisic acid (Smith et al. 1968).

Abscisic acid

As Osborne was starting her work on these abscission stimulators from senescent leaves, investigations on abscission stimulators from cotton fruits were beginning in the United States. The great asset of cotton fruits as experimental material is the tons of material available for extraction if you live in a cotton-growing state. Liu and Carns (1961) isolated a crystalline material, called Abscisin I, from mature cotton fruit-walls (the fruit from which seeds and fibers had been removed). The yield was 0.1 mg/kg of dried fruits. This material speeded abscission of debladed leaves of cotton explants. Switching to young cotton fruits, a different abscission-speeding substance was isolated and identified (Ohkuma et al. 1963; Addicott et al. 1964; Ohkuma et al. 1965). Initially called Abscisin II, it is now generally known as abscisic acid (ABA) (Figure 8-10). The yield was 40 μg/kg dried cotton fruit. Abscisic acid speeded abscission of cotyledonary explants of cotton, with as little as 1 μg per petiole inducing close to 100% abscission in 2 days, at which time there was zero abscission in the controls (Addicott et al. 1964). Similar abscission-speeding effects were reported from ABA application to debladed leaves or explants of *Coleus*, *Phaseolus*, or *Citrus* (Figure 8-11; Smith et al. 1968). ABA also countered IAA's stimulation of elongation in the *Avena* coleoptile straight-growth assay, with a concentration of 0.3 mg ABA/liter reducing by more than half the elongation that 0.1 mg IAA/liter produced (Addicott et al. 1964). The levels of endogenous ABA rose sharply in cotton fruit at about the time early fruit abscission occurred in the intact plant, falling again soon thereafter (Davis & Addicott 1972). This parallel variation between ABA content and natural abscission time increased the likelihood that ABA was functioning as an endogenous regulator in cotton fruits.

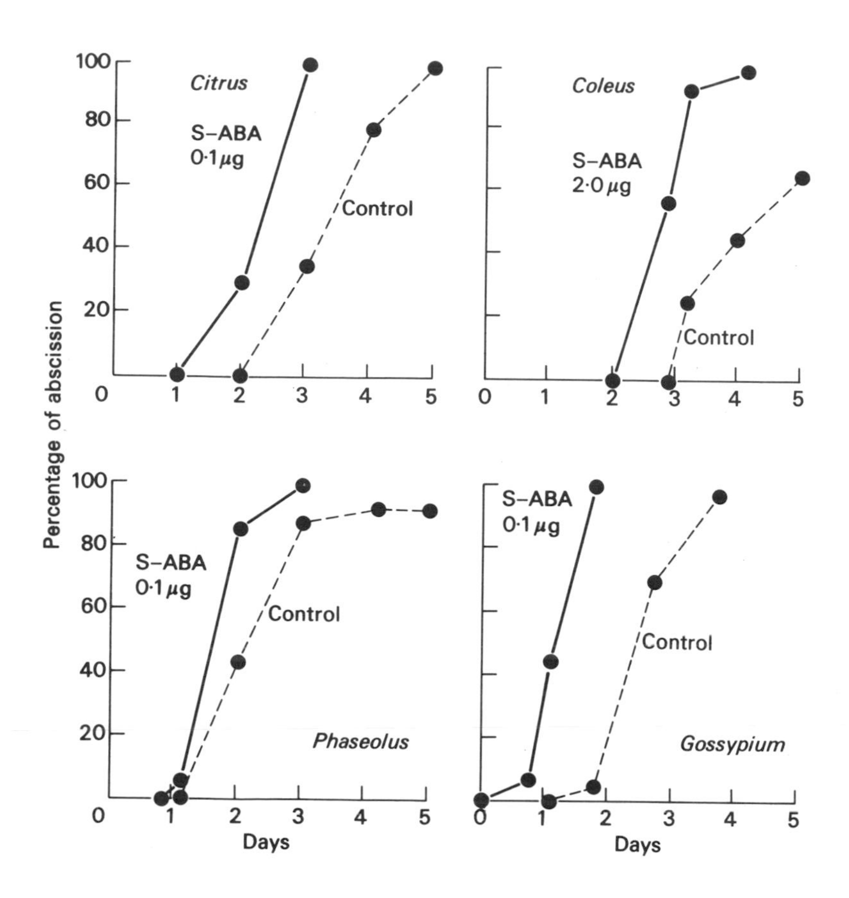

Figure 8-11. The abscission-speeding effect of abscisic acid (ABA) on explants of the four genera listed (Smith et al. 1968).

Since 1964, almost all subsequent research has been done with ABA rather than Abscisin I. The greater interest in ABA has been fostered by reports from other laboratories that ABA was present and active in their material. Several groups, Wareing's in particular, had been working on factors controlling dormancy in woody plants. Short days initiated dormancy in a number of these plants (e.g., Wareing 1954) and chromatographed extracts of shoot tissue revealed a parallel increase in the growth inhibition produced in coleoptile sections by eluates of the chromatogram zones that had been dubbed the *β-inhibitor* by Kefford (see Figure 3-3). Hemberg had suggested as long ago as 1949 that dormancy might be induced by the production of inhibitors, so botanists were quick to suggest that β-inhibitor inhibited the growth of woody shoots (as well as inhibiting the elongations of coleoptiles, the activity that provided its name). After persuading the famous organic chemist, Cornforth, to work on the project, Wareing's group found that ABA was in their *Acer* extracts, as well as in several other plant parts (Cornforth et al. 1965). Cornforth et al. (1965) soon thereafter described the synthesis of ABA. A substance speeding the abscission of young flowers of *Lupinus* and having antiauxin effect (Van Steveninck 1959) was also identified as ABA by Cornforth et al. (1966) and Koshimizu et al. (1966).

As soon as ABA was available, even if only in the tiny amounts one could obtain as gifts, it was tested in various systems previously believed to involve inhibitions. The flowering of several long-day plants was inhibited by added ABA (Evans 1966; El-Antably et al. 1967; Cathey 1968; Kinet et al. 1975), fitting the view that ABA might be a floral inhibitor made in SD leaves. A variety of other inhibitory responses were recorded in other systems (see review by Addicott & Lyon 1969).

Abscisic acid: senescence factor or β-inhibitor? Because of ABA's inhibitory and antiauxin effects, it is natural to ask if senescence factor is ABA. Böttger (1970a, b) confirmed on *Coleus* debladed leaf explants the abscission-speeding effect of ABA that Smith et al. (1968) had reported and added that ABA apparently was present in extracts of 6 kg of *Coleus* shoots (judging by an optical rotatory dispersion curve from the purified extract that was like that published by Milborrow [1968] for authentic [+]ABA). To see if ABA was present in the hypothesized higher amounts in diffusate from senescent leaves, Böttger considered the β-inhibitor activity on his chromatograms as if it were all due to ABA. Using the *Avena* straight-growth bioassay, on these assumptions, he found no ^{Rf}ABA^{Rf} in diffusate from young leaves (#2), a very small amount from grown leaves (#4), and a considerable amount from senescent (yellow) leaves. Böttger concluded that the big increase in ^{Rf}ABA^{Rf} in senescent leaves was what induced abscission (1970b).

Chang and Jacobs (1973) approached the ABA–senescence-factor problem from a different angle. They looked for the basis of action of senescence factor and asked if ABA could duplicate it. Sections cut from naturally senescent *Coleus* petioles were placed on petioles of younger explants as in Figure 8-9, so that senescence factor could diffuse for 22–24 hours into the petiole on the explant. This treatment gave the expected speeding of abscission of the younger petiole, of course. But, in addition, when sections were cut from explant petioles that had just been treated with senescence factor

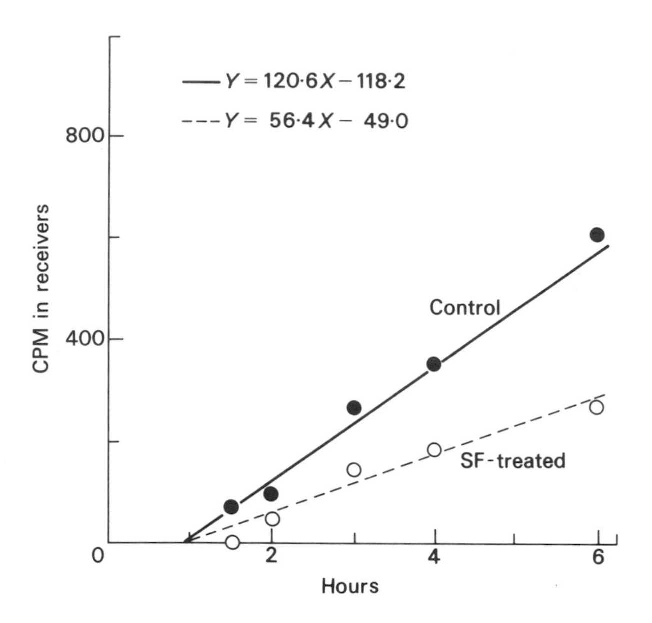

Figure 8-12. The time course of basipetal ^{14}C-IAA movement through *Coleus* petiole sections as affected by pretreatment with senescence factor (SF) (Chang & Jacobs 1973).

and their ability to transport ^{14}C-IAA was determined, results as shown in Figure 8-12 were obtained. Senescence factor decreased IAA transport in the direction leaf blade to abscission layer. This decreased basipetal movement was a reflection solely of decreased flux; there was no change in the velocity of 5.6 mm/hour (as measured by the time intercept). If the transport sections were extracted immediately after the 4-hour ^{14}C-IAA transport tests were over, the extract chromatographed, and the location of radioactivity measured on the chromatogram, the pretreatment with senescence factor was found to have increased the conjugation of the added ^{14}C-IAA with aspartic acid, judging by Rf values (Figure 8-13). The more senescence factor added, the less ^{14}C-IAA was transported basipetally and the more conjugation with aspartic acid occurred. The obvious conclusion was that senescence factor speeded abscission by decreasing the flow of IAA down the petiole from the leaf blade and that the decrease of transport resulted from increased conjugation with aspartic acid.

If ABA were added to the young petioles of the explants, it speeded abscission and decreased ^{14}C-IAA transport to an extent

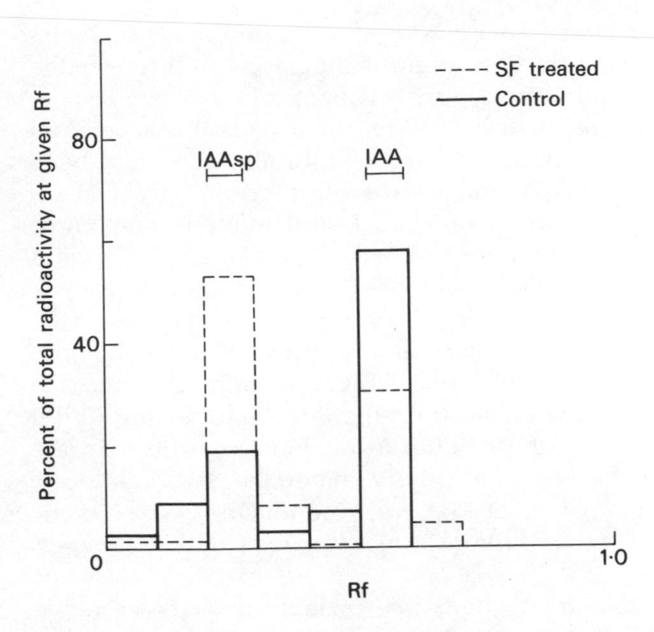

Figure 8-13. Decrease in free ^{14}C-IAA and increase in ^{14}C-IAAsp resulting from pretreatment with senescence factor of transport sections of *Coleus* petioles (Chang & Jacobs 1973).

that was statistically highly significant (Chang & Jacobs 1973). Increasing amounts of ABA progressively decreased IAA transport, paralleling the effects of adding increasing amounts of senescence factor, with 0.04 μg ($\pm$)-ABA having an effect equivalent to the diffusate from one senescent petiole. The decrease in ^{14}C-IAA movement brought about by prior treatment with ABA was also accompanied by an increase in ^{14}C-indoleacetylaspartate. The close similarity in the action of ABA and senescence factor on abscission, IAA transport, and IAA conjugation supported the hypothesis that senescence factor includes ABA or related compounds. And the abscission-speeding effects of both ABA and senescence factor could be explained by their inhibition of IAA movement.

A sardonic joke in experimental laboratories is that one sign of a skillful experimenter is knowing when to stop experimenting in order to write the research paper: one more experiment can cause the whole house of cards to tumble. Unfortunately for the hypothesis that ABA was the sole functional ingredient in senescence factor and in β-inhibitor zones, quite a few more experiments were done that did not support that hypothesis.

Doubts were first raised with respect to β-inhibitor. As I emphasized in discussing the use of chromatography to discover the chemical nature of endogenous auxins (Chapter 3), any one zone of a chromatogram is apt to include more than one substance. And even before ABA was discovered, the β-inhibitor zone had been shown to contain other compounds with inhibitory activity in auxin bioassays. Salicylic acid, cinnamic acid, and related compounds were reported to be in that zone (Köves 1957; Varga 1957), with some possiblity that scopoletin might be there too (e.g., Housley & Taylor 1958). Holst (1971) tried to substitute ABA for the β-inhibition from potato peelings in inhibition of the coleoptile straight-growth assay but found that ABA could inhibit elongation only to 20%, whereas the eluate from the β-inhibitor could inhibit elongation completely (100%). This result led her to look in the β-inhibitor eluate for other inhibitory materials. Two inhibitory zones other than that typical of ABA were found. One gave phenolic color reactions and had a Rf like that of salicylic acid; the other was uncharacterized.

The β-inhibitor zone from woody trees, which showed reasonable increases and decreases in activity in relation to SD-induced dormancy, was then reinvestigated by Lenton et al. (1972). They used gas-liquid chromatography to identify and measure the endogenous ABA. Contrary to their expectations, endogenous ABA did not increase in amount parallel to the previously reported increase in β-inhibition upon transfer of trees from LD to SD; in fact, in all three species ABA decreased under SD!

Spinach was one of the LDPs whose flowering was inhibited by added ABA, according to the first reports supporting the views that SD may be inhibiting flowering in LDPs by increasing the endogenous ABA levels. However, the levels of endogenous ABA (estimated by optical rotatory dispersion) did not decrease when spinach plants were shifted from SD to LD; in fact, ABA increased (Zeevaart 1971b) even though β-inhibition decreased (Wareing & El-Antably 1970).

Hence, for dormancy in both potato tuber and photoperiodically sensitive trees, as well as for flowering in LDPs, the initial hopes that ABA would explain the inhibitory effects have been dampened. What about ABA and senescence factor? Osborne et al. (1972) concluded that ABA and senescence factor are different compounds, with senescence factor affecting abscission indirectly by stimulating the production of ethylene. I find much of their data unconvincing. (For example, they claim separation of ABA and senescence factor, but all their chromatograms show the zone said to contain ABA immediately adjacent to the senescence factor zone. One would like to see at least a valley of activity between the two zones said to contain two different substances.)

The research group at Hamburg, which was interested in ABA not only through Böttger's work on abscission but also from Dörffling's long-term interest in the role of inhibitors in controlling the growth of lateral shoots, undertook a group effort to solve the senescence factor problem (Dörffling et al. 1978). They found three zones with abscission-speeding activity on chromatograms of extracts of *Coleus* petioles. One of these zones they identified as ABA by Rf on chromatograms and by retention time in gas chromatography. However, the amounts of ABA did not increase as leaves became senescent (the *concentration*, in fact, decreased). A second zone speeded abscission because it contained xanthoxin (a substance chemically similar to ABA, but with only 0.01 times the abscission-speeding activity). The active material(s) in the third active zone was not identified. Contrary to expectations from Osborne et al., ethylene production was not increased by material from any one of the three abscission-stimulating zones. Dörffling et al. concluded that there was a more or less constant level of abscission stimulators present in developing leaves but that the stimulators could only act when the level of IAA decreased sufficiently.

Summary

To summarize the ABA situation, although ABA added to plants inhibits various processes, no one has so far demonstrated that any of the naturally occurring inhibitory effects can be quantitatively replaced by substituting the endogenous levels of ABA – or, in fact, by substituting the endogenous levels of any combination of inhibitory chemicals. There are several likely explanations: (1) there may be endogenous inhibitors that are still unknown; (2) the increase in β-inhibition from chromatograms of extracts of the trees may be due to a decrease in gibberellins in that zone rather than an increase in ABA (as Lenton et al. 1972 suggested); (3) inhibition of development is much less specific than stimulation, so we should be especially wary of giving credence to the hypothesis that a substance is an endogenous inhibitor merely because it inhibits when added from the outside. Valuable though it is in any case, exact substitution with demonstration of exact replacement should be the general practice when dealing with inhibitors.

In answer to the question "What makes leaves abscise?" we currently have to say that we are not sure about the relative contribution of the various factors known to speed abscission. IAA is clearly the paramount preventer of abscission: as long as it moves down the petiole, abscission does not occur. But when the IAA available on the distal side of an abscission layer declines sufficiently, then abscission-speeding factors can operate. If the leaf is so old that its color has started to fade, the senescence factor (whether that is ABA, some other substance, or a mixture of substances) moves down the petiole and delivers the *coup de grâce* to the declining trickle of IAA, decreasing the basipetal flow of IAA while (or, perhaps, "by") increasing its conjugation with aspartic acid. The abscission-speeding effect of IAA available proximally from younger leaves presumably reinforces the effect of senescence factor.

9

Movement of hormones

Movement is ineluctably tied to the definition of hormones: hormones are substances, active in minute amounts, produced in one location and active in another location. In angiosperms this movement of hormones coordinates development. The characteristics of hormonal movement and the regulation of it are, therefore, of prime importance in understanding plant development. In this chapter I shall discuss the experiments and conclusions that have led to our current broader understanding of hormonal movement, its polarity, and its regulation in shoots. (Hormone movement in roots is discussed in the next chapter.)

Investigations of the last decade or two have wrought drastic changes in our views about hormone movement in higher plants. Some of the changes are diagrammed in Figure 9-1. In place of the belief that one hormone only, auxin, moves with strong polarity in plant tissues, investigators now have established the polar movement of other classes of hormones. And the movement of auxin itself, earlier thought to be remarkably unvarying, has since been shown to change with age, gravity, illumination, and growth, as well as from the effects of other hormones. These new findings in hormone movement naturally suggest changes in our view of the manner in which the development of higher plants is coordinated.

Auxin movement: early views and research techniques

The classical view of hormone movement was as follows (Went & Thimann 1937): auxin was the only class of hormone believed to move with polarity and it was thought to move with strong basipetal polarity (i.e., from the shoot tip toward the shoot base and from the root tip toward the root base). The rate of movement (about 12 mm/hour) was thought to be several orders of magnitude greater than diffusion could account for (Went 1928, van der Weij 1932) and

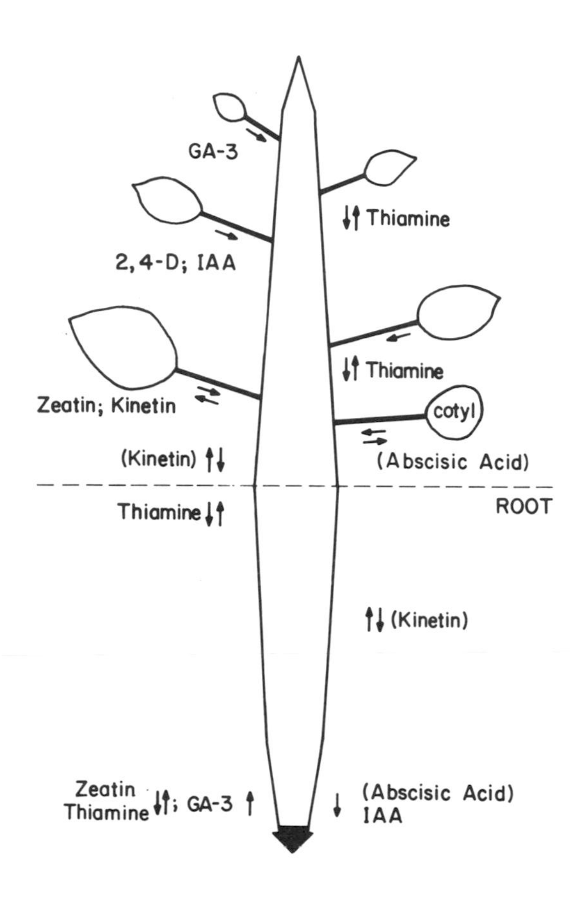

Figure 9-1. Changes since 1937 in information on the polarity, or lack of it, of various plant hormones in shoots and roots (Jacobs 1978b). (Arrows of equal length but in opposite directions indicate movement without polarity.) Note that no one species has been investigated using all these hormones. Also, these results are based on studies of only a single hormone added to one end of an excised transport section. The addition of a second hormone often affects the movement of the first (see text). Names in parentheses refer to results from other laboratories than Jacobs's.

to be unaffected by various external and internal factors. Unfortunately, most of the classical work was based on young seedlings grown in the dark, a rather narrow base for generalizing about the world of independently growing green plants! An additional handicap of the early work was that measurements of auxin were solely by bioassay, and with the primitive statistics then available the investigators could feel sure only about large differences in apparent auxin levels.

The classical technique for studying the polarity of auxin movement was surprisingly simple (Went 1928; van der Weij 1932). A section several millimeters long was cut from the shoot organ to be studied (usually a coleoptile). A block of 1.5% agar containing auxin (the donor block) was placed on one cut end and a plain agar receiver block was placed on the other (Figure 9-2). After a few hours the receiver block was removed and tested for auxin content with a bioassay. If the donor had been on the cut nearer to the original shoot or coleoptile tip, the basal receiver would typically show sizeable auxin activity. If the donor had been on the cut nearer the original root end of the coleoptile, there was typically little or no auxin activity detectable in the apical receiver. Gravity had no discernible effect on auxin movement. This was, therefore, the first evidence that the tissues of plants were polar in a way that was reflected in the polar movement of a plant hormone (Went 1928).

The major early investigator of auxin movement was van der Weij (1932, 1934). His results were the basis for the classical view of auxin movement. Unfortunately, the lack of precision of his data (often for reasons beyond his control) meant that the conclusions were largely based on guesses (Jacobs 1961). As in most such cases, sometimes the guesses could be confirmed by later work with more precise methods and sometimes they could not. The two conclusions of van der Weij that have been confirmed repeatedly by others are that auxin moves with strong basipetal polarity in coleoptiles and that the rate of movement is very roughly 10 mm/hour. He introduced the method of estimating rate of auxin movement that is illustrated in Figure 9-2. Van der Weij felt that the time course of auxin movement into a basal receiver followed reasonably closely a straight-line relation. On that assumption, he drew such lines and extrapolated them back to the time axis. That time intercept was presumably the time when the main front of auxin, added in the donor, started to arrive in the receiver agar. From the time intercept and the length of the transport section the rate of movement was then calculated. For the example in Figure 9-2 auxin moved through the 3-mm section in 0.3 hour for a calculated rate of 10 mm/hour. Van der Weij (1932) pointed out that the rate of auxin movement, estimated by him in this way as roughly 10 mm/hour, was paralleled by the rate of basipetal movement of tropistic curves in intact coleoptiles (Dolk 1930; DuBuy & Nuernbergk 1930), giving credence to the applicability to the intact plant of results from transport sec-

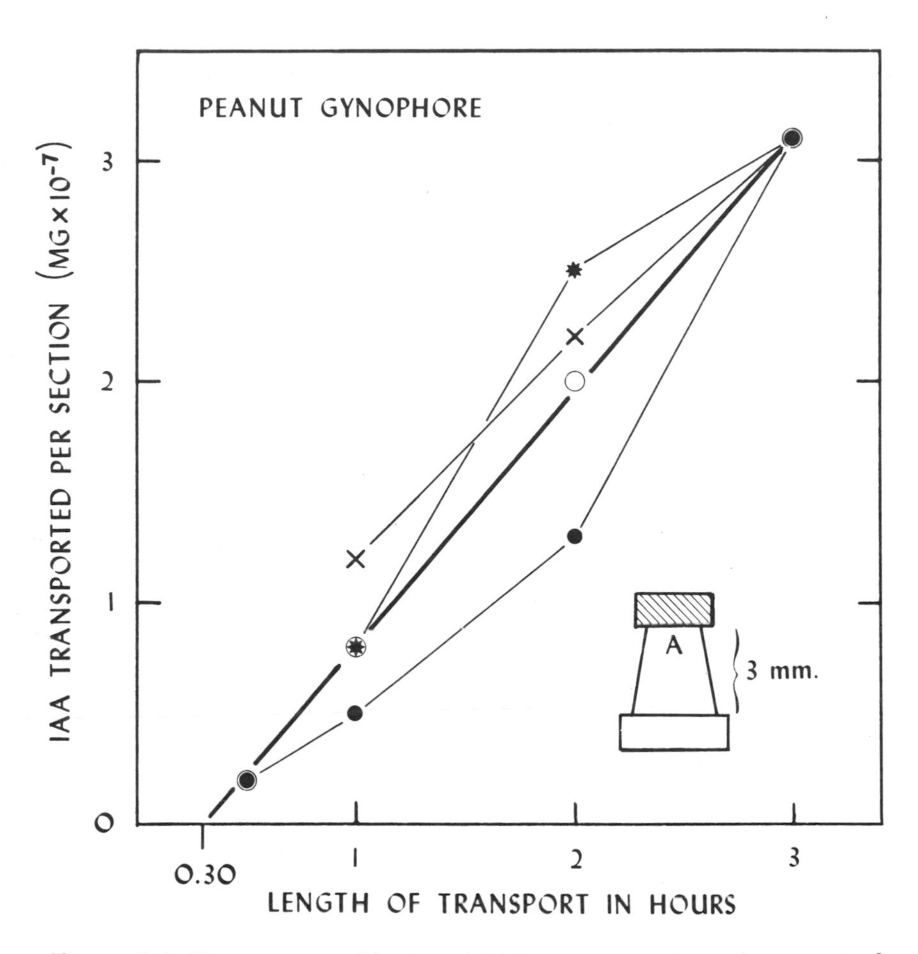

Figure 9-2. Time course of basipetal IAA movement through an excised section of the *Arachis* gynophore. The auxin in the basal receiver blocks was assayed with a calibrated *Avena* curve test. Results from three different days are shown by asterisks, crosses, and solid circles. Increasing sample size by pooling the results of the three experiments (open circles) revealed a linear accumulation of auxin in the basal receivers. The heavy line is the best-fitting straight line by regression calculations (Jacobs 1951, 1961).

tions. The strong basipetal polarity of movement of auxin through sections and of the phototropic curve in intact coleoptiles is another parallel that supports both the involvement of auxin in phototropism and the validity of using excised sections to study auxin movement.

Most other conclusions of van der Weij that have been critically checked by later workers have not been confirmed; and I think that in view of the limitations of technique under which the early investigators labored, it is not sensible to believe any early conclusions that have not yet been confirmed with quantitative reproducible data.

Improvements in techniques

The very different current views about hormone movement have resulted from questions about the generality of the classic view as well as from improvements in techniques. The first technical improvement was calibration of the *Avena* curve bioassay with synthetic indole-3-acetic acid (IAA); this allowed assays run on different days to be compared validly. The usefulness of this is demonstrated in Figure 9-2 (Jacobs 1951, 1961). The results from three transports and bioassays done on different days are shown with the three different symbols that are connected with thin lines. The pooled average of all three experiments is shown with open circles connected with the solid line. It is clear that it would be misleading to use only one experiment to try to determine the linear relationship shown by the pooled data, yet data from single experiments were all that workers had available to them before IAA calibration was introduced.

After IAA calibration made possible more accurate estimates of auxin levels, it became worthwhile to improve the data treatment. Instead of the subjective method of drawing the straight lines by eye as van der Weij had done (and as many others still do), the best-fitting straight line could be calculated by linear regression techniques, with the time intercept then calculated more accurately from the linear equation than from graphical extrapolation. Selected data of van der Weij were so treated by Gregory and Hancock (1955), and Jacobs adopted it for treatment of his own data (1961 and later papers). The solid line in Figure 9-2 is based on such linear regression calculations.

By far the most important technical advance, however, came directly from one of mankind's continuing efforts to find more efficient ways to kill his fellows, namely, the radioactive compounds that became available as a byproduct of making atomic bombs. Once IAA labeled with radioactive ^{14}C could be purchased from the Radiochemical Center at Amersham, studies of hormone movement were based increasingly on counts of radioactivity with occasional confirmation using bioassays. There were several abortive attempts in the 1950s to use radioactive auxins, but the first thorough treatments were in the doctoral theses of Goldsmith (1959) and Hertel

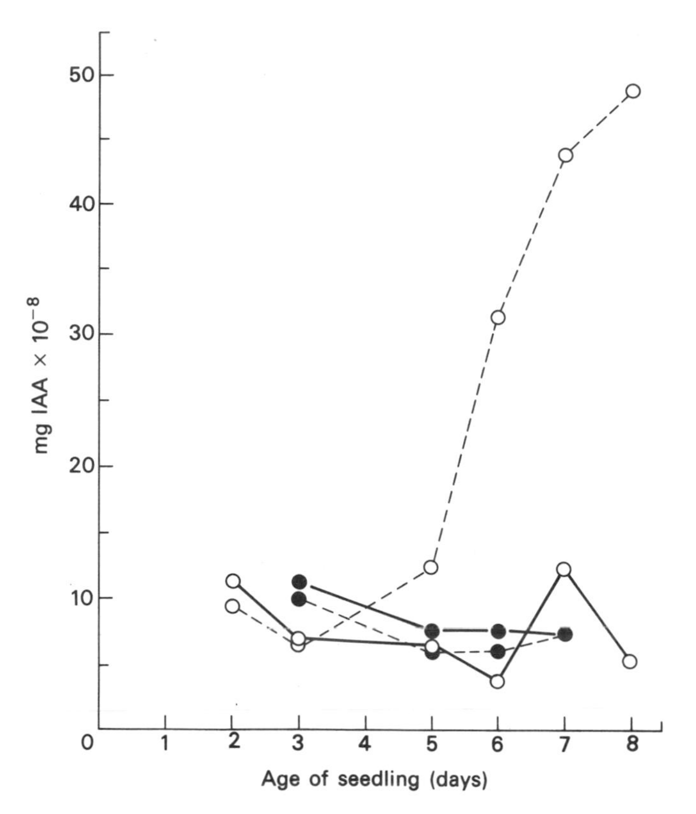

Figure 9-3. The development from day 2 to day 8 of basipetally polar IAA movement through the tip of the growing *Phaseolus* hypocotyl, as assayed with the *Avena* curve test (Jacobs 1950a). The amounts of endogenous auxin diffusing from control sections are shown by dots. The amounts collected when synthetic IAA (2 mg/liter) was added at the opposite end are shown by open circles (dashed line for basipetal movement, solid line for acropetal movement).

(1962), who used thin-window Geiger counters, and of Naqvi (1963), who adapted liquid scintillation counting for transport studies. The two main cautions in using ^{14}C-IAA are to be sure that the labeled material added to the plant is essentially pure IAA and that the radioactivity counted after the plant has acted on the added hormone is still all with the added compound. (A critical description of the techniques for using radioactively labeled hormones is given in the appendix to this chapter.) There has been remarkable unanimity in the reports that the radioactivity extracted from basal receiver blocks is still with IAA (judging by Rf) if the transport has not lasted more than 6–8 hours.

Changes in auxin movement with developmental stage

The classic view that IAA always moves with unvarying speed and intensity and with strict basipetal polarity has been shown to be incorrect in many instances. Polar movement of IAA was found to develop gradually in the tip zone of the young bean hypocotyl, no polar movement being apparent 3 days after germination but full basipetal polarity being present after 8 days (Figure 9-3; Jacobs 1950a; Smith & Jacobs 1968). The gradual development of polar IAA movement in the tip zone apparently acted to control growth in the zones below. The zone just below the tip was limited in its growth by available IAA and sucrose, judging by the growth of excised sections (Jacobs 1950b; Klein & Weisel 1964). In the intact seedling the necessary auxin came from the shoot above the tip region and had to pass through the tip to reach the fast-growing region below. There was a high and statistically significant correlation between the growth of the hypocotyl during days 3–8 and the amount of IAA that could be transported basipetally through the tip during that same period ($r = 0.94$). This quantitative parallel variation supported the hypothesis that it was the development of basipetal movement in the tip that controlled the elongation of the auxin-starved regions below (Jacobs 1950b).

The 8-day bean seedling had a long hypocotyl of quite uniform diameter and anatomy. However, it was far from uniform along its length in ability to move IAA (Figure 9-4). Sections cut from regions progressively nearer the shoot-root transition at the hypocotyl base moved progressively less IAA basipetally in a given time (Jacobs 1950a). All regions showed complete absence of acropetal IAA movement (i.e., no more auxin activity was found in apical receivers when IAA-containing donors were on the basal end than were found in control receivers that merely collected the small amounts of endogenous auxin from the untreated sections). These bioassay findings were confirmed using ^{14}C-IAA (Smith & Jacobs 1968). Such gradients in ability to move IAA have since been found along the axes of several organs (e.g., *Helianthus* seedling stems by Reiff & von Guttenberg 1961 and Leopold & Lam 1962; *Fritillaria* flower stalks by Kaldewey 1965; *Nicotiana* stems by Sheldrake 1973), and even some data in the classic literature on coleoptiles indicate such a situation (see Fig. 4 in Jacobs 1961).

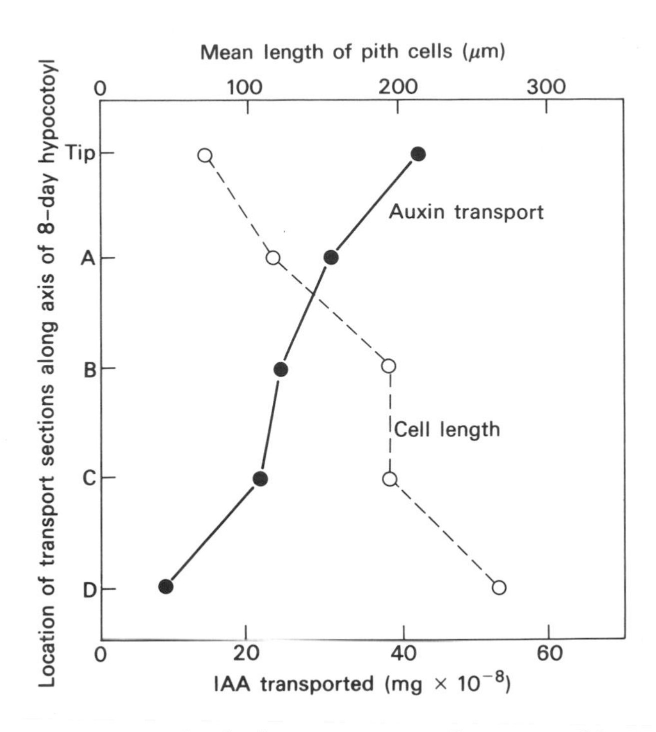

Figure 9-4. The declining amounts of IAA moved basipetally through sections cut from progressively more basal regions of the 8-day hypocotyl of *Phaseolus*. IAA at 2 mg/liter was added in the apical donors, and auxin in receivers was assayed with the *Avena* curve test. The average cell length increases as the ability to transport IAA declines (Jacobs 1950a).

IAA movement was found to be not always strictly basipetal. When young elongating internodes of greenhouse-grown *Coleus* were tested, instead of dark-grown coleoptiles of *Avena*, one-third as much IAA moved acropetally as basipetally (Jacobs 1952, 1954). These first results using the curvature bioassay were confirmed using liquid scintillation counting of ^{14}C-IAA (Naqvi & Gordon 1965). Evidence that the acropetal movement was not an artefact of the excised transport sections but was operative in the more intact plant came from parallel studies on xylem regeneration in such internodes (Jacobs 1956). There is extensive quantitative evidence that IAA is the limiting factor for xylem regeneration in *Coleus* (see Chapter 4). When *Coleus* plants were treated so that IAA would be available to the regenerating young internode only from below, the number of xylem strands regenerated in a week was one-third of the number regenerated in treatments providing IAA only from above (Figure 4-12). That is, there was quantitative parallel variation between IAA polarity in the excised sections (as measured by curvature bioassay or ^{14}C counting) and in the more intact plant (as measured by xylem regeneration). Leopold and dela Fuente have confirmed the presence of sizeable acropetal movement of ^{14}C-IAA through *Coleus* stem sections (1967).

After the fact it does not seem unreasonable that coleoptiles (seedling organs of simple structure, ephemeral life-span, and limited growth, specialized solely for orienting the shoot toward the light) should show different properties of hormone movement from histologically more complex, longer-lived, young internodes that normally have materials for the development of younger leaves and stem moving up through them from the older leaves, stem, and root system below.

Having found that young bean hypocotyls gradually develop polar movement of IAA, investigation of progressively older leaves revealed that polar movement gradually declined with age. Petiolar sections cut from progressively older primary leaves of *Phaseolus* showed increasing amounts of acropetal ^{14}C-IAA movement, coupled with a general decrease in the amount of basipetal movement (McCready & Jacobs 1963b; Leopold & de la Fuente 1968). Expressing these changes as the polarity ratio (amount of acropetal movement divided by the amount of basipetal) revealed a decline with age that was paralleled by a decline in the elongation of the sections (Figure 9-5). McCready and Jacobs (1963b) found such a relation also when the auxin-type weed killer 2,4-dichlorophenoxyacetic acid (2,4-D) was tested similarly in bean leaves. A decline with age of leaves in polar ^{14}C-IAA movement was also found in *Coleus* (Werblin & Jacobs 1967; Veen & Jacobs 1969a, see Figure 8-5), but in that organism the change with age was manifest only in decreased basipetal movement; even sections from very old *Coleus* leaves showed no significant acropetal movement of IAA. (The *Coleus* studies used leaves at six successive nodes as representatives of various ages; the *Phaseolus* studies compared leaves at one node that were from seedlings of various ages.)

The only other genus for which such data are available is *Xanthium*, famous as one of the few plants that can be induced to flower with only one short-day cycle. Sections cut from older petioles of *Xanthium* moved less ^{14}C-IAA basipetally and more acropetally than sections from younger petioles (Jacobs 1978a). With *Coleus* in the large Labiatae family, *Phaseolus* in the still larger Leguminosae,

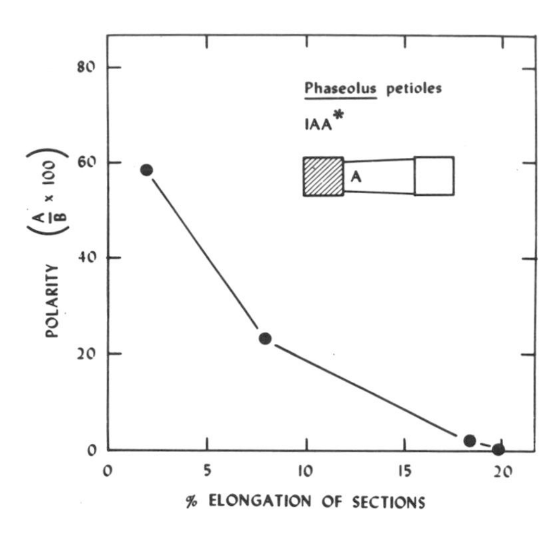

Figure 9-5. The decline in polarity of IAA movement (shown by an increase in the polar ratio) through progressively older *Phaseolus* petioles is associated with a progressive decline in the elongation of the older petioles. Data from youngest petioles tested are at the lower right (McCready & Jacobs 1963b).

and *Xanthium* in the Compositae, a third major family of angiosperms, the results on IAA movement through petioles stand some chance of representing the usual situation in angiosperms.

Aside from ageing, the onset of reproduction marks the most strikingly obvious change of state during plant development. Accordingly, the photoperiodic sensitivity of *Xanthium* was exploited to see if photoinduction changed parameters of IAA movement, as would be expected from the drastic changes in developmental physiology that accompany flowering. Surprisingly, there was no significant change in either basipetal or acropetal movement of IAA through petioles (Jacobs 1978a), even after as many as 5–7 short-day cycles, by which time the shoot apex of *Xanthium* would have been already converted into making reproductive organs (Kirk et al. 1967).

Changes in auxin movement resulting from changes in the environment

The classic literature said that as one increased the concentration of auxin in the donor agar, more and more auxin came into the basal receiver (van der Weij 1932; Went 1937). There was no sign of a

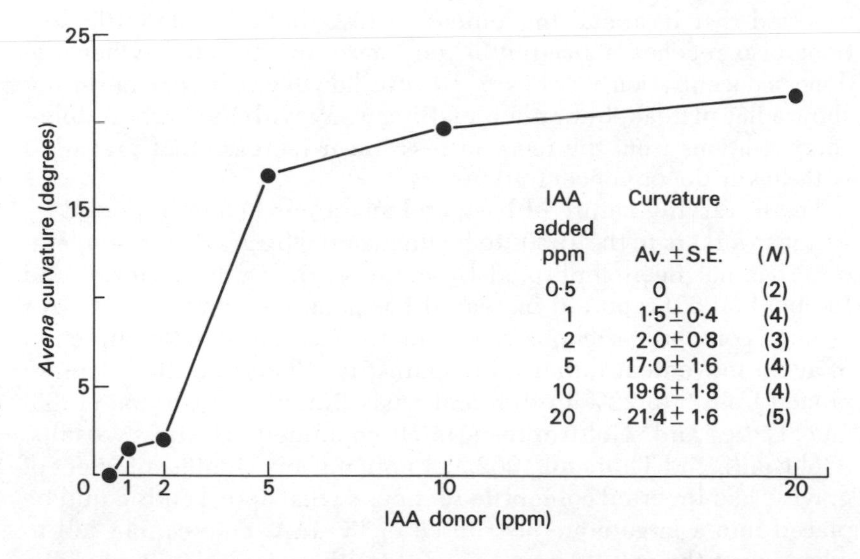

Figure 9-6. Evidence for a plateau in basipetal movement of IAA through young *Coleus* internode sections as the donor concentration is increased. Receivers were assayed with the *Avena* curve test (Scott & Jacobs 1963).

plateau, which is surprising in terms of other transport systems in the cell biology literature. Examining this relation, therefore, in young *Coleus* internodes of the developmental stage I had used for the xylem regeneration and 3 : 1 polarity studies, we found a clear plateau (Figure 9-6; Jacobs 1961 and Scott & Jacobs 1963 using the *Avena* curve assay, Naqvi 1963 using ^{14}C-IAA, and Thompson & Jacobs 1966 counting the number of xylem cells that regenerated as a bioassay). When donor concentrations of IAA were in the 2–5 mg/ liter range an absolute plateau of basipetal movement was reached: no increase in transport occurred even with 10 times as much in the donors. Jacobs (1961) pointed out the importance of such a strict maximum in the basipetal movement of IAA for controlling normal development – if the plateau occurs in the normal plant. This upper limit on how much IAA can move down through the young internode would be able to act as an upper-limit control on hormonal flow, preventing excessive amounts of auxin produced by the leaves from disrupting the normal correlations with which auxin is involved. We do not know how general such a plateau or saturation effect is. Naqvi (1963) did not find it in the older internode 5 of *Coleus*, using the same techniques that had revealed it in internode 2. Goldsmith and Thimann (1962) and Hertel and Leopold (1963) reported that basipetal movement of IAA through coleoptile sections also reaches a maximum, or "starts to saturate," when the donor concentration reaches 5–10 μm; however, their data do not show a flat plateau, but rather a falling away with the highest donor concentrations from the more or less linear increase that prevailed with lower donor concentrations.

The unvarying nature of basipetal auxin movement, reported by various workers in the 1930s to be unchanged by gravity or temperature, has not been confirmed by more recent studies. Hertel and Leopold (1963) reported increased basipetal movement of ^{14}C-IAA through coleoptile sections when gravity was acting in the direction of auxin movement rather than against it, saying that the increase ranged from 6 to 22% in different trials. Naqvi and Gordon (1966) and Little and Goldsmith (1967) confirmed Hertel's results. (Goldsmith and Thimann 1962, not noticing any significant effect of gravity, had inverted coleoptile sections so that apical ends could be placed into a large aqueous source of ^{14}C-IAA. One cannot tell to what extent the results from use of this otherwise useful technique were confounded by gravity effects.) Illuminating transport sections increased the IAA movement, as Scott and Wilkins (1969) first showed for *Zea* root sections and Koevenig and Jacobs (1972) demonstrated for *Coleus* stem sections. One of the anomalies of van der Weij's conclusions was that temperature had no effect on the rate of auxin movement in coleoptiles, but Gregory and Hancock (1955) demonstrated that linear regressions fitted to van der Weij's data did indicate an increase in both rate and slope as the temperature was increased from 2.5°C to 22°C. This fitted their own conclusion from their studies on diffusible auxin in woody apple sections. However, their technique, which involved studying the rate at which auxin "disappeared" from apical zones of stem cuttings, confounded temperature effects on auxin metabolism with those on auxin movement. The same caution applies to Kaldewey's use (1965) of the technique to arrive at the same conclusion, namely, that the rate of auxin transport, whether endogenous or added, increases with temperature to about 30°C and declines above that. Reverting to a van der Weij type determination but using ^{14}C-IAA instead of bioassays, Hertel and Leopold (1963) reported only a small effect of a 10° temperature increase (from 14° to 24°) on the rate of IAA movement through coleoptile sections, but a sizeable increase in the slope (their Fig. 4). Naqvi (1963) used the more objective techniques of regression-fitting and statistical testing of the significance of different intercepts for his studies on young internodes of *Coleus:* he found that increasing the temperature 10° significantly increased the rate of basipetal movement (from 2.8 to 4.8 mm/hour) but did not change the slope (his Fig. 11b). Several investigators had considered a $Q_{10} = 2$ as evidence that the process involved metabolism. (Q_{10} designates the rate of a process at one temperature divided by the rate of 10°C lower.) A $Q_{10} = 1$, by contrast, was considered evidence that the process was physical (as in diffusion). Chang and Jacobs (1972) showed that this assumption was not valid: although they confirmed that the Q_{10} was close to 2 (1.86) for ^{14}C-IAA movement through live sections of *Coleus*, they demonstrated that the Q_{10} of IAA movement through killed petioles or even through cylinders of agar of similar dimensions was also about 2 (2.10 and 1.94, respectively).

Van der Weij concluded that polar auxin movement could even continue against a gradient (as by placing auxin in receivers as well as in donors). In addition to the general problems of that period with accurate assaying and with artefacts from movement of auxin in surface films of water (see Went & White's 1939 critique of earlier work such as Went 1937), van der Weij considered loss from apical donors as evidence of basipetal movement into unassayed receivers that contained auxin at the start. We now know that this is an incorrect assumption: starting with McCready and Jacobs (1963a) and confirmed by many later studies, it has been shown that substantial loss from donors occurs irrespective of the concomitant occurrence of polar movement through the sections. We have recently investigated this question in young *Coleus* stems, prodded by unexpected results from our studies in xylem regeneration. If leaves were present only above the regenerating young internode, the amount of xylem regeneration was larger – usually to a statistically significant degree – than if all the leaves below the internode were also present (Aloni & Jacobs 1977a). In view of all the evidence that IAA is the limiting factor for xylem regeneration in this internode, the results suggested that somehow the leaves below were preventing as much IAA produced by the leaves above from moving down to the regenerating area. The most obvious guess as to mechanism was that, contrary to the classical view, the IAA produced by the leaves below decreased the basipetal gradient and thereby decreased basipetal IAA movement. Results from excised young *Coleus* internodes confirmed this hypothesis. Providing IAA solutions to the base of the sections, as well as to the apex, reduced the number of xylem strands that regenerated midway along the internode. And direct tests of ^{14}C-IAA movement through transport sections cut from such internodes confirmed that a drastic decrease in basipetal movement of ^{14}C-IAA occurred when unlabeled IAA was in the basal receivers (251 cpm versus 1 cpm; Jacobs & Aloni 1978).

The endogenous path of auxin movement

In what tissues does IAA move? A glance back at Figure 1-2 indicates that the coleoptile is a relatively simple organ in terms of heterogeneity of tissue (although it is scarcely all parenchyma except for the two small vascular strands, as some physiologists have considered it). Although there was some unresolved dispute in the early literature as to whether more auxin was transported basipetally when the donor blocks were placed over a side of the coleoptile that contained a vascular strand, no specific evidence was forthcoming for many years. Hertel and Leopold (1963) reported that ^{14}C-IAA could move through pieces of the coleoptile from which the vascular strands had been excised. The first specific evidence about cell type came from time-course studies on older *Coleus* stems: cylinders cut from the center of the stem with a cork-borer and shown by histological sections to consist solely of pith parenchyma cells transported ^{14}C-IAA in the classical, strongly polar manner (Figure 9-7, Jacobs & McCready 1967). Sheldrake, apparently independently, confirmed these findings using *Nicotiana* pith (1973).

Although the studies with pith cores demonstrate that excised pith parenchyma can transport IAA in classically polar fashion, they do not, of course, ensure that pith is the path for IAA movement in the intact plants. Some indication that IAA added from outside can move in phloem came from studies in which IAA was added on the leaf surface or through a flap of leaf-midrib to an otherwise intact *Phaseolus* plant (Little & Blackman 1963). No direct determination was made of the rate at which IAA moved through the plant, but assuming that curvatures of the *Phaseolus* stem indicated the recent arrival of IAA led them to estimates of 200–240 mm/hour – closer to the rate at which photosynthates have been estimated to move (Table 14.1 in Canny 1973) than to the rate of IAA movement in coleoptiles. This suggested that the IAA might be moving in the phloem, as the photosynthates were believed to do. More specific evidence that auxins can move in sieve tubes came from aphid-feeding studies (see Chapter 7 for technique). After methylene-labeled IAA was added to a leaf of *Vicia faba* – unfortunately at a high concentration and as a radiochemically impure mixture – aphids feeding on sieve-tube contents either above or below the treated leaf were found to have ^{14}C-IAA in their honeydew, judging by location of ^{14}C on

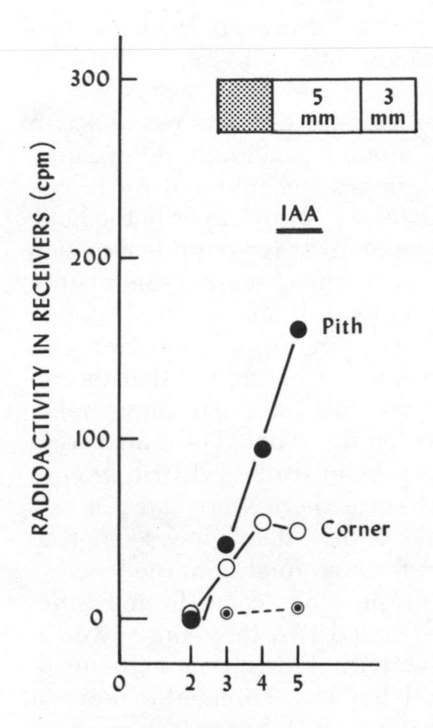

Figure 9-7. Time course of ^{14}C-IAA movement through cylinders cut from pith parenchyma or the corner of a *Coleus* stem (the "corner" comprised all cell types). Data for basipetal movement are connected by solid lines, those for acropetal movement by the dashed line (Jacobs & McCready 1967).

chromatographed extracts of honeydew as well as by the coleoptile section bioassay (Eschrich 1968).

These results raise an intriguing problem. If sieve tubes are the major or sole path for IAA movement in the intact plant, then the rate of movement of IAA in the intact plant would be expected to be much higher than that observed through the classical transport sections, because phloem physiologists have found that excision in most species renders the sieve tubes near the cuts nonfunctional. That is, excised sections might be expected to not show any sieve-tube-dependent movements (Eschrich 1968). The nice, quantitative agreement in coleoptiles between the rate of movement of the phototropic curve in the intact plant and the rate of IAA movement in excised sections one might explain away, according to this hypothesis, as reflecting the unusual paucity of sieve tubes in the coleoptile (see Figure 1-2). In such a specialized organ, we might argue, the fast, essentially nonpolar, movement that can occur through sieve tubes is not needed, and the slower, polar movement typical of IAA movement through other living cells can predominate.

However, if blocking of sieve tubes near the cuts produces an artefact in transport sections (and no one has actually determined that such is the case), there is also an artefact inherent in the approach of those who add labeled auxins to mature leaves in the hope of understanding normal auxin movement. It is young leaves that are typically the main producers of auxin (Figure 8-2); mature leaves produce little or none. Hence, even if adding ^{14}C-IAA to a mature leaf results in the label being swept around the plant with the photosynthate in the sieve tubes, it can be argued that the results have nothing to tell us about the "normal auxin movement" that comes from the youngest leaves on the shoot. This latter argument made little impression on those who studied distribution of label added to mature leaves, and since there were more investigators in that camp, and vocal ones at that, their view seemed to prevail more through decibels than through quality of evidence.

The first attempt to resolve these questions came from Bonnemain (1971). He added methylene-labeled IAA to young leaves or old leaves, then followed the radioactivity with autoradiographs of transverse sections of the stem. The label was estimated to move at about 10 mm/hour when added to a young leaf, but at 160 mm/hour when added to an old leaf. Furthermore, label from IAA added to the young leaf appeared most heavily in the cambial zone of the autoradiographs rather than in the sieve-tube area. By a different experimental pathway, and using *Pisum* rather than *Vicia*, Morris and Kadir (1972) arrived at essentially the same conclusion as Bonnemain. Avoiding all cuts, they confirmed that some of the ^{14}C-IAA added to mature leaves could be recovered from phloem-feeding aphids stationed above or below the application site. They also added a drop of carboxy-labeled IAA to the apical bud, on the expectation that it would supplement the normal auxin flow from the young leaves. This apically applied IAA did not move in sieve tubes, judging by its nonappearance in the aphids, but moved basipetally nevertheless (judging by the ^{14}C that could be extracted from the stem below). ^{14}C-sucrose moved like ^{14}C-IAA if both were added to that mature leaf, but when ^{14}C-sucrose was added to the apical bud there was no sign that it moved basipetally in the same 4.5 hours that moved many counts of ^{14}C-IAA down into the stem. A further indication that IAA from young leaves moved in a different system was that tri-iodobenzoic acid (TIBA) blocked its movement but not the movement of ^{14}C added as ^{14}C-IAA to a mature leaf

(Morris et al. 1973; the latter portion confirmed by Goldsmith et al. 1974).

In summary, there seems little doubt that, if ^{14}C-IAA is added to fully grown leaves, some of it at least can move in the sieve tubes at much faster rates than the classical 10 mm/hour through coleoptilar sections. However, there is substantial evidence that fully grown leaves produce little or no IAA. Hence, fast movement of IAA in sieve tubes is apparently "pharmacological," not biological. In contrast, Bonnemain and Morris present evidence that ^{14}C-IAA moving in unwounded plants from normal endogenous sites of IAA production (viz., the young leaves) moves at the slow rate previously observed through excised transport sections and that such movement is not in sieve tubes. This indicates that the earlier objection to use of excised sections (on the grounds that sections cannot be normal, because one would expect sieve tubes near the cuts to be nonfunctional) is not pertinent.

Nevertheless, we should not have illusions about the certainty of these conclusions. No one has measured total flow of endogenous IAA in stems and then accounted for its quantitatively by the observed flow in various cell types. Autoradiographs present many interpretive problems. Are the radioactive molecules still with the original chemical? Were the radioactive chemicals actually *moving* through the blackened cells of the autoradiograph or were they merely being stored there (analogous to guessing on the basis of an airplane photograph if all the cars seen in open parking lots of a big city indicated that parking lots were the main channel of car movement)? Does the amount of radioactivity sufficient to make a autoradiograph correspond in any meaningful way to the amount detected in agar receivers on the transport sections? These questions are all unanswered. But keeping these cautions in mind, current evidence supports the view that endogenous IAA does not move normally in sieve tubes.

The kinetics of auxin movement

The kinetics of basipetal IAA movement through transport sections have been studied often since van der Weij's time, with the usual wide range of quality in the reports. The problem of adequate sample size, already discussed in relation to the *Avena* bioassays of Figure 9-2, can be easily solved using ^{14}C-IAA. McCready's technique (1958, 1963) of pooling 16 receiver blocks of agar, melting them on one planchet, and then drying them down to a thin agar film for counting provided much better sampling. The use of this tech-

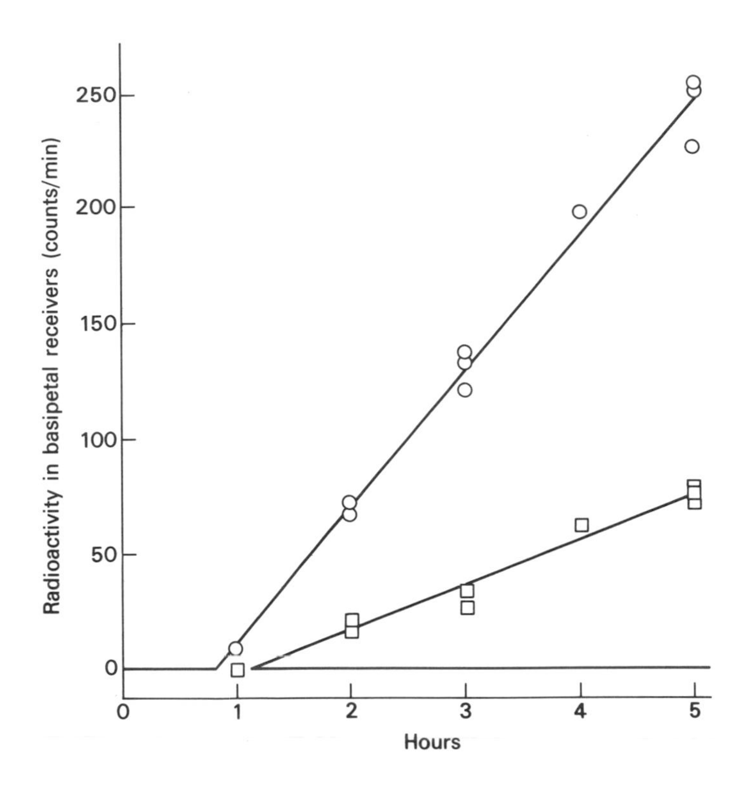

Figure 9-8. The time course of movement of ^{14}C-IAA into basal receivers on transport sections cut from *Phaseolus* petioles. Donor blocks of agar contained 5 (squares) or 50 (circles) μM IAA (McCready & Jacobs 1963a).

nique, along with the greater precision possible with radioisotope counting, is illustrated in Figure 9-8, where data from three experiments are shown, analogous to Figure 9-2. The best-fitting lines were fitted by least-squares technique to these data from young bean petioles and the results were analyzed statistically (McCready & Jacobs 1963a). The ^{14}C later shown to be still essentially all with IAA increased in the basal receivers in linear fashion for at least 5 hours, whether donors contained 5 or 50 μm IAA. The tenfold difference in initial IAA concentrations in the donors changed the slope of the line threefold but not its intercept. (A similar result was also reported by Hertel for *Zea* coleoptiles, although without statistical analysis [Fig. 6 of Hertel 1962].) The time intercepts of the two lines, calculated from the regression equations, were not significantly different by statistical test and averaged 0.96 hour, for a calculated rate of basipetal movement of 5.7 mm/hour.

Collections of receivers at 8 hours suggested a falling off from linearity and 12-hour collections showed no net increase in ^{14}C over 8 hours. A large absolute decline in counts in the receivers was found at 24 hours no matter which of the three concentrations had been in the original donors (Fig. 3 of McCready & Jacobs 1963a). Obviously, it is not sensible to collect only at 20–24 hours, as some workers have done, if one wants to study the primary process of hormone movement. An additional stricture against these long-term studies comes from Naqvi's (1963) studies of ^{14}C-IAA movement through *Coleus* stem sections: using liquid scintillation counting, he reported the same sort of large decline from 12 to 24 hours in receiver counts, but added that of the ^{14}C remaining in the receivers, a progressively smaller percentage was still with ^{Rf}IAA^{Rf} (81% at 8 hours, 75% at 12 hours, only 61% at 24).

Restricting ourselves to the early, and presumably more meaningful, hours of section transport, therefore, we find that Naqvi (1963) also fitted linear regression lines to his ^{14}C-IAA data from *Coleus* internodes 2 and 5 (young elongating and older with cambium, respectively) and from the equations calculated rates of 3–6 mm/hour for basipetal IAA movement. The data in his Figures 6a and 6b show fairly close fits to the straight lines, considering that his sample number was 2/datum, instead of the 16–48 of McCready and Jacobs. The six sections per sample and three replicates per point used in Naqvi and Gordon's (1966) study of gravity effects in *Zea* coleoptiles resulted in almost perfect matching of the regression line (Figure 9-9). The calculated rate of basipetal movement was 17.7 mm/hour at 25°C for that corn hybrid, 11–12 mm/hour for "Snowcross" variety. All the counts in the receivers in these early hours were still with IAA, judging by Rf after paper chromatography.

The kinetics of ^{14}C-IAA movement through sections from *Coleus* petioles of various ages was studied similarly by Werblin and Jacobs (1967) and data from their study were given in Figure 8-5. The rate of 4.8 mm/hour for basipetal movement did not change significantly as the petioles aged (calculated from their raw data, unpublished); only the slope decreased.

Data from *Phaseolus* hypocotyl sections also showed an exact fit to a straight line for the 1.5- to 3-hour collections (Smith & Jacobs 1968) for a calculated velocity of 4.0 mm/hour.

Hertel and Leopold (1963) used such large samples (20/data point) that their data fitted convincingly on their straight lines, even though regression calculations were not used. By extrapolating their lines "drawn by eye" (their Figure 5), they estimated rates of

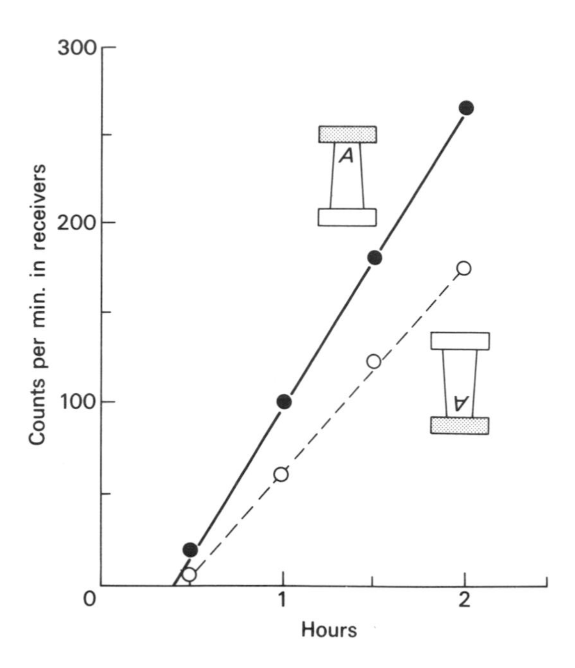

Figure 9-9. The effect of inversion with respect to gravity on the time course of basipetal ^{14}C-IAA movement through *Zea* coleoptile sections (Naqvi & Gordon 1966).

basipetal movement at 14 mm/hour through coleoptiles of *Zea* that were 4 days old, 10 mm/hour through those 3 days old, and 7 mm/hour through *Helianthus* epicotyls.

The general conclusion we can draw from these papers is that IAA accumulation in basal receivers follows a linear relation for several hours. (Of course, it is not unexpected that a close look at the region of the extrapolated intercept shows a brief period of accelerating counts before the linear accumulation is established: such transitions have been noted by Hertel & Leopold [1963] with *Helianthus* epicotyl, Thornton & Thimann [1967] for coleoptiles, and Smith & Jacobs [1968] for *Phaseolus* hypocotyl.)

Other pieces of evidence tend to confirm conclusions from the large-sample, quantitatively analyzed studies previously described. Some of these are based on extrapolations from the first two points of an apparently nonlinear accumulation in the receivers (e.g., Leopold & Lam 1961; Leopold & de la Fuente 1967; Greenwood & Goldsmith 1970; Koevenig & Sillix 1973). Others are based on entirely different techniques, such as measuring the rate at which a pulse of IAA either comes out into receivers that are frequently changed (van der Weij 1932; Hertel & Leopold 1963; Hertel & Flory 1968) or moves down within transport sections that are cut up into slices for extraction at successive times (Goldsmith 1967; Newman 1970; Bayer 1972).

The first calculations of best-fitting straight lines were not done because any theory of polar transport required them. They were looked on as merely a convenient way to describe the data, and as a nonbiased and more accurate way to calculate the time intercept and hence the rate of basipetal movement.

Investigation of differences between acropetal and basipetal movement in shoots

The idea that polarity of IAA movement might result from differential secretion at the upper and lower ends of the individual cells was suggested by Hertel and Leopold (1963). Leopold and Hall (1966) creatively elaborated on this, devising a mathematical model that demonstrated that a very small difference in the amount of IAA secreted at two opposite ends of a single cell would be amplified by passage through 100 or so such cells arranged in a linear file into a sizeable polarity of the overall movement. (A close relation between the number of cells in a transport section and the polarity of IAA movement had been pointed out earlier for *Phaseolus* hypocotyl sections [Fig. 9-4, from Jacobs 1950a, 1961].) Leopold and Hall calculated that the 3 : 1 ratio of basipetal to acropetal IAA movement that Jacobs had observed through sections from young stems of *Coleus* could result after enough time from the 187 pith cells along a 5-mm section even if those cells had individual polarities so low that 50.15% of the auxin in each cell would be secreted from the basal end and 49.85% from the apical. The simple form of the mathematical model resulted in concave predicted curves for the arrival of IAA in receivers, but adding a factor for the immobilization of some IAA in each cell could change the predicted arrival curves to "nearly linear," said Leopold and Hall. With more immobilization, they said, the curves would become increasingly convex.

In a companion paper dela Fuente and Leopold (1966) reported that actual IAA arrival curves, fitted by eye, were linear, concave, or convex, depending on the organ and species; they felt that their actual results thus fitted the mathematical model. We cannot compare validly their arrival curves with those mentioned earlier, because dela Fuente and Leopold gave arrival data as "cpm/g fresh weight" instead of "cpm." But aside from that difficulty, "linear, concave, and convex" cover fairly well the possible shapes of simple curves – and there was no basis provided for expecting the bean petiole, for instance, to have a linear curve as contrasted to the convex one reported for *Helianthus*. More satisfying was the evidence from extraction of transverse slices of transport sections that the polarity actually did increase as IAA transversed more cells, as the mathematical model predicted. They found sizeable counts in the apical receivers and from extrapolations of the arrival curves pointed out that the intercepts and therefore the rates were apparently the same for movement in the two directions. The next year identical rates of acropetal and basipetal movement of ^{14}C-IAA were demonstrated, through older bean petioles, with full statistical analysis (Figure 9-10, Jacobs 1967): in both directions the arrival curves were linear.

This work of Leopold and co-workers and of Jacobs focused attention not only on the kinetics of acropetal movement but also on its nature. There had been a general feeling among researchers on IAA movement that basipetal movement was fundamentally different from acropetal, with the latter considered to be mostly diffusional (e.g., McCready 1963; Goldsmith 1966). Leopold and Hall had suggested that acropetal movement might be due to secretion at the apical end of each cell, rather than merely diffusion. The similarity, except in slope, of Jacobs' acropetal and basipetal arrival curves supported the hypothesis that the processes were not fundamentally different in the two directions. The distinctions assumed to exist between IAA movement through living tissue and through nonliving "diffusion" systems were shown to be, at least in part, not valid by the research of Chang and Jacobs (1972), discussed earlier in this chapter.

Chemical specificity of polar auxin movement

The chemical specificity of polar auxin movement has been the subject of much research since the mid-1930s. The general result has been that the more growth activity a synthetic compound showed in one of the auxin bioassays, the more polar its movement was likely to be, although Thimann as early as 1935 (1935b) pointed out that some compounds could show little auxin activity in a bioassay requiring transport (such as the *Avena* curve assay), but substantial amounts in bioassays in which transport was of much less consequence (such as the *Avena* section growth assay). Using a modified *Avena* curvature test, Went and White (1939) estimated that the rate of basipetal movement through *Avena* coleoptile sections of

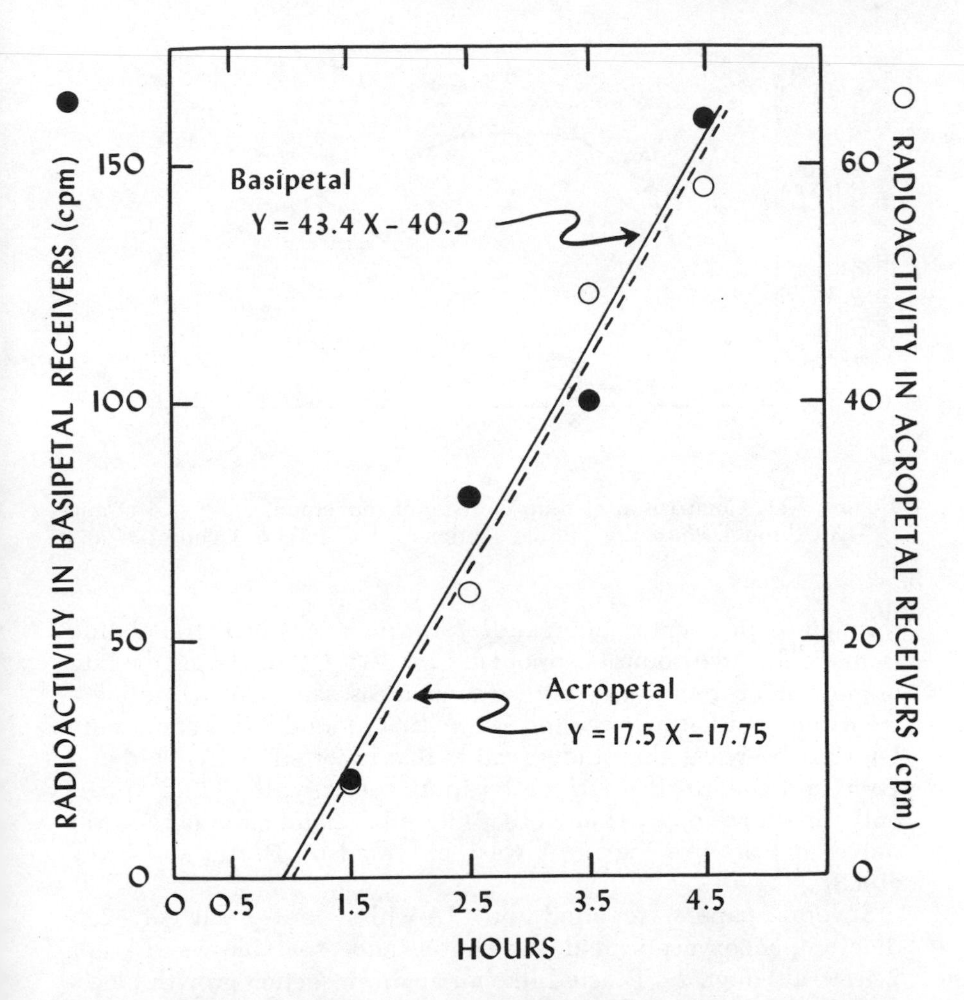

Figure 9-10. The time course of ^{14}C-IAA movement in the basipetal and acropetal directions through old *Phaseolus* petiole sections (Jacobs 1967). The lines are the best-fitting straight lines to the data; their equations are shown on the graph. The *X*-intercepts were not significantly different by statistical test, so the calculated velocity of movement in the two directions was not different either.

IAA, indole-3-butyric acid (IBA), anthracene acetic acid, and naphthalene acetic acid (NAA) was 9.0, 6.6, 5.4, and 3.8 mm/hour, respectively. Indolepropionic acid and *cis*-cinnamic acid apparently did not move in amounts sufficient to be detected in the three or so hours of their test. Rates in mm/hour of 7.5 (IAA), 6.7 (NAA), and 3.2 (IBA) were estimated for the basipetally polar movement

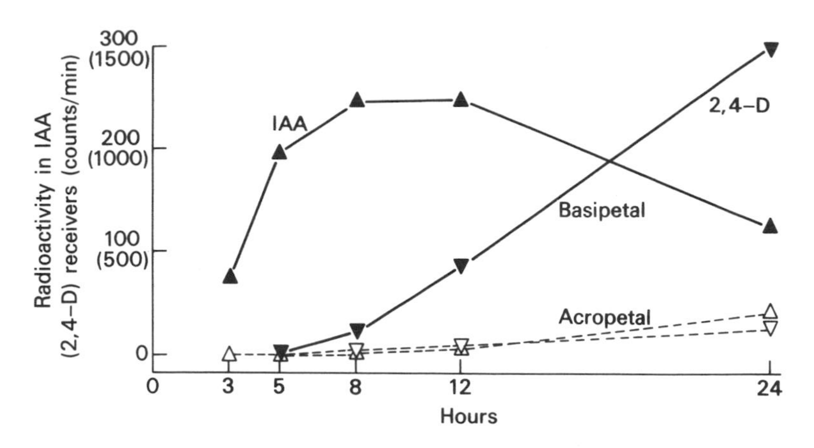

Figure 9-11. Comparison of time courses of movement of ^{14}C-2,4-D and ^{14}C-IAA through *Phaseolus* petiole sections (McCready & Jacobs 1963a).

through sections cut from green *Helianthus* stems by extrapolating from the first two points (Leopold & Lam 1961). Calculating the rate of movement from linear regression analysis using *Coleus* petiole–stem combinations, Gorter and Veen (1966) found a rate of 6.9 mm/hour for ^{14}C-NAA, almost identical to that reported by Leopold and Lam, and confirmed its strong basipetal polarity, the latter apparently first reported by Haupt (1956) for other plant material. Slower movement of NAA than IAA was confirmed by Hertel and Flory (1968).

Several papers reported no IAA-like movement of 2,4-dichlorophenoxyacetic acid (the famous and profitable weed killer 2,4-D), although 2,4-D acted like an auxin in section growth bioassays. Suspecting that this was because different measuring systems were being used for 2,4-D or because investigators were tacitly assuming that the rate of movement of 2,4-D must be approximately the same as that of IAA, McCready (1963) and McCready and Jacobs (1963a) tested ^{14}C-IAA and ^{14}C-2,4-D with the same transport system and for longer periods. The weed killer was then found to move much like IAA, in basipetally polar fashion but at a slower rate (Figure 9-11). Rates of basipetal movement of 2,4-D through bean petioles were 0.6 to 1.0 mm/hour from linear regression equations fitted to the data (McCready 1963). The same rate of 1.0 mm/hour was calculated by Naqvi (1963) from regression equations for ^{14}C-2,4-D basipetal movement through *Coleus* stem sections. He added statistical evidence that the rates were identical, only the slopes changing, when the donor concentration of 2,4-D was changed from 0.2 to 2.0 mg/liter – thus paralleling nicely the data for IAA in Figure 9-8. The basipetal polarity of 2,4-D movement and the rate of movement have been confirmed (Jacobs & McCready 1967 for polarity and rate in tissue cylinders cut from *Coleus* stems; Hertel & Flory 1968 for polarity and rate in *Zea* coleoptiles; Harel 1969 for polarity in bean petioles; Abrol & Audus 1973 for polarity and rate in *Helianthus* hypocotyls). The close similarity between 2,4-D and IAA is even manifested in the same sort of declining polarity with decreasing elongation in ageing bean petioles (McCready & Jacobs 1963b; Fig. 3 of Jacobs 1967). Gravity decreased the basipetal movement of ^{14}C-2,4-D in coleoptiles, as it does that of ^{14}C-IAA, but the effect on 2,4-D was greater (1540 cpm in the receivers after 2 hours in the normal position, 825 cpm for inverted section [Hertel & Flory 1968].)

The closeness of the relation between auxin activity and strongly polar movement was made even more convincing by studies of compounds closely related chemically but with large differences in auxin activity. The highly active, auxin-type weed killer 2,4,5-trichlorophenoxyacetic acid was much more polar than the weakly active 2,4,6-trichlorophenyoxyacetic acid (Jacobs 1967). With 20 μM of ^{14}C-labeled compound in the donors, the active auxin gave 2376 cpm/34 as the number of counts in basal as contrasted to apical receivers, whereas the inactive chemical gave 360/144 in the same 24-hour period (Jacobs 1968, recalculated to correct for the different specific activities). Shifting the side chain by one position in the auxin-active 1-NAA to 2-NAA removed most of the auxin activity as well as the basipetal movement through short coleoptile sections (as estimated by fluorescence of extracts of receivers [Hertel et al. 1969]). The (+)- form of 3-indole-2-methylacetic acid, which is active as an auxin, showed more basipetal movement than its optical isomer, which is less active as an auxin (Hertel et al. 1969). Similar results were found by Veen (1972), studying carboxy-labeled 1-NAA, 2-NAA, and decahydronaphthaleneacetic acid (1-NAA with saturated rings). The last two compounds showed no auxin activity in two bioassays and no polarity in their movement – only 1-NAA moved with polarity.

The relation between elongation and polar movement of auxin

The closeness of specificities for polar movement and for activity in bioassays involving growth suggests that both occur at the same site – a view that has been pushed by Hertel and co-workers and that has led them into their work on auxin-binding sites in the cell (see the section in Chapter 5 on the biochemical basis of auxin action).

The relation between polar movement of IAA and elongation has often been commented on. The correlation found in developing bean hypocotyls (Jacobs 1950a, b) has already been discussed, as has the close relation found in *Phaseolus* petioles between decreasing polarity and declining elongation (Figure 9-5). Scott and Briggs (1960) followed both the basipetally diffusible auxin (measured directly with *Avena* curve assay) and the auxin activity obtained with cold ether extraction, and related them to the elongation from different zones of epicotyls of green pea seedlings. There was not a close relation between either of the auxin determinations and the distribution of elongation along the pea epicotyl (their Figs. 2 and 3), although diffusible auxin showed a somewhat better match to the pattern of elongation than did the extractable auxin. With either extraction or diffusion the yields of auxin declined much farther down the epicotyl than did elongation. Similarly, Leopold and Lam (1962) confirmed and extended the report of Reiff and von Guttenberg (1961) that there was a declining gradient of basipetal auxin transport down the stem of sunflower seedlings (*Avena* curve assay being used for both papers). The later authors added that although elongation per 2-mm zone also declined down the stem, its decline was more precipitous than that of auxin transport. They concluded that growth was apparently not needed for auxin transport, because essentially no elongation occurred in basal sections that could still transport auxin. (They also pointed out that Scott and Briggs may have been measuring declining ability to transport rather than a decline in amounts of endogenous diffusible auxin.)

The elongation rate of the leaf of the fern *Osmunda* showed a close relation to the leaf's production of basipetally diffusible auxin – which may, here also, have been more a function of ability to move auxin than to produce it. The two parameters, measured over a 32-day period, gave curves of similar shape even in details, with the curve for diffusible auxin lagging a few days after the elongation curve (Fig. 1 of Steeves & Briggs 1960).

In addition to the preceding cases, in which a decline in elongation is associated with a decline in basipetal movement of auxin, there are several cases in the literature where ageing (presumably accompanied by a decline in elongation) is associated with declining polarity manifested partly by increasing acropetal movement (e.g., old *Coleus* stems and petioles [Leopold 1964], the basal portions of old hypocotyls [Smith & Jacobs 1968], old stamen filaments of *Cleome* [Koevenig & Sillix 1973]).

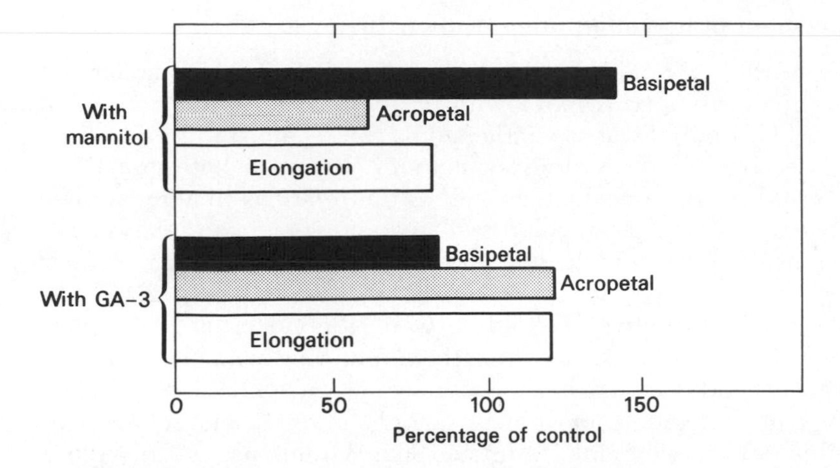

Figure 9-12. The effect of mannitol in decreasing the elongation of *Phaseolus* petiole sections and increasing the polarity of movement of ^{14}C-2,4-D and the opposite effect of GA-3 on both elongation and polarity (data from McCready & Jacobs 1967, figure unpublished).

Because the progressively older primary leaves of *Phaseolus* showed such a clear decrease of polarity for both IAA and 2,4-D with the decreasing elongation that accompanied ageing (Figure 9-5), McCready and Jacobs (1967) used that system to test the hypothesis that elongation was required for polar auxin movement. Sections from young petioles, which showed the typical strong basipetal polarity of ^{14}C-2,4-D movement, were treated with 0.2 M mannitol – just enough to stop their elongation when the mannitol was supplied in donor and receiver blocks. Not only was polar movement of 2,4-D not decreased by the osmoticum, but it increased 41% in the basipetal direction and decreased a like amount in the acropetal (Figure 9-12). In other words, approximately stopping elongation drastically increased polarity. Similarly, using 20 μM gibberellic acid in the donors to increase the elongation of the transport sections resulted in a decrease of polarity, which was a reflection of both a decrease in basipetal and a corresponding increase in acropetal movement of 2,4-D (Figure 9-12).

The conclusion from all these papers is that although the developmental stage that shows the maximal polarity or basipetal movement of auxin is likely to be the stage with the fastest elongation rate, concomitant elongation is not required for strongly polar basipetal movement of auxin.

Movement of hormones other than auxins

One of the long-held dogmas in the study of plant hormones has been that only auxins move with basipetal polarity through plant tissue. This belief was strengthened by the recent studies discussed earlier, showing the stereospecificity of auxins for both growth effects and polar movement, as well as the relations in a given plant between the strength of polarity of auxin movement and the amount of elongation. Even Picloram, a herbicide much used in the Vietnam war, showed auxin activity in bioassays coupled with 2,4-D-like polarity and kinetics (Kefford & Caso 1966, Horton & Fletcher 1968). The belief was only strengthened further when tryptophan, a favorite candidate as the endogenous precursor of IAA, showed either no detectable movement or only a small amount with no visible polarity (Schrank & Murrie 1962; Whitehouse & Zalik 1967). So firmly held was the belief that "only auxin moves with polarity" that some investigators jumped to the assumption that any developmental effect occurring below a stem position but not above it must have been due to the basipetally polar movement of auxin.

In most of the 1930s no other plant hormones were known, so one cannot fault the early investigators for not testing them. Went (1939) did, in fact, test the polarity of movement of sodium, bromine, and phosphate ions and found them to be nonpolar in *Avena* coleoptiles and *Helianthus* hypocotyls. However, by 1962, when McCready and I were doing our first study of 2,4-D and IAA movement under identical conditions, several new classes of hormone were known. It seemed sensible to test the polarity of their movement, too, under conditions that were standard for the auxin transport tests. My working hypothesis was that polar movement would not be restricted to the auxins but would be a more general characteristic of plant hormones (Jacobs 1972). The reason for thinking so was that the ability to move hormones in polar fashion would be expected to be so valuable to the higher plants in coordinating their development that one could expect that strong selection pressure would have been exerted in favor of it during evolution (Jacobs & Pruett 1972). Accordingly, we embarked on a program to test for movement and polarity of movement of cytokinins, gibberellins, thiamine, and other presumed hormones. So that the results would be as directly comparable as possible with other auxin studies, we used the same experimental arrangement.

How does one start such a program? From many earlier statements in this book you are aware that the investigator will want to use radioactively labeled compounds, if it is at all possible, rather than to plow through bioassays. Unfortunately, in the years immediately after a new hormone is discovered one cannot buy the hormone with a radioactive label. Demand is too low for the commercial radioisotope suppliers to take a chance. The only feasible source for a physiologist who cannot synthesize the compound himself is a friendly colleague, who has usually carried out the synthesis himself and who provides the chemical as a straight donation. As one who has been the lucky recipient of several such gifts (^{14}C-IAA from B. Stowe, ^{14}C-zeatin from E. Sondheimer, ^{3}H-GA-1 from L. Rappaport), I feel this generosity is too little known and is given too little credit. In the normal publication process, a footnote thanks the donor "for the generous gift." In the actual research, the labeled hormone can save several years of research over similar investigations forced to use bioassays.

Gibberellins

Several authors reported that gibberellin (GA) could move basipetally through stem sections (Kentzer & Libbert 1961, in a qualitative study of *Helianthus* using direct bioassay of 3-hour receivers; and Cohen et al. 1966, who found significant activity in 12-hour receivers from *Pisum;* in both studies GA-3 was added in the donors). GA-1 or GA-5 was added in donors to 15-mm-long *Pisum* sections and after 20 hours gibberellin activity was found in the receivers from both treatments (Jones & Lang 1968). They tested receiver-extracts directly with the dwarf-corn bioassay (Jones, personal communication), so there was, of course, no evidence as to what gibberellins were in the receivers. (Jones and Lang ran their transport experiment because they had initially thought that GA-1 might be mobile but not GA-5: this had been suggested by their observation that gibberellin activity was found at the Rfs typical of both these gibberellins when extracts of shoots were chromatographed, but only ^{Rf}GA-1^{Rf} was found when diffusates of the apical bud were chromatographed. Cleland and Zeevaart 1970 similarly found ^{Rf}GA-5^{Rf} in extracts but not in diffusates of *Silene* shoots.)

In testing for polarity of movement of gibberellic acid, we followed several principles. First, only physiological doses or concentrations should be added. (Kato's report [1958] of movement of GA-3 without polarity was based on the addition of extremely large concentrations in the donors; these unphysiological levels were made mandatory, apparently, by his use of the relatively insensitive ultraviolet absorption to measure GA in the receivers. There is no reason to doubt the accuracy of the data, but much reason to doubt its applicability to the normal plant.) Second, by analogy with the extensive auxin results, we looked for polar GA movement in organs at a developmental stage known to be responsive to added GA-3. For our *Coleus* clone, young petiole 3 was selected as a developmental stage that elongated more and abscised faster under the influence of GA-3 (Jacobs & Kirk 1966). Third, because lack of labeled GA forced us, for the first year, to measure GA with bioassays, we selected the barley endosperm test from the various bioassays (see the techniques section of Chapter 7) as being the one that seemed to measure GA by the biologically most direct method and was also one of the most sensitive. (The high sensitivity was important because we were adding low physiological amounts in the donors and could thus expect much less to appear in the receivers.) Fourth, transport sections at least 5 mm long were used because it is well established for auxin movement that polarity declines as section length decreases (e.g., dela Fuente & Leopold 1966). (The brief report of Hertel et al. 1969 that radioactivity added as ^{3}H-GA-1 moves without polarity through *Zea* coleoptile sections should be viewed with some skepticism on this account [the sections were only 2 mm long] as well as because there was apparently no evidence that the radioactivity in the receivers was still with GA-1.) A flow sheet of the methods used is given in Figure 7-12.

GA was found to move with basipetal polarity through sections of these young petioles (Jacobs & Kaldewey 1970; Jacobs & Pruett 1972). The time course of basipetal movement (Figure 9-13) was remarkably similar to that of IAA, showing a linear accumulation in the receivers for the first 5–6 hours with a later decline. (The heavy line in the figure is the best-fitting line by least-squares regression analysis.) By using the equation to determine the time intercept, as earlier described for IAA, we could calculate the rate of movement of the main front of gibberellin. The rate was 1.4–1.8 mm/hour in different experiments, the same order of magnitude as that of IAA through green tissue, although a bit slower. Gibberellin movement was like that of IAA through such petioles in that 1% of the activity in the original donors was found in the receivers at 5–6 hours (Jacobs & Pruett 1972) (i.e., with 1 μg GA-3 in the original donors, gibberellin activity equivalent to 10 ng of GA-3 was found in the receivers by bioassay). The most striking difference between auxins and GA-3 in their polar movement through petioles is that not all the gibberellin activity in the receivers could be attributed to unchanged GA-3. Auxins come out into receivers apparently unchanged, but when GA-3 receivers were extracted, thin-layer chromatograms run, the GA-3 Rf zone eluted and assayed with the

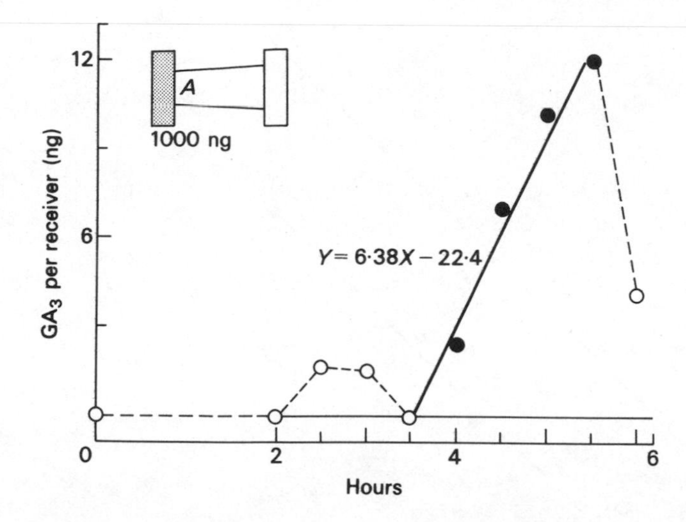

Figure 9-13. Time course of gibberellin accumulation in basal receivers on sections cut from young *Coleus* petioles (Jacobs & Pruett 1972). GA-3 was added in the donor blocks, and GA activity in the receivers was assayed with barley endosperm, then calculated as "GA-3." The equation is that of the heavy line and the solid circles.

barley endosperm test, the amount of ^{Rf}GA-3^{Rf} was only 20–30% of the total gibberellin activity in directly assayed basal receivers – even *after* corrections had been made for the 20% loss found by calibrating the entire procedure.

By the different route of using autoradiographs of tissue collected every 0.5 hour, G. H. Palmer (1972) arrived at the similar estimate of 2.5 mm/hour for ^{14}C-GA-3 movement in the embryo of barley, assuming that the ^{14}C was still with GA-3.

GA-1 labeled with ^{3}H was reported to move with polarity through 40-mm sections of bean stem and into receivers (Phillips & Hartung 1974). There were, however, significantly more counts in the apical than the basal receivers after the 16 hours – polarity that was opposite in direction to that found for GA-3 in *Coleus* petioles. Phillips and Hartung interpreted their polarity as the result of GA-1 moving toward elongating regions, said to be near the top of the long sections.

Thiamine

Thiamine, which is one of the B complex of vitamins for some of us animals, is a likely candidate as a root growth hormone in higher

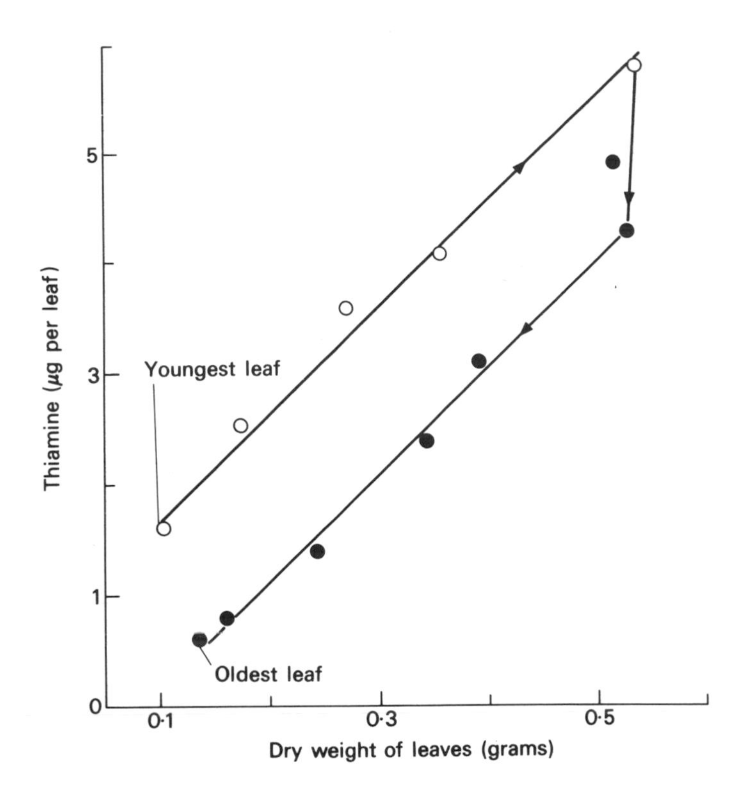

Figure 9-14. The content of thiamine in tomato leaves as plotted against the dry weight of the leaves. The arrows indicate the sequence of leaves from the youngest to the oldest (data from table of Bonner 1942, figure unpublished).

plants. When excised roots of several species were grown in aseptic culture, thiamine had to be added at hormonal levels to the medium (along with minerals, sucrose, and other predictable substances) to achieve continued growth (Bonner 1937, 1940; Robbins & Bartley 1937; White 1937). In the intact plant most thiamine is found in the leaves. If one could assume that the needs of the roots when excised accurately reflected their needs on the intact plant, it was a reasonable guess that thiamine acted as a hormone by moving from the leaves down to the roots. (Although the highest concentration of thiamine is found in the younger leaves of tomato, the older leaves were considered to be the major site of synthesis on the basis of data showing an accumulation of thiamine below a piece of stem killed above the older leaves and above a piece killed below them [Bonner 1942]. An intriguing double-linear relation is obtained [Figure 9-14, unpublished] by plotting the data from Bonner's paper.)

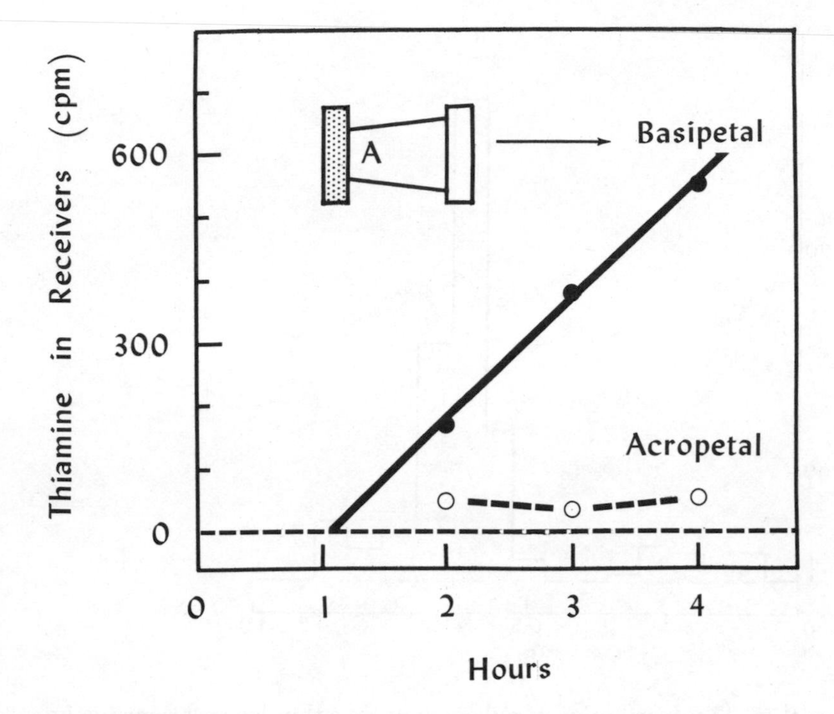

Figure 9-15. The time course of ^{14}C-thiamine movement in the basipetal and acropetal directions through tomato petiole sections (Kruszewski & Jacobs 1974).

Applying the same principles and techniques used for our GA-3 studies, we investigated the polarity of ^{14}C-thiamine movement through petioles of older leaves (the presumed endogenous sources), and we selected tomato (*Lycopersicum esculentum*) as a species shown by several early investigators to need thiamine for growth of excised roots in vitro. Strongly polar movement was found in the predicted leaf-blade-to-stem direction, that is, from the site of synthesis toward the site of presumed use in roots (Kruszewski & Jacobs 1974). The time course showed a linear accumulation in the basal receivers (Figure 9-15), paralleling that observed for auxins and gibberellin. All the radioactivity in the receivers was still with thiamine, judging by its Rf (Figure 9-16). The rate of basipetal movement, calculated in the standard way from the time intercept of the extrapolated best-fitting straight line, was 3.4 and 4.7 mm/hour in the two experiments – close to that of auxins and gibberellin through petiolar sections. The most striking difference from the auxins and gibberellin was the large amount of thiamine moved in

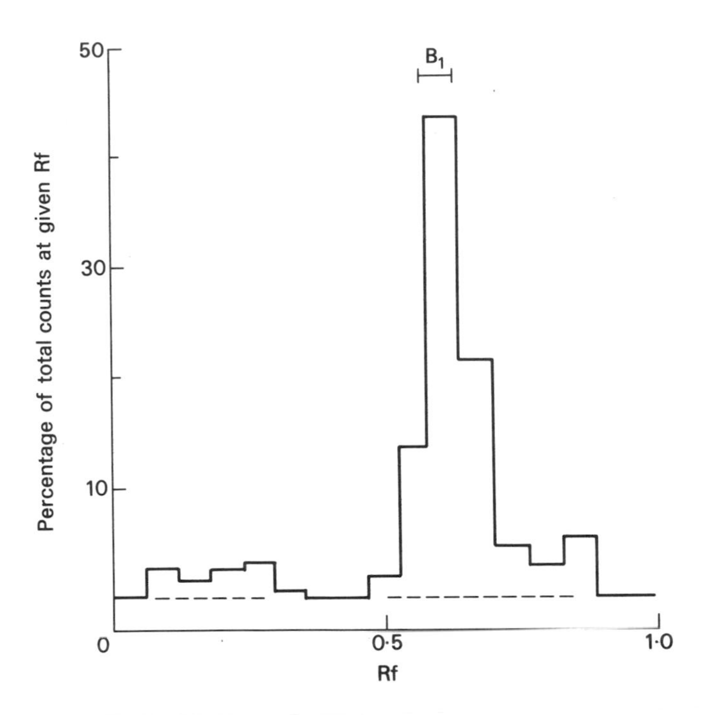

Figure 9-16. The distribution of radioactivity on a thin-layer chromatogram of an extract of basal receiver blocks used for ^{14}C-thiamine transport. The location of marker thiamine (B_1) is shown above (Kruszewski & Jacobs 1974).

these few hours; as much as 25% of the thiamine gone from the donor blocks was in the receivers by 4 hours, in contrast to the 5% or less of IAA and GA moved through petiolar sections of *Coleus* in that same time (Jacobs & DeMuth 1977). The ability of green petiolar sections of tomato to move thiamine approaches, in fact, the legendary ability of etiolated coleoptiles to move large amounts of IAA.

When transport sections were taken from the stem, whether above or below the node to which such older leaves were attached, there was significant movement of ^{14}C-thiamine but no polarity (Table 2 of Jacobs & DeMuth 1977). All the radioactivity being measured was still with thiamine, judging by its location on chromatograms. Considered in the light of the evidence from *Coleus*, these results suggest that green stems are typically not as polar with respect to hormone movement as are the petioles. (In *Coleus*, you will remember, strong polarity of IAA movement in petiole 2 was associated with the weaker 3 : 1 polarity in the adjacent stem, and GA-3 movement through sections from the young stem was weak enough [Greenblatt & Jacobs, unpublished] to force us to switch our studies to the young petioles, where more strongly polar GA movement was found.) The evidence from Bonner's early work indicating that thiamine moves both up the stem to the younger leaves and down the stem to the roots fits our evidence from transport sections that there is no polarity of thiamine movement in the stem.

The discoveries of polar movement of some gibberellins and of thiamine are too recent for the necessary evidence to have accumulated that such polarity, observed in sections, has quantitative parallels with phenomena controlled by those hormones in the intact plant. Only for gibberellic acid is such information starting to appear. Aloni (1976) has shown that the differentiation of phloem fibers in older portions of stems of the Princeton clone of *Coleus* is controlled by the leaves and that the leaf effect shows strong basipetal polarity. For complete replacement, both GA-3 and IAA are required as substitutes for the leaves, suggesting that both hormones move with polarity (Aloni 1979). Couillerot & Bonnemain (1975) added GA-3 labeled with ^{14}C to leaves of intact tomato plants, and followed the time course of distribution of the label. For the first three hours only a tiny fraction (0.02%) of the added label left the leaf to which it had been added. The label moved only basipetally during that time. Both findings fit expectations from the experiments with excised transport sections.

Cytokinins

The movement of cytokinins, as another class of growth regulators, has been subject to much uncertainty. Some physiologists thought cytokinins moved little if at all in the plant – if so, cytokinins would not be hormones by definition. However, this belief was based largely on indirect experiments in which cytokinin effects were not seen at a distance from the site of application, a procedure that carries much less weight than direct determinations. Osborne and Black (1964, Black & Osborne 1965) were the first to study cytokinin movement in the standard transport arrangement, using a gift from van Overbeek of the synthetic cytokinin ^{14}C-benzyladenine. After movement through *Phaseolus* petiole sections, the radioactivity in the agar receivers was extracted in ethanol, concentrated, then placed on filter paper for electrophoresis. From autoradiographs of the electrophoresed material, Black and Osborne concluded that "most, possibly all" of the ^{14}C was with unchanged benzyladenine.

The photograph they present shows only one band of radioactivity, but there is no way for us to tell what percent of the total radioactivity could be at other locations and still not show up in the autoradiograph. (This point is important in view of Fox & Weis's contrary results discussed later.) ^{14}C-benzyladenine showed basipetal polarity of movement in 22–24 hours, whether sections were freshly cut or were from pieces of the seedling that had been artificially aged for 16–18 hours. The polar ratios were 6 or 7 to 1. The rate of basipetal movement was estimated as 0.73 mm/hour, by extrapolating from three points that looked quite linear in the time course. (No regression calculations or statistical tests of significance were given.) The weak cytokinin ^{14}C-adenine, was added to *Phaseolus* sections for comparison: 7 μM in donors resulted in no counts in receivers, 70 μM gave only traces. With 364 μM in donors 508 cpm were found in basal receivers after 23 hours compared to 132 in apical receivers, but Black and Osborne stated that none of the counts were still with adenine – all were in another band after electrophoresis. This band, at the Rf of neither adenine nor adenosine, showed no kinin activity in tobacco pith bioassay (Osborne et al. 1968). Libbert and Kanter (1966), using *Helianthus* hypocotyls and color tests on thin-layer chromatograms, confirmed that adenine did not move through transport sections into receivers.

Fox and Weis (1965) used methods that would be expected to give more reliable data (e.g., a liquid scintillation counter rather than a 5% efficient Geiger tube, and benzyladenine of higher specific activity) but confirmed neither the polarity nor the chromatographic purity of ^{14}C-benzyladenine movement. For all four genera tested there was sizeable movement in both directions of ^{14}C into 24-hour receivers, but also in all four there was substantial radioactivity at zones other than that typical of benzyladenine (about 30% for receivers from *Phaseolus* petioles, up to 40% for *Avena* coleoptiles). Osborne and McCready's (1965) "answering" article to Fox and Weis did not resolve the conflict. It ignored the latter's evidence that much of the ^{14}C in receivers was with compounds other than benzyladenine, pointed out that different varieties of *Phaseolus* had been used (always a possibility for explaining differing results), and provided data interpreted by them as showing polar movement through green and etiolated *Pisum* stem segments and *Phaseolus* petioles, as well as movement without "clear evidence of polarity" through *Xanthium* petioles. Unfortunately, neither pair of authors provided statistical tests of significance for the reality of the difference between counts in apical and basal receivers or between receiver counts and background – the latter being particularly needed for judging the reality of Osborne and McCready's 5–7 counts above background for *Xanthium* or 8–38 for *Pisum*.

Guern and Hugon (1966), who injected labeled benzyladenine into the stem base of etiolated *Cicer* seedlings, found more ^{14}C above the injection site than below, but stated that after various intervals "almost all" of the radioactivity extracted from the apical buds was with a substance different from benzyladenine. No evidence was presented as to whether benzyladenine moved acropetally and then the new substance was formed or whether conversion to the new substance occurred prior to its acropetal movement. The new substance was 6-benzyladenosine, thought Guern et al. (1968), basing their conclusion on its Rf and the fact that treatment with nucleoside phosphorylase caused the Rf of the labeled material to change to the Rf of 6-benzyladenine.

Veen and Jacobs (1969b) wished to test the transport of cytokinins as McCready and Jacobs (1963a) had tested that of the auxins IAA and 2,4-D. With the movement, much less the polarity, of benzyladenine being so much in question, they turned to the artificial kinetin and the endogenous (though weak) adenine as counterparts among the cytokinins for 2,4-D and IAA, respectively. Evidence that kinetin could move in stem tissue had already been provided by Seth et al. (1966), who added ^{14}C-kinetin to the top of a decapitated shoot and found a few counts at the Rf of kinetin when stem pieces 40 mm below were extracted and chromatographed. Using the standard *Coleus* petiole sections, Veen and Jacobs found that both adenine and kinetin moved through the sections but completely without polarity. The time course of ^{14}C-kinetin movement in the two directions through older petiole 7 is shown in Figure 9-17 (Jacobs 1972). The rate of movement, calculated from the time intercept of the regression equations, was 1.1 mm/hour in both directions, similar to that found for 2,4-D. All the counts in the 24-hour receivers, whether apical or basal, were still with kinetin, judging by thin-layer chromatography (Fig. 4 of Veen & Jacobs). Kinetin's movement paralleled that of 2,4-D in another way too: about 7% of the hormone leaving the donors was found in the receivers at 24 hours. In short, the artificial cytokinin was like the artificial auxin in the kinetics measured, except that (1) the kinetin was completely nonpolar, (2) older petioles moved more kinetin than younger ones (Figure 9-18, from Jacobs 1972) even after correction for differences in uptake (Veen and Jacobs 1969b), whereas the reverse was true for 2,4-D. ^{14}C-adenine was like ^{14}C-IAA in showing a maximal number of counts in receivers at earlier hours, with a sizeable decline in

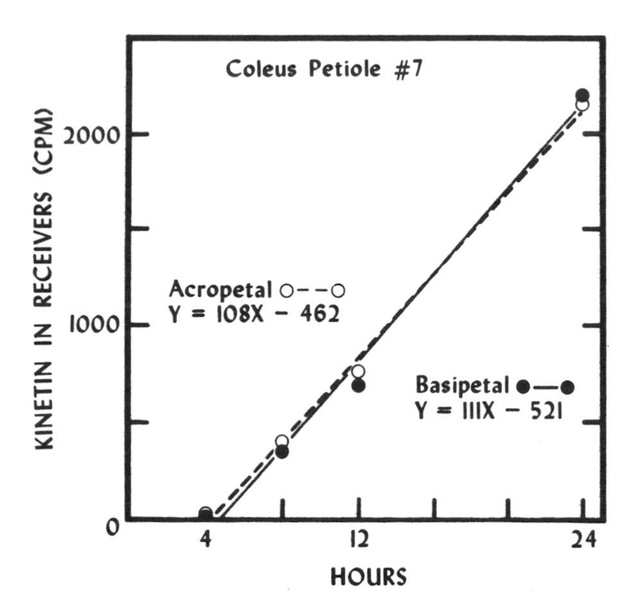

Figure 9-17. The complete absence of polarity of movement of ^{14}C-kinetin through *Coleus* petiole sections, as shown by linear regression equations fitted to the data (raw data from Veen & Jacobs 1969b, regression calculations and figure from Jacobs 1972).

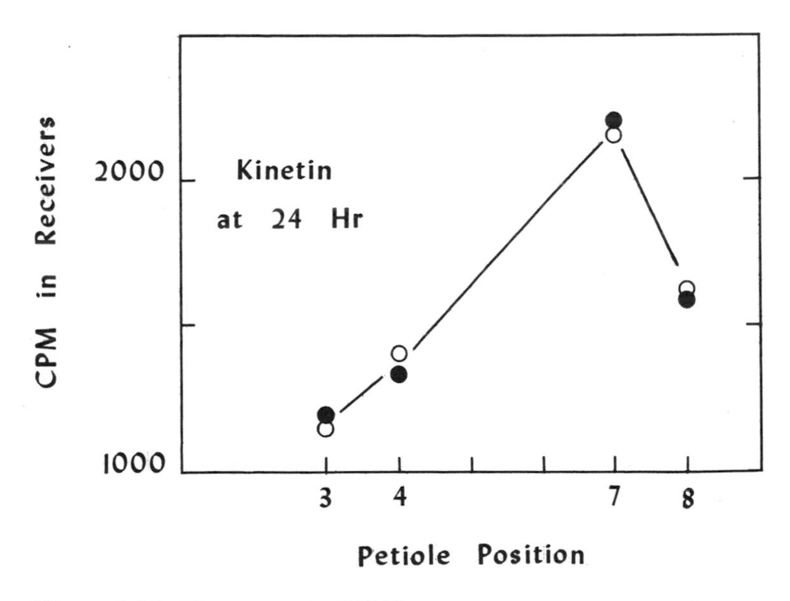

Figure 9-18. The amount of ^{14}C-kinetin in receivers on *Coleus* petioles of various ages (no. 3 being the youngest shown, no. 8 the oldest). Counts in basal receivers (solid circles) and apical receivers (open circles) are shown (Jacobs 1972).

receiver counts by 24 hours, and in showing faster uptake from the donor than its artificial counterpart. Adenine differed from IAA, however, not only in the lack of polarity, but also in the fact that even as early as 6 hours the adenine receivers contained a second radioactive substance, judging by the distribution of counts on the thin-layer chromatograms (Veen & Jacobs).

No statistically significant difference was found between counts in apical and basal 24-hour receivers when ^{14}C-benzyladenine was in donors placed on sections from *Coleus* no. 2 petioles or internodes, *Cleome* floral organs, or *Avena* coleoptiles (Koevenig 1973). Koevenig emphasized, as had Veen and Jacobs, the large variance associated with these ^{14}C-cytokinin counts, and pointed out that without tests of significance one could easily be misled into thinking the differences in average counts were real. The time course of ^{14}C accumulation in *Avena* receivers, described by linear regressions, was essentially identical in the two directions. The rate of movement, based on the linear part of the time course, was 1.4 mm/hour – close to that reported by Veen and Jacobs for kinetin. (Unfortunately, in view of the earlier conflicting results on the purity of labeled material in ^{14}C-benzyladenine receivers, no report was given as to what chemicals the ^{14}C was with in Koevenig's receivers.)

The absence of polarity in the movement of ^{14}C-kinetin as well as its chromatographic purity in 20-hour receivers were confirmed by Radin and Loomis (1974), using sections from hypocotyls and cotyledonary petioles of 3-day-old *Raphanus* seedlings. The significance of differences was assessed by Duncan's multiple-range test following analysis of variance. From the occurrence of substantial counts at 4 hours in receivers on their 9-mm sections, they concluded that the rate of movement must be somewhat more than 2 mm/hour. If hypocotyls from younger seedlings were used, however, kinetin moved with basipetal polarity through the sections (no statistical tests were described on this point).

In planning a comparison of artificial and endogenous cytokinins, Veen and Jacobs (1969b) had used adenine as their example of endogenous compound, mainly because it could be purchased with a ^{14}C label. However, adenine's weak activity as a cytokinin (0.001 that of kinetin) leaves one uncertain whether its transport properties are related to its endogenous nature – the main interest – or to its weak activity. Zeatin, by contrast, is both endogenous and highly active, about 100 times more active than kinetin in cultures of carrot phloem tissue (Letham 1967a) or in the soybean callus bioassay (Jacobs 1976). Zeatin moved in both directions through *Coleus* petiole sections but with no discernible polarity. Direct assay of receivers in the soybean callus bioassay as well as liquid scintillation counting of receivers after adding ^{14}C-zeatin in donors confirmed this. The amounts moved into the receivers were low but statistical tests confirmed that they were significantly above background, as well as not significantly different in the apical and basal receivers (Jacobs 1976).

The fastest rate of movement observed so far through the classical transport sections was that found by Sherwin and Gordon (1974) for the movement of tritiated cyclic adenosine 3′,5′-monophosphate (cyclic AMP, the famous "second-messenger" of mammalian hormone physiology). They reported that cyclic AMP accumulated in receivers at a linear rate, starting just a few minutes after it was added to 7-mm sections from corn coleoptiles. Regression calculations showed a calculated rate of basipetal movement of 183 mm/hour. Acropetal movement, also linear, averaged 79 mm/hour. The difference in rates was statistically significant – the first such case observed. (See the results with IAA in Figure 9-10 and with benzyladenine by Koevenig.) The slopes differed also, being unexpectedly greater for acropetal movement. Gordon et al. (1973) had already shown that most of the radioactivity was still with cyclic ^{Rf}AMP^{Rf} after 30 minutes. In most parameters, therefore, the movement of cyclic AMP was unlike that of cytokinins. Whether cyclic AMP functions as a hormone in vascular plants is not yet known. In fact, even its presence in plants is contested.

Abscisic acid

The movement of abscisic acid (ABA) has been followed in only a few cases. Dörffling & Böttger (1968), aided by a gift of racemic, synthetic abscisic acid from Cornforth but also using native abscisic acid they had isolated from tomato fruits, first reported on movement through *Coleus* stem and petiole sections. Because no radioactively labeled ABA was available then, they devised a new bioassay: inhibition of growth of inverted coleoptile sections was resorted to. (This new bioassay seems a weak link in their chain of evidence. Inhibition is notoriously unspecific in general and no evidence about specificity was given by them. They did show [their Fig. 1] that increasing the (±)-ABA dose up to 0.02 μg inhibited elongation more and more, up to a maximal useable inhibition of about 12%. [Many of the values given in their tables are higher than 12%.] In addition, they mentioned the relatively high variability of their results.) Evidence that ABA did actually move through petiole sections was obtained by extracting receiver blocks, running the extract

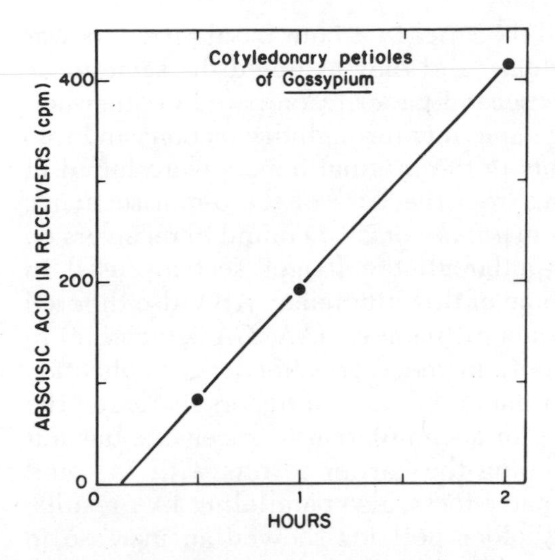

Figure 9-19. Time course of movement of abscisic acid through 3-mm long sections of cotton petioles (raw data from Ingersoll & Smith 1971, figure and regression line from Jacobs 1978b).

on thin-layer chromatograms, and determining the inhibiting activity of various zones. They stated that only the quenching zone corresponding to the Rf of (±)-ABA showed inhibitory activity in two bioassays. The rate of basipetal movement through older petioles, as estimated by direct inhibition-bioassay of receiver blocks, was 24–36 mm/hour; the most convincing data showed a reasonably linear time course of accumulation in receivers on 9-mm long petiole sections, with none of the values over 12% inhibition, and with the time intercept about 15 minutes. Basipetally polar movement occurred through young petioles and internodes, but the older no. 5 petioles and internodes showed so much acropetal movement that polarity was weak or perhaps absent, they felt. (The polarity conclusions are vitiated, unfortunately, by the duration of these experiments. From their Fig. 2 the 2- and 4-hour durations would place the amount of inhibitor far above the useable response range of the bioassay, that maximum having been passed after only 1 hour.)

Through short sections from cotyledonary petioles of *Gossypium* (the genus from which ABA was first isolated) the movement of ^{14}C-ABA was nonpolar but fast (Ingersoll & Smith 1971). Accumulation in receivers looked linear for 2–3 hours. Regression lines indicated a rate of 22–23 mm/hour (Figure 9-19). (Their previous paper

[1970] had reported that all the label in 4-hour basal receivers was ^{Rf}ABA^{Rf} in two-solvent systems, and they assumed the same to be true for all receivers in this second paper.) Compared to other hormones, ABA moved in large amounts through the sections and into receivers (70% of the counts in the original donors were found in 24-hour receivers collecting from the base of the 3-mm sections). Compare this with only 7% of *uptake* of 2,4-D found in receivers on bean petioles (Fig. 9-11) – although the longer sections used in beans would account for some of this difference. ABA also differed from other endogenous hormones (namely, IAA, GA-3, adenine) in showing no decline in activity in receivers after longer collection times. Doubling the concentration of ABA in donors increased the slope of the regression line for accumulation in receivers, but left the rate unchanged (paralleling the earlier results with IAA and 2,4-D). There were strong age effects, also paralleling IAA results: sections from progressively older petioles showed an increase in movement at 14 days compared to 10, with a progressive decline thereafter. They felt that counts in the apical receivers were essentially the same as those in the basal receivers at all four ages. Ingersoll and Smith calculated regression lines for their time-course studies but said nothing about statistical tests of significance on the acropetal versus basipetal data.

After ^{14}C-ABA became commercially available, Veen used it to reinvestigate its movement in *Coleus* petioles (1975). He found no movement of ^{14}C into either apical or basal receivers, even after 8 hours. Extraction of the four quarters of the 10-mm transport sections revealed longitudinal distribution patterns of ^{14}C that were nonpolar; that is, essentially the same pattern was found whether basipetal or acropetal movement was being followed. The pattern was like that of ^{14}C-mannitol and unlike the polar pattern from ^{14}C-NAA run at the same time. Veen's results on *Coleus* petioles from measuring ^{14}C are very different from those of Dörffling and Böttger from measuring inhibition. I calculate that the rate of ^{14}C movement would be less than 1.3 mm/hour in Veen's petiole sections. (Veen did not present evidence that the ^{14}C he counted was still with ABA, nor did he check to see whether "inhibitory activity" could be found in his receivers even though no ^{14}C from ABA had reached there.)

Dörffling et al. (1973) used their coleoptile-inhibition bioassay to estimate ABA movement through 6-mm sections from *Pisum* internodes. The results indicated a rate of only 2.4–3.0 mm/hour, with no discernible polarity.

At present there is no convincing evidence as to why such different results were obtained in these three laboratories. Veen (1975) discusses most of the possible explanations, but does not discuss the main weakness of his paper, namely, the assumption that the ^{14}C he counted was always and solely with ABA.

However, despite these tenfold differences in estimated rate of movement, there is movement of ABA in and through transport sections. Is there evidence of ABA movement in the intact plant? As with most other known hormones, ABA has been found by bioassay from aphids' honeydew, and some of it, at least, was presumably in the sieve tubes of willow (Hoad 1967).

ABA labeled with ^{14}C was added to the abraded oldest leaf on *Phaseolus* seedlings by Hocking et al. (1972 in a brief note in *Nature*). One day later they extracted pieces of the plant and found some of the ^{14}C still with ABA they said – the percent of ^{14}C extracted being unspecified – but with label at other zones of the chromatograms, too. The label that had been added as ABA was extracted from all areas of the plant. Shindy et al. (1973) published the first detailed paper on ^{14}C-ABA movement in intact plants, adding ABA of high specific activity to the cotyledon or the first (oldest) true leaf of cotton seedlings. Six hours after adding the ^{14}C-ABA most of the ^{14}C was below the site of application (judging by autoradiographs), although after 4 days ^{14}C was all over the seedling. However, even after 8 days most of the ^{14}C not still in the treated leaf was in the roots (9.3% of the total recovered) and, very surprisingly, the ^{14}C in the roots was still all with ABA, judging by Rf in thin-layer chromatograms, gas–liquid chromatography, either alone or combined with mass spectroscopy. Also of interest was their observation that the added ABA never caused abscission, even though it obviously had to pass the abscission zone of the treated leaf. As authors of both papers pointed out, the pattern of ^{14}C distribution they observed after adding ^{14}C-ABA to these basal leaves was similar to that found by others for various other metabolites added to exporting leaves. But whether such leaves typically export *endogenous* ABA is still a question.

Dörffling and co-workers borrowed the approach that Morris and Kadir (1972) had used with IAA to study ^{14}C-ABA added to intact *Pisum* seedlings (Dörffling et al. 1973, Bellandi & Dörffling 1974). If ^{14}C-ABA was applied as a droplet in 45% ethanol between the young leaves of the apical bud, label was found only a short distance down the stem even after 24 hours. None reached the roots. ^{14}C-IAA added in the same way resulted in counts in the root 2–4 hours later, confirming Morris et al. (1969). But if ^{14}C-ABA was injected into the internode just below the apical bud, instead of being merely placed on the surface, counts were found in the extracts of roots at 4 hours (indicating a rate of movement of more than 10 mm/hour, although the estimate was admittedly rough). Radioactivity from ^{14}C-ABA added to an old leaf at the shoot base showed a distribution like that reported by Shindy et al. for cotton. The only evidence about what compounds the ^{14}C was with in all these extracts was provided in a statement that after "application" (presumably to leaf surface) of ^{14}C-ABA, thin-layer chromatograms of ethanol extracts of stem showed only one radioactive zone, which was at the Rf of ABA. The interval between applying and collecting was not stated.

In summary, investigations of hormonal movement of ABA are in too early a stage for any quantitative parallel variations to have been worked out. Apparently no one has looked for ABA effects in the intact plant that move with rates matching the 22+ mm/hour reported by Ingersoll and Smith. In fact, it is still uncertain what organs in what developmental stage are normally producing ABA as a hormone (which may well not be the same as what organs provide ABA when extracted).

Interactions of hormones

The study of hormone movement through classical transport sections represents isolation not only of the section from the rest of the plant but also of the single added hormone from all the other hormones that would presumably be interacting with it in the intact organism. A beginning has been made in studying hormone movement as affected by the presence of other hormones.

Adenosine triphosphate (the famous "energy-rich" ATP) increased basipetal IAA movement through *Helianthus* hypocotyl sections, judging by direct bioassay of auxin in receivers (Reiff & von Guttenberg 1961; Libbert et al. 1961). Adenine, adenosine, or adenosine monophosphate, added at similar mM levels in the donors, gave similar increases (Libbert et al. 1961). Benzyladenine at 23 μM, or kinetin at 23 μM, also increased the basipetal movement of ^{14}C-IAA (Black & Osborne 1965; McCready et al. 1965, respectively), although the stimulation was only seen if the excised bean petioles were "aged" for many hours. McCready et al. found that more of the auxin-type herbicide 2,4,5-trichlorophenoxyacetic acid also moved into basal receivers after 12 hours' treatment with kinetin. (Although one might be skeptical about the reality of the differences between the low counts McCready et al. reported for

^{14}C-IAA, they said that all 34 experiments showed more counts from kinetin treatment.) Kinetin affected IAA movement as had the other adenine derivatives that the Rostock group had tested on *Helianthus* (Leike 1967). For *Coleus*, at least, kinetin increased basipetal ^{14}C-IAA movement to a statistically significant degree even when tested on freshly cut petiole sections (Chang 1971; Fig. 9 in Jacobs 1978b), in contrast to McCready et al.'s results with freshly cut *Phaseolus*.

A similar effect of kinetin was found in more nearly intact bean seedlings (Davies et al. 1966). They substituted ^{14}C-IAA for the top of the shoot, with or without kinetin added. Kinetin caused more ^{14}C to be in 24-hour extracts of all sections of the stem below the application site, with almost all the radioactivity said to run to the Rf of IAA. A nice addition was their evidence that this increased basal IAA was functional: kinetin plus IAA substituted for the shoot tip inhibited the lateral buds for many days longer than did IAA alone. Davies et al. wondered if kinetin were really increasing "transport" of IAA or only increasing uptake (because the addition of kinetin did not result in less IAA near the application site with more farther away). McCready et al. (1965) had noted that kinetin usually increased uptake of IAA along with its effect on IAA movement, but they did not think the effect on uptake was the primary one. Basipetal movement through *Pinus* stem sections of ^{14}C added as ^{14}C-IAA was increased by pretreatment with either kinetin or benzyladenine. The increased xylem differentiation resulting in decapitated stems (see Chapter 4) supported the view that the label was still with IAA (Hejnowicz & Tomaszewski 1969). Bayer (1973) also found more ^{14}C-IAA movement into basal receivers along with more uptake when tobacco stem sections were pretreated with kinetin. That auxin uptake can be stimulated by kinetin without concomitant "transport" was shown in leaf discs of *Phaseolus*, using 2,4-D as the auxin (Sargent 1968). We might derive a hint as to the mechanism of kinetin action in these cases from Lau & Yang's fine paper (1973) presenting evidence that kinetin increased ^{14}C-IAA uptake into *Phaseolus* hypocotyl sections, decreased the conjugation of IAA with aspartate, and thereby resulted in a sizeable increase in free IAA in the sections.

In addition to cytokinins increasing IAA movement, IAA can increase the basipetal movement of cytokinins. Black and Osborne (1965) reported a doubling of counts in basal receivers when IAA was added with ^{14}C-benzyladenine to *Phaseolus* petiole sections (e.g., the 153 cpm in 22-hour basal receivers was increased to 323 cpm when IAA at 11 μM was in the donors). Counts in apical receivers were not changed from their low values. Fox and Weis (1965) also saw changes in receiver counts from adding IAA along with ^{14}C-benzyladenine but the counts decreased in their *Phaseolus* petioles rather than increased, increased in *Avena* coleoptile, and probably decreased in *Pisum* epicotyl (neither of these papers gave statistical tests of significance). Seth et al. (1966), substituting hormones for the top of the *Phaseolus* shoot as described earlier, found that IAA increased the basipetal movement of ^{14}C-kinetin down the stem, as estimated by the number of counts running to the general Rf of kinetin when stem extracts were chromatographed. Particularly clear-cut results were provided by Radin and Loomis (1974): in radish petioles and hypocotyls, the nonpolar movement of ^{14}C-kinetin through transport sections was changed to polar movement by the addition of IAA. The effect of IAA was to increase the basipetal movement to a statistically significant degree. (Their valuable results with roots of radish are described in Chapter 10.) Hypocotyl sections cut from younger seedlings showed polar movement of kinetin without added IAA – a finding they guessed was owing to more endogenous IAA in the younger hypocotyl. (No measurements of endogenous auxin were made.)

Gibberellic acid also affects the movement of IAA. Kuraishi and Muir (1964b), whose primary aim was to find the basis for the large increase in endogenous diffusible auxin that resulted from prior treatment with gibberellic acid, concluded that in *Pisum* GA-3 was not acting by increasing IAA movement. When 2 μM of IAA was in the donors, the addition of GA-3 changed the average curve in the *Avena* bioassay not the slightest bit. And although GA-3 added with 20 or 200 μM of IAA did cause a statistically significant increase of 2–7° in the average *Avena* curve, they concluded that the effect was too small to account for the very large increases in endogenous auxin that they (1964a) and others (e.g., Phillips et al. 1959) had observed from GA treatment.

The first specific evidence that GA-3 increased IAA movement in a developmentally functional way came from Jacobs and Case (1965). They were investigating why IAA substituted for the apical buds did not replace it for more than a few days in terms of inhibiting outgrowth of lateral buds. By the PESIGS rules, an obvious guess would be because the apical bud is producing, in addition to IAA, something else required for such inhibition. Gibberellic acid was a likely candidate as the additional component. When GA-3 was substituted along with ^{14}C-IAA for the apical bud of *Pisum*, the lateral buds were inhibited for a longer period and much more ^{Rf14}C-

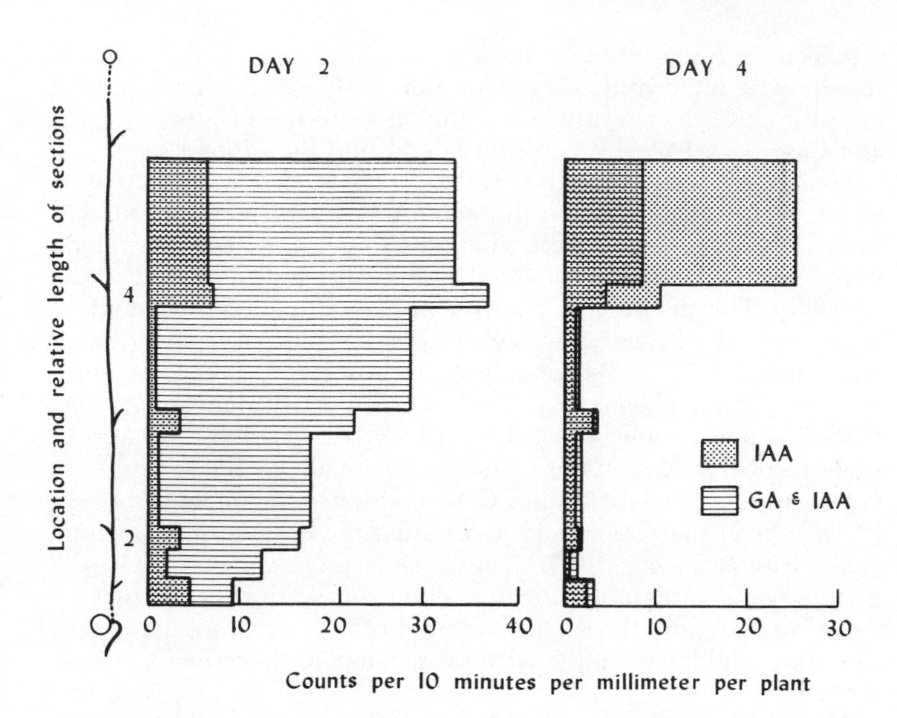

Figure 9-20. The effect of GA-3 (substituted, along with ^{14}C-IAA for the top of the *Pisum* shoot) in causing more ^{14}C-IAA to be present two days later far down the stem from the point of application. This effect of GA-3 has disappeared by day 4 (right side; Jacobs & Case 1965).

IAA^{Rf} was extractable from stem pieces far below the application site and in the region of the inhibited lateral buds (Figure 9-20).

The authors suggested that GA-3 increased the movement of IAA, although they realized their observations were equally explicable if GA-3 somehow prevented IAA from being metabolized to an inactive and nonlabeled form. Basipetal movement of ^{14}C added as IAA to transport sections of *Helianthus* stems was found to increase by 100% if the intact plants had been sprayed with GA-3 24 hours before (Palmer & Halsall 1969), and while this GA effect slowly declined over the next 100 or so hours, there was a concomitant slow increase in stem elongation from the prior treatment with GA-3. GA-3 given as a pretreatment to *Pinus* stem sections caused much more basipetal movement of label from either IAA or ^{14}C-NAA (Hejnowicz & Tomaszewski 1969). Evidence that the radioactive label was still with the auxin came from parallel experiments in which IAA plus GA-3 substituted for the apical bud completely replaced the bud's effect in causing new xylem cells to differentiate in otherwise intact pine trees. (IAA alone only partially replaced the apical bud, so their results were analogous to the results of Jacobs and Case with IAA and GA-3 on lateral bud inhibition.)

Auxin movement through petioles, as contrasted to stems, seems to be decreased by gibberellic acid. When transport sections cut from *Phaseolus* petioles were treated with GA-3 in the donors along with ^{14}C-2,4-D, less label moved basipetally and more moved acropetally. This decrease in polarity was accompanied by about two times as much elongation as occurred in the control sections treated only with the 2,4-D (McCready & Jacobs 1967). *Coleus* petioles showed similar phenomena. GA-3 increased the elongation of debladed young petioles treated with IAA but speeded their abscission (Jacobs & Kirk 1966). Pretreating intact young leaves with GA-3 caused more label from ^{14}C-IAA added to the intact leaf blade to be kept in the petiole and away from the basal abscission zone (Kaldewey & Jacobs 1975). The greater retention of IAA in the petiole could explain the greater elongation of the petiole that results from GA, and the smaller amount of IAA reaching the abscission zone could explain the faster abscission, as described in Chapter 8.

Abscisic acid decreases the basipetal movement of IAA, as discussed in Chapter 8, on senescence and abscission. According to an abstract by Kaldewey, Weis, and Wakhloo in the 1969 Seattle Botanical Congress, ABA decreased ^{14}C-IAA-2 movement through coleoptile sections. Chang showed that pretreatment with ABA decreased subsequent basipetal movement of IAA by decreasing the flux but not the velocity, increased conjugation of IAA with aspartate, and – as was well known before – speeded petiolar abscission (Chang 1971; Chang & Jacobs 1973). Senescence factor from senescing *Coleus* leaves had the same set of effects. Chang and Jacobs proposed as a working hypothesis that ABA's primary effect in these experiments was to induce increased conjugation of IAA, thus removing it from the pool of IAA available for basipetal movement, and indirectly causing abscission by preventing IAA from getting down through the petiole to the abscission layer at its base. Such an effect would be analogous to the effect of GA-3 on IAA movement, petiole elongation, and abscission, as discussed earlier. Whether the primary action of ABA was to increase IAA conjugation or to decrease basipetal IAA movement could not be determined from Chang's data. But in either event, the decreased IAA movement to the abscission layer seemed able to explain the abscission-stimulating action for which ABA was named. Naqvi, apparently independently, confirmed part of these results in reporting that label added as IAA to *Zea* coleoptiles moved down the coleoptile under the influence of ABA with essentially the same velocity as the controls (as measured by intercept) but with lowered flux (Naqvi 1972; Naqvi & Engvild 1974).

As an example of hormonal "interactions" between hormones of the same class, the discovery by Leopold and Lam (1962) is important. They started from their observation that the longer they waited to cut out a transport section after decapitating a *Helianthus* seedling, the less the basipetal movement of ^{14}C-IAA through that section. However, when they waited as long as 7–11 days (when the lateral buds above the tested zone had started to elongate because of their release from apical dominance), the ^{14}C-IAA movement was restored to a level even higher than the zero-day controls. Suspecting that endogenous auxin from the elongating lateral buds might be the controlling factor, they substituted synthetic auxin (NAA) for the main shoot apex and found that NAA could replace most of the "transport-maintaining effect of the shoot apex" (their Fig. 4). They suggested that apical meristems, through their production of auxin, might in just such a way maintain a polar gradient in the tissues proximal to them, and thereby help determine the polar development of tissues. IAA or NAA similarly increased the amount of ^{14}C added as ^{14}C-2,4-D that moved basipetally through *Zea* coleoptiles (Hertel & Flory 1968; Rayle et al. 1969).

In summary, various hormones have been found to affect the movement of other hormones in all cases so far tested. (Several hormones, such as ABA, thiamine, and the gibberellins, are so recently recognized as hormones that no one has yet tested the effect of other hormones on their movement.) The study of hormonal effects on hormone movement is in such an early stage, however, that little is known about the mechanisms even in the relatively simple "isolation" provided by transport sections – the main exception being the evidence that ABA and senescence factor decrease IAA movement by increasing IAA conjugation with aspartic acid. For most of the other cases decreased hormone levels in the receivers might be the result of increased destruction, increased conjugation, decreased uptake, a decreased ability to move the hormone in polar fashion – or any combination of these.

Techniques – radioisotopes in research on hormones

Molecules labeled with radioisotopes provide one of the most powerful tools available to chemists, biochemists, and physiologists. But nowhere are they more valuable than in the study of plant hormone movement and metabolism. This is because the hormones are present in such minute amounts in the plant that few techniques can measure endogenous levels: bioassays and fluorescence measurements provide two (as discussed for auxins in the technique section of Chapter 1), radioisotope labeling provides another – although when using radioisotopes, one must add the hormone from outside.

The radioactive label used most often in research on plant hormones has been carbon-14 (^{14}C), with tritium (^{3}H) being a poor second in number of users. Both radioisotopes are unusually safe, their radiation being of such low energy (maxima of 0.155 and 0.018 MeV, respectively) that they penetrate only tiny distances. (Tritium, in fact, is the preferred label for autoradiographs because its extremely low energy allows optimal localization.) Both ^{14}C and ^{3}H have long radioactive half-lives, so one does not have the problems with them that one does with the short-lived phosphorus-32, with its much more penetrating radiation.

Counting of ^{14}C-labeled hormones has almost all been done with Geiger counters (e.g., Goldsmith 1959, Hertel 1962, McCready & Jacobs 1963a, b) or with liquid scintillation counters (e.g., Naqvi 1963 and most later publications from Jacobs' laboratory). The scintillation counters are currently favored, not only because of their much greater efficiency (about 90% of ^{14}C's disintegrations per minute being counted, as contrasted to a maximum of about 40% with a windowless gas-flow Geiger counter), but also because manufacturers have provided more automatic features on the liquid scintillation counters. For counting ^{3}H Geiger counters are so inefficient that scintillation counters have a virtual monopoly on tritium counting. One of the greatest assets of scintillation counters with several channels is their ability to count ^{3}H and ^{14}C in the same sample at one time; however, this asset has been almost completely unexploited so far in plant hormone research.

Hence, it is now easy to count ^{14}C and ^{3}H. But the major problem with the use of radioactively labeled hormones is telling what compound those "counted radioisotopes" are with at the time of counting. If the ^{14}C is no longer with IAA, you do not want to be counting ^{14}C and referring to it as ^{14}C-IAA. This problem starts when your first order of ^{14}C-labeled hormone arrives. Check it for chromatographic purity within a few days or the shipper will not refund your money. (In 15 years we have had only one sample arrive with noticeable radioactive contaminant, but that was from a regular supplier.)

Stock solutions of labeled hormones should be checked for radiochemical purity once per month if the stock is being used regularly. (From self-radiolysis, radiochemicals are apt to decompose faster than nonlabeled chemicals.) Our experience has been that ^{14}C-IAA stock solutions typically maintain radiochemical purity for at least 6 months if refrigerated.

Obviously, hormone that has been exposed to plant tissue is much more likely to have been metabolized than when it was sitting in solution in a glass container in the refrigerator. However, the general experience with ^{14}C-IAA in the classical transport experiment has been that all the counts in receiver blocks of agar are still with IAA, judging by Rf on a chromatogram – as long as one does not run the transport for more than 6–8 hours. (Progressively more counts are clearly not with IAA in receivers collected at 12 and 24 hours [Naqvi 1963, see Fig. 6 of Jacobs 1967].) We have found the same apparent chromatographic purity in receivers collecting thiamine (Kruszewski & Jacobs 1974, Jacobs & DeMuth 1977), 2,4-D (Jacobs & McCready 1967), and kinetin (Veen & Jacobs 1969b). So many reports of "only ^{Rf}IAA^{Rf} in the receivers" have been published that investigators have tended to become careless: they tacitly assume that, if control transports show nothing but ^{Rf}IAA^{Rf} in receivers, any sort of treatment subsequently applied will not change the radiochemical purity of material in the receiver blocks either. Such an assumption seems unjustified. For some hormones other than auxins activity has been demonstrated in chromatographed extracts of receivers at Rf zones other than that typical of the substance added in the donor (e.g., benzyladenine, adenine, and gibberellic acid, as described earlier). Clearly, one should check the radiochemical purity of labeled compounds in receiver blocks, even when using a hormone as apparently stable as IAA.

Extracts of tissues exposed to ^{14}C-IAA typically provide a sharp contrast to extracts of receiver blocks. After only a few hours, while the ^{14}C in the basal receivers is still all ^{Rf}IAA^{Rf}, other labeled zones appear on chromatograms of green petiole extracts (Veen & Jacobs 1969a). The same was true of ^{14}C-NAA (Veen 1966) and ^{14}C-kinetin (Veen & Jacobs 1969b). For tissue extracts, therefore, it seems a waste of time to report total ^{14}C counts without providing at least minimal chromatographic evidence as to what compounds the label is with.

These problems with the radiochemical purity of ^{14}C-labeled material are exaggerated when tritium labeling is used instead. Tritium is notorious for its lability. It can exchange with the hydrogen of surrounding water. It can even exchange within organic molecules (the cryptically named N.I.H. shift, first noticed at National Institutes of Health, refers to such exchange within amino acids). These are not theoretical problems only; in a recent doctoral thesis, which gave rise to several later publications, the stock solution of ^{3}H-IAA had about 50% of the ^{3}H at Rfs other than that of IAA! It is ingenous, at best, to add such stuff in the donors and then count the ^{3}H as though it were all IAA. Such problems doubtless provide the main reason tritium has not been used so often as ^{14}C in plant hormone transport studies.

Thin-layer or paper chromatography is routinely used to check radiochemical purity. I prefer paper, solely because it has been in use long enough for more artefacts to have been discovered. Unfortunately, most researchers still use only isopropanol-ammonia-water as their solvent, despite the evidence that the ammonia converts more than half of any IAA-glucose present to other products (see Chapter 3). The usual chromatogram scanner uses Geiger tubes for counting and is therefore not sensitive enough to detect hormonal impurities. Counting zones of the chromatogram in a liquid scintillation counter allows impurities to show up that are not seen in a Geiger scan. We have found that a ^{14}C-IAA stock with such an impurity showed *acropetal* movement of the ^{14}C, whereas after the stock had been purified sufficiently to show only one ^{14}C zone in the scintillation counter the ^{14}C movement was strictly polar in a basipetal direction. When one remembers that only about 2–10% of the IAA leaving a donor block actually gets transported through green stem or petiole sections in 5–6 hours, it is clear that a relatively small percentage of radiochemical impurity in the donor IAA can be drastically misleading.

Even counting 20 zones of a paper chromatogram is not sufficient to detect all radiochemical impurities. Veen recommended adding an autoradiograph of a second paper chromatogram, and convinced me by showing that a zone that was "pure ^{14}C-adenine^{Rf}" by 20-zone scintillation counting actually had two radioactive spots – one nestled up close to, but separate from, the other.

Hence, as the minimum check for radiochemical impurities, I recommend using scintillation counting of 20 zones of a paper chromatogram, preferably with a second chromatogram being checked with autoradiography. To check the *chemical* concentration of IAA in your stock solution (which will be in the mg/liter range), use ultraviolet absorption.

A final caution: the papers of many scientists contain the tacit assumption that *all* the counts in tissues have been extracted with the solvent used. Evidence for this assumption is seldom given. Be on the lookout for this flaw, particularly if it affects the conclusions.

For background reading on radioisotopes the book by Francis et al. (1959) is excellent despite its age. It provides a judicious mix of theory and practice, with liquid scintillation counting being the only area for which later books must be consulted (e.g., Neame & Homewood 1974). Technical details of scintillation counting of ^{14}C-IAA in transport experiments and on paper chromatograms are covered in Naqvi's thesis (1963). I would add the caution that glass scintillation vials should be used instead of plastic ones. We have found (unpublished) that plastic vials give extremely erratic counts with ^{14}C-IAA; apparently, the IAA is adsorbed on the plastic, drastically lowering the counts.

A quite different use of radioactively labeled IAA was recently reported by Pengelly and Meins (1977). They adapted the radioimmunoassay technique, already well worked out for measuring levels of animal hormones, for the measurement of IAA in extracts of plant tissues. The technique seemed reasonably specific (judging by the dozen or so chemicals tested) and could detect even a few ng of IAA. (This sensitivity is in the same low range as the fluorometric method of IAA assay developed by Knegt and Bruinsma [1973].)

10

Roots and hormones

During the last decade there has been an explosion of information about the hormonal physiology of roots. A glance back at Figure 9-1, which summarizes current evidence about polarity of hormone movement, gives some indication of the resulting changes in viewpoint. Perhaps most surprising is the drastic change from the 1937 view that auxin moved basipetally from the root tip to the view (based on evidence amassed in 1967 and later) that auxin moved with strong polarity in the opposite direction. How could such a reversal of opinion occur? To some extent it was a function of the small number of people investigating roots. But the main reason, I believe, was the stultifying effect of the dead hand of dogma. The Cholodny–Went auxin theory of tropisms explained tropisms as being the result of auxin moving basipetally from shoot tip or root tip, being redistributed laterally under the influence of gravity or light, and thereby causing differential growth. The extensive and quantitative evidence supporting this theory with respect to phototropism and geotropism in shoots was so strong (see Chapter 2) that people were willing to believe it for geotropism in roots even though the evidence was, by comparison, meager and painfully lacking in quantitative detail. Furthermore, the theory itself was so simple and all-explaining that it was natural to want to believe it. All these factors combined to discourage a critical assessment of the role of hormones in root geotropism.

Early work on auxin polarity in roots

Cholodny clearly enunciated the basic theory for geotropism in 1924 and 1926. Because coleoptile tips of *Zea* could cause positive geotropism in decapitated *Zea* roots, which otherwise did not react to gravity, he concluded that the same substance caused positive geotropism in roots and negative geotropism in shoots (1924). And because *Zea* coleoptile tips also restored negative geotropism to *Lupinus* hypocotyl sections, he concluded that the opposite geotropic reactions of roots and shoots did not result from roots receiving inhibitory levels of growth hormone (auxin), whereas shoots were provided with lower stimulatory levels, and that the opposite responses of the two organs were a function of their innate properties (1926). He guessed that under the influence of gravity the growth hormone went to the lower side of a horizontal organ. (It was this same basic idea that Went took up to explain the action of light in causing redistribution of auxin in the phototropism of coleoptiles [1926, 1928], hence, the adoption by Went and Thimann [1937] of the designation "the Cholodny–Went theory" of tropisms.)

However clearly enunciated Cholodny's theory was, his evidence in support of the theory was shockingly scanty by current standards, 50 years later. He asserted that *Zea* coleoptile tips not only restored a completely normal geotropic curve but also inhibited the elongation of decapitated *Zea* roots (1924) – the latter being a point of obvious importance in support of his theory. But no absolute values were given, much less any measure of variability. We were told only that if the elongation of decapitated roots as measured by a millimeter ruler was considered to be 100, then the coleoptile tip reduced it to 64. Similarly, he stated (1926) that decapitating *Lupinus* roots increased elongation by 12% in 4–5 hours, whereas replacing the root tip decreased it by about 5% in 8 hours. It is only sensible to be suspicious of results given solely as percentages, but even more so when differences claimed are as small as these. One's suspicions deepen on seeing the few absolute values included in the 1926 paper; one learns that Cholodny was using his millimeter ruler to measure total elongations of about 1 mm in the control sections of *Lupinus* hypocotyl, and that the addition of 4–6 *Zea* coleoptile tips to the hypocotyl raised the hypocotyl increment to 2.5–3 mm.

In the ensuing years this problem of inadequate data (and typically no statistical tests of significance) continued to plague the literature on the auxin physiology of roots. Various inconsistencies in results were brushed aside. Although diffusing root tips onto agar blocks did provide auxin, judging by the *Avena* curve assay, sugar had to be added to the agar to obtain curvatures of reasonable size (Boysen-Jensen 1933b; Cholodny 1934) or, in van Raalte's case (1937), to obtain any detectable curvature. This requirement for sugar was explained away by guesses about the need to maintain the delicate excised root tip. This assumed delicacy was also in-

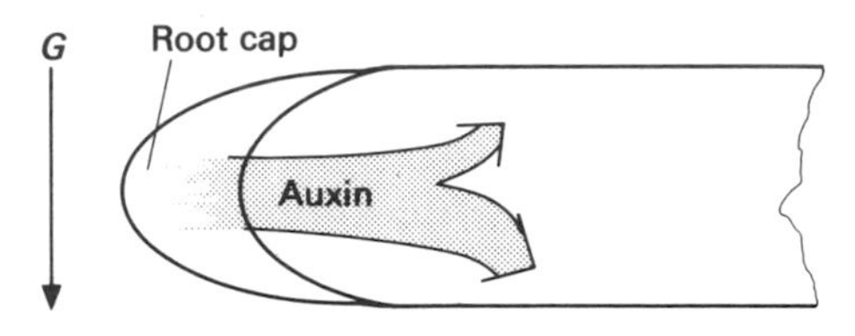

Figure 10-1. The Cholodny–Went theory of auxin and root geotropism. Auxin moving from the root tip toward the root base was partially diverted, under the influence of gravity (*G*), to the lower side of the root. The greater concentration of auxin on the lower side of the root caused greater inhibition of elongation there, thereby resulting in a positive geotropic curvature.

voked by Cholodny to explain why excised root tips, in contrast to coleoptile tips, did not restore geotropic curvature to decapitated roots (1924) or to hypocotyl sections (1926). The view of the 1930s to 1950s was that auxin was produced in root tips, moved from there toward the root base, and was present at levels that inhibited the elongation of roots (Went & Thimann 1937; Leopold 1955). When a root was placed horizontally, the basipetally moving stream of auxin was believed to be diverted to the lower side where it inhibited elongation even more strongly, thus causing the positive geotropic curve by differential inhibition (Figure 10-1).

Gorter (1932) and Faber (1936) used transport sections and the *Avena* curve assay to test for polarity of auxin movement through roots. Both concluded that although auxin activity was found in the receiver agar, there was little or no polarity. Two reports in the older literature – both ignored by Cholodny–Went adherents – were particularly difficult to reconcile with the Cholodny–Went view that the direction of auxin movement was polarized from the root tip back toward the root base: one was the report that xylem in roots regenerated acropetally (toward the root tip) (Simon 1908); the other was that cambium reactivation also occurred in an acropetal direction in roots (e.g., Brown 1935). Research on shoots provided extensive evidence that indole-3-acetic acid (IAA) was controlling these phenomena (see Chapter 4); so, unless some other factor suddenly became limiting as cambium reactivation spread down from shoot to root, the polarity of both these developmental phenomena suggested that IAA moved toward the root tip rather than away from it.

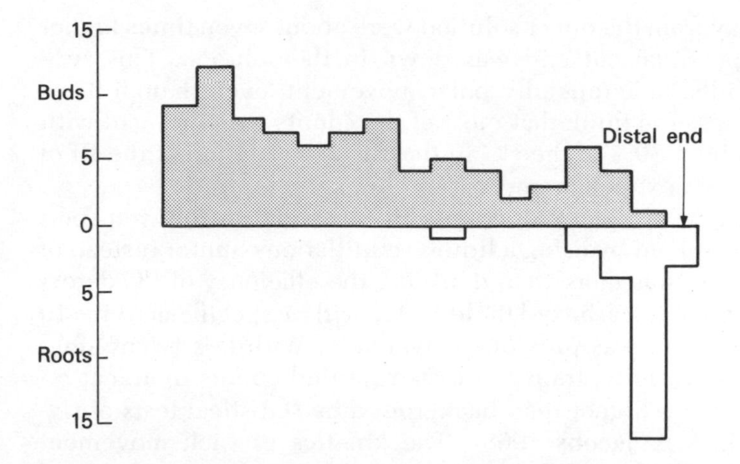

Figure 10-2. The distribution of buds and roots regenerating on a 15-mm section of *Convolvulus* root grown in sterile nutrient culture (after Bonnett & Torrey 1965b). Note the strong polarity of root regeneration.

Increasing evidence for auxin movement toward the root tip

A final impetus to reinvestigate IAA movement through classical transport sections of roots came from the papers of Bonnett and Torrey on the regeneration of roots and buds from sections cut from *Convolvulus arvensis* roots. These roots were derived from a clone that had been grown for many years in aseptic organ culture. Lateral roots grew out near the distal (root tip) end of cultured sections 10 or more mm long, regeneration of roots manifesting a strong polarity compared to regeneration of buds (Figure 10-2). IAA added to the sections increased the number of roots that grew out in 6 weeks of culture (about three primordia per 10-mm length were already present upon sectioning [1966]), leading Bonnett and Torrey (1965b) to conclude, as had earlier workers, that in the untreated sections endogenous auxin moved to the distal end of the excised section and there caused the polar pattern of root regeneration. Testing this hypothesis with transport sections and carboxy-labeled IAA, Bonnett and Torrey (1965a) unfortunately did not get enough ^{14}C into the receivers to count and were therefore driven to the expedient of counting ^{14}C from dried and ground tissue slices. When root sections were placed upright, with either their distal or proximal cut surface in an ^{14}C-IAA solution for 2.5–4 hours, counts from slices farthest away from the donor solution were about seven times higher when the proximal cut end was down in the solution. This supported the idea of acropetally polar movement, even though there was good reason to think that most of the counts were not still with IAA (only about 30% of the ^{14}C in the slices was found at the Rf of IAA after ether extraction and paper chromatography).

The problem of detecting counts in receivers on root transport sections was solved by using a liquid scintillation counter instead of a Geiger tube (thus more than doubling the efficiency of ^{14}C detection) and by using carboxy-labeled IAA with a specific activity 10 times higher than was previously available. With this twenty-fold increase in sensitivity, transport tests revealed counts in apical receivers that were higher than background by statistical tests of significance (Kirk & Jacobs 1968). The kinetics of such movement through *Lens* root sections were much like earlier results from shoots (Figure 10-3). For instance, there was very strong polarity (33:1 at 6 hours, with the differences being statistically significant at the 2% level). Receivers collecting in the shoot-toward-root-tip direction showed a linear increase in counts from 3 to 6 hours. The equation for the best-fitting straight line to the 3–6 hour data, when used to calculate the time intercept, indicated that the main front started to come through the 5.1-mm root section after 2.34 hours, for a rate of 2.2 mm/hour. This is the same order of magnitude as the rate of IAA movement through various shoot structures (other than coleoptiles) and is essentially the same as that found by the same techniques for tissue cylinders cut from more mature internodes of *Coleus* (Jacobs & McCready 1967). Losses of counts from donor blocks also followed expectations from shoots, as did the decline in receiver counts after a 6- to 8-hour maximum. All the counts in receivers before this decline were still with IAA, judging by Rf–another parallel with IAA movement in shoots. The chief difference from results with shoots was that only about one-tenth as much of the IAA lost from the donors appeared in the apical receivers by 6 hours. Similar results were obtained with *Phaseolus* roots. This lowered ability of roots, in contrast to shoots, to move IAA is apparently why earlier workers with their less sensitive detection systems did not discover this strongly polar IAA movement with receivers. Kirk and Jacobs also showed that counts of radioactivity in tissue slices manifest less strong polarity than do counts of receivers, which seems reasonable in view of the chromatographic purity of hormone in receivers, as contrasted with the heterogeneity of labeled products in tissue slices. The big difference from earlier views was, of course, that the polarity was acropetal rather than

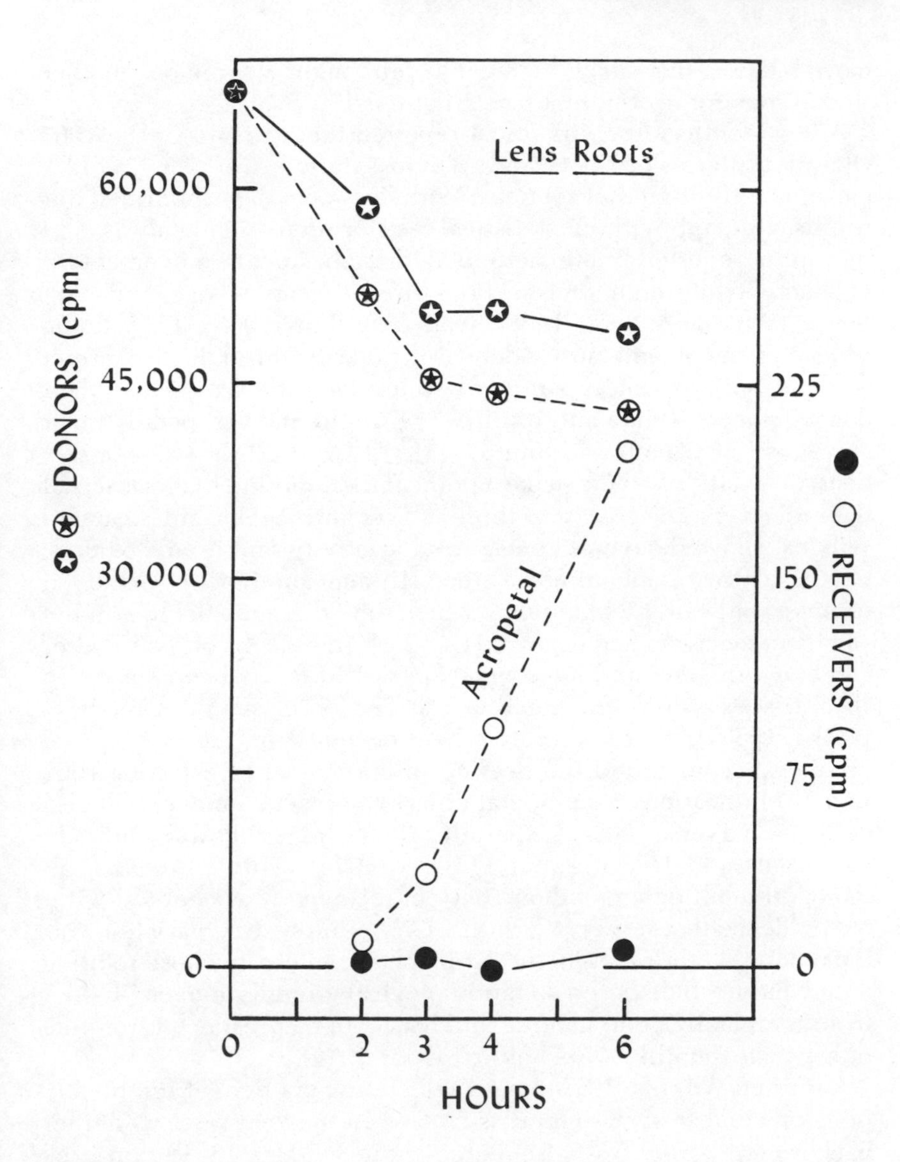

Figure 10-3. The time course and polarity of movement of ^{14}C-IAA through transport sections cut from *Lens* roots and into agar receivers (Kirk & Jacobs 1968).

basipetal (i.e., toward the root tip rather than away from it). Considering the whole plant, the new results suggested a unified polarity of the organism—polar movement of IAA was typically from shoot tip toward root tip instead of movement in which one IAA stream moved down the shoot to meet in an unknown region another stream moving up the root (see Figure 9-1).

A few months after Kirk and I reported these results at the 1967 Ottawa conference, Scott and Wilkins started work on ^{14}C-IAA movement through root sections. With *Zea* roots they confirmed our results, although without statistical tests or regression analysis, and they also reported more acropetal than basipetal movement of radioactivity through root sections from *Avena,* wheat, and sunflower (Wilkins & Scott 1968; Scott & Wilkins 1968). They found little movement and no evidence of polarity through *Pisum* root sections, but this was apparently because too little IAA was in their donors, since Hillman and Phillips (1970) did find acropetally polar movement of ^{14}C after adding 57 μM ^{14}C-IAA to "Alaska" pea sections. (The latter authors make a point of their finding that extracts of their receivers, in contrast to those of Kirk and Jacobs and Scott and Wilkins, showed several zones of radioactivity on chromatograms, but the contrast is meaningless since Hillman and Phillips extracted receivers only after 24 hours, by which time the same decline in net counts in receivers had occurred that Kirk and Jacobs observed after 6–8 hours in *Lens* and *Phaseolus* roots. And it has been shown in shoot tissues, where this decline was first observed [McCready & Jacobs 1963a], that increasing heterogeneity of the radioactive products accompanied the decline in net counts [see Jacobs 1967 Fig. 6].) Indications of acropetal polarity of ^{14}C-IAA movement also came from Iversen and Aasheim (1970) using sunflower and cabbage roots and Hartung and Phillips (1974) using *Phaseolus,* although in both papers radioactivity in receivers was counted without evidence that it was still with IAA. An interesting variation was the report that the *Selaginella's* rhizophore, an organ whose rootlike nature has been disputed on morphological grounds, moves ^{14}C-IAA in an acropetally polar manner and hence, in that respect, is rootlike rather than shootlike (Wochok & Sussex 1974).

Cane and Wilkins (1970) found a declining gradient of acropetally polar movement of ^{14}C added as IAA when they cut sections farther back from *Zea* root tips, although even in sections 13–19 mm back there was still some acropetal polarity judging by their averages. Batra et al. (1975) confirmed the declining gradient for *Zea,* but Hillman and Phillips (1970) found little, if any, such difference along *Pisum* roots. (None of the three "gradient" papers reported checking to be sure that all the ^{14}C in receivers from these more basal sections was still with IAA or had auxin activity.)

Dogmas are not relinquished easily, even in science, and a common scientific ploy against results that seem to overthrow a dogma is to attempt to explain away the results. Davies and Mitchell (1972) thought there were two possibilities for doing so with the reports of acropetal polarity in roots: they raised once more the possibility that cut sections might act differently from similar regions in the intact organism, and they also suggested that the IAA concentrations added in donors by their immediate predecessors to roots might be too high (resuscitating the view from Cholodny's time in the 1920s that roots might be much more sensitive than shoots to a given concentration of IAA). Accordingly, they applied ^{3}H-IAA of 85–90% purity to the intact root surface of *Phaseolus,* adding only 0.002–0.02 μM IAA (rather than the 17 μM Kirk and Jacobs had used to match earlier studies on shoots). However, they obtained essentially the same results: label piled up in the root tip and this acropetally moving ^{3}H was almost all IAA by Rf. And although label was found toward the root base, it was in a sharply declining gradient and much of it ran to zones on chromatograms other than that typical of IAA. The rate of tritium movement was roughly 3–7 mm/hour.

Auxin movement across the shoot–root transition zone

With this polarity of IAA movement toward the root tip confirmed in sections cut from roots of several genera, and even reported for IAA added to the surface of roots on intact seedlings, the question remained of whether IAA can move down the shoot, across the shoot–root transition, and then down the root axis to its tip. (The shoot–root transition is the portion of the longitudinal axis where the vascular strands change their location and arrangement from that typical of the shoot to that typical of the root. The rearrangements can be complex and the transition region may be in the hypocotyl or epicotyl, depending on the species [see Esau 1953].) Early work with excised sections suggested that IAA movement from shoot to root ceases at the transition region. For instance, the transition region of *Phaseolus* seedlings is at the base of the hypocotyl, and sections cut from this region showed no movement of added IAA in either direction (Jacobs 1950a; Smith & Jacobs 1968 – see Figure 9-4). Similarly, Scott and Briggs (1962) obtained essentially no diffusible auxin from the base of *Pisum* epicotyls, which includes the lower part of the transition region, although sizeable amounts diffused from shoots cut above the transition region. However, the few available studies of IAA movement in intact seedlings do not confirm the inference that IAA movement down the shoot ceases at the transition region.

Morris et al., in a fine paper (1969), considered this problem. They added 0.9 μg ^{14}C-IAA (0.25 μCi) in a droplet to the apical buds of intact *Pisum* seedlings and followed the distribution of label both down the plants and to different radioactively labeled compounds (as evidenced by the location of ^{14}C on paper chromatograms). They found that label moved down the stem at a rate of 11 mm/hour and moved into the root system where total counts increased up till 12 hours, the longest time checked. Ethanol extracts of the roots at various times after adding ^{14}C-IAA to the shoot apex revealed, upon chromatography, that ^{Rf}IAA^{Rf} increased in the roots for from 3 to 6 hours and then remained at about the same level at 9 and 12 hours, whereas label at the Rf of indole-3-acetylaspartic acid increased throughout the 3- to 12-hour period. Particularly interesting was the discovery that more label was at the Rf of indole-3-aldehyde than at ^{Rf}IAA^{Rf} in extracts of both roots and the most basal internode. Radioactivity at the aldehyde Rf increased to a maximum at 6 hours in roots and at an earlier hour in the basal internode just above the roots. Sections from the upper part of the root were placed root tip end down onto agar and diffused for 4 hours, after which the receivers were extracted for chromatography. In contrast to receivers on the basal end of sections from the upper internode, which showed essentially all counts at the Rf of IAA (as expected from the earlier literature), receivers from the root sections showed the most radioactivity at the Rf of indole-3-aldehyde, with much smaller amounts still with ^{Rf}IAA^{Rf}. Also supporting the view that the ^{RF}indole-3-aldehyde^{Rf} was moving from shoot base to root base was the observation that in the basal internode ^{14}C at this Rf of tissue extracts declined precipitously after reaching its maximum there at about 4.5 hours, and that diffusates from the bottom end of this same internode contained ^{Rf}indole-3-aldehyde^{Rf} as well as ^{Rf}IAA^{Rf}. Their conclusion was that both IAA and indole-3-aldehyde moved from the stem into the root. There was no indication that the indoleacetylaspartic acid moved. They did not follow distribution within the root system. Davies and Mitchell (1972), using intact *Phaseolus* seedlings, confirmed that label from IAA added to the shoot tip could move down the shoot and into the root, but they added a bit of localization in the root by noting that acidic chloroform extracts of the most distal 50 mm of the root showed about one-third or more of the ^{3}H was still with ^{Rf}IAA^{Rf}, even 24 hours after ^{3}H-IAA had been added to the shoot tip. (Unfortunately for our interest in localization, they did not give the total length of these 5-day roots, but they do state elsewhere that 3- to 4-day seedlings had roots 50–70 mm long.)

Further evidence that the shoot-toward-root-tip polarity of IAA movement was operative in intact plants came from Kendall et al. (1971). They found that very little ^{14}C-IAA added to a solution bathing the roots of sterile intact *Pisum* seedlings moved up into the shoot, and the few counts found there did not get beyond the first internode.

Endogenous auxin of roots

All these investigators of IAA "transport" added IAA from the outside. But what is the endogenous auxin of roots? Kefford's (1955) chromatograms of extracts of *Vicia*, *Pisum*, and *Zea* roots had shown auxin activity at the Rf of IAA (Figure 3-3), although in each case there was also auxin activity at a zone nearer the origin. Similar chromatographic evidence for the presence of IAA in extracts of roots has been reported by many workers since then, notably by Audus and co-workers. Two-way chromatograms of ether-soluble substances from *Vicia faba* roots gave Lahiri and Audus (1960) evidence for four auxins (Figure 10-4), one of them running to the Rf of IAA in these two solvents and giving a positive color test with Ehrlich's reagent (indicating the presence of an indole ring). Four zones of significant coleoptilar inhibition were also present, one apparently being the β-inhibitor zone that Kefford discovered. (One of the auxin zones and three of the inhibitory zones showed up only on the two-way chromatograms, an additional caution [in case one is still needed] against putting too much faith in the completeness of separations on the usual one-way chromatograms.) More recently Greenwood et al. (1973) asserted, without presenting data on this point, that "almost all" the auxin activity that they found in their final acid-ether, after extracting *Zea* roots, ran to the Rf of IAA. There was a 60% loss of IAA in their preparations. Without correcting for loss, the authors found that extracts of stelar tissue showed auxin activity equivalent to 142 μg IAA/kg fresh weight at the Rf of IAA. Extracts of cortical tissue, by contrast, gave only 1 μg IAA/kg, indicating a striking concentration in the stele. More secure identification of IAA as an endogenous auxin of *Zea* roots came from the use of gas chromatography and mass spectrometry (Elliott & Greenwood 1974), a finding confirmed in a paper submitted several weeks later by Bridges et al. (1973). The latter authors also estimated the amounts of IAA recovered from various regions, and reported 10 times as high a concentration of IAA present in stelar tissues as in the cortex (53.3 μg/kg fresh weight in the stele, as contrasted to only 4.8 in the cortex). Root tips gave 27 μg/kg. (These

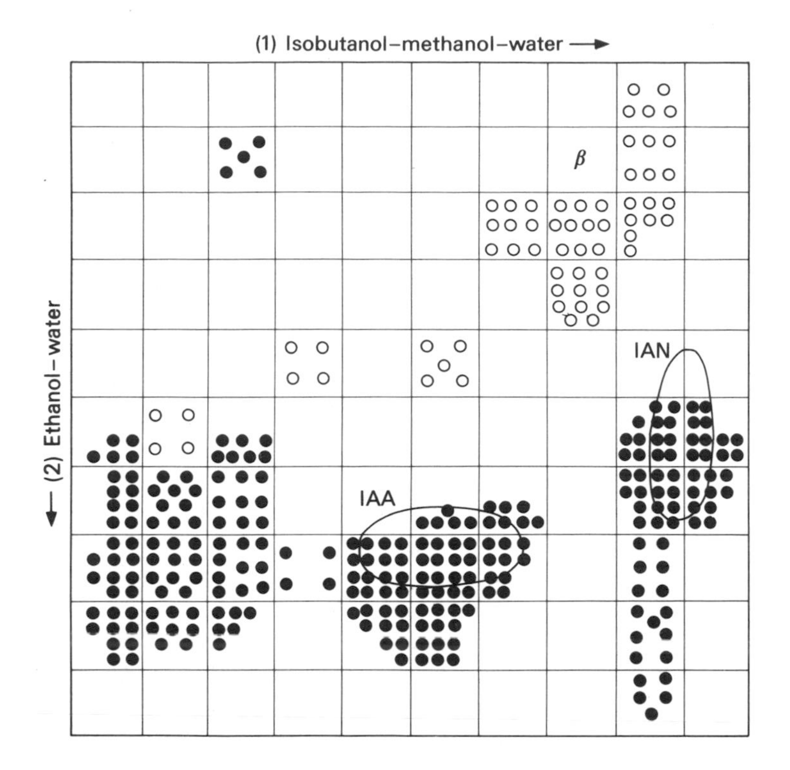

Figure 10-4. Distribution of zones of statistically significant growth promotion (black points) and inhibition (open circles) on a two-way chromatogram of the ether-soluble acid material from ethanol extracts of *Vicia* roots. The 100 squares were assayed with the *Avena* section test. The locations to which marker IAA and IAN ran on control chromatograms are designated by the ellipses. The number of points in each square increases with the strength of the response (Lahiri & Audus 1960).

values are without any correction for losses in preparation and chromatography.) Furthermore, the stele seems to be the major pathway for the polar acropetal movement of added IAA, judging by the collection of ^{3}H-IAA in receivers after application of the labeled IAA solely to the stele or cortex by means of a micropipette (Shaw & Wilkins 1974).

However, although there is convincing evidence from mass spectrometry for the presence of IAA in extracts of *Zea* roots, we are far from knowing whether IAA is the only endogenous auxin in roots, or even the major auxin, and the chromatographic evidence (e.g., Figure 10-4) would suggest it is not. The addition of radioactively labeled IAA to roots results in a relatively fast appearance of label in other compounds. Morris et al.'s (1969) data showing more label at the Rf of indole-3-aldehyde than of IAA have already been cited. Iversen et al. (1971) injected ^{3}H-IAA into the stem–root transition of intact *Phaseolus* seedlings and found some pileup of label at the root tip after 5 hours, but thin-layer chromatography of methanol extracts of the root tip of the roots showed no peak of radioactivity of ^{Rf}IAA^{Rf} and two large peaks elsewhere. And even ^{3}H-IAA, which had only moved the few millimeters represented by excised sections of *Zea* root, was completely converted to other compounds if the stele had been removed from the sections (Shaw & Wilkins 1974). If the stele had been left in the sections, some ^{3}H in the total extract still ran to ^{Rf}IAA^{Rf} after 8 hours, but as much or more was at two other zones.

Obviously we very much need a quantitative assessment of total auxin activity in roots, with an exact balance sheet telling how much of that total can be explained by the endogenous IAA, indoleacetaldehyde, indoleacrylic acid (Hofinger et al. 1970), chloroindoleacetic acid (Gandar & Nitsch 1967), phenylacetic acid (Wightman 1977), and other auxins.

The functions of auxin in roots

What are the functions of auxin(s) in the root? Synthetic auxins added to roots typically inhibit elongation (Went & Thimann 1937; Aberg 1957). (Sporadic reports have been published, claiming some stimulation of root elongation when very low levels of IAA are added to roots, but I agree with the critical assessment by Aberg [1957] that the claimed stimulations are unconvincing. They have usually been given only as averages with no statistical evidence that the differences were anything but sampling error.) Auxin added to the shoots of intact plants can also inhibit root elongation. For instance, Eliasson (1961) added various auxins to the uncut surface of *Pisum* hypocotyls and followed the time course of the resulting inhibition of root elongation. The kinetics of root inhibition with IAA and 2,4-dichlorophenoxyacetic acid (2,4-D) looked remarkably like a mirror image of Figure 9-11, which shows the kinetics of movement of IAA and 2,4-D through bean petioles (McCready & Jacobs 1963a). Inhibition by IAA increased until 6 hours, then decreased again until about 18 hours. Comparison of the two figures suggests that movement of the two auxins into the pea root is the cause of the inhibition of root elongation, with the difference between IAA and 2,4-D in the timing of inhibition and its cessation being a function of their known different rates of movement and of metabolism.

The potential role of IAA, present in high concentration and moving acropetally in the stele, in developmental phenomena that are initiated in the stele is obvious. This is particularly so for the initiation of lateral roots, which start to differentiate with specific relation to the vascular tissue in the stele (Esau 1953). Xylem cells of the vascular tissue differentiate acropetally in the developing root, paralleling the polarity of IAA movement, and the differentiation of xylem cells is stimulated by IAA (see Chapter 4).

Abscisic acid and the role of the root cap in geotropism

If the root tip is not a source of auxin, in inhibitory concentrations, that moves basipetally and is redistributed under the influence of gravity, as Cholodny proposed (Fig. 10-1), what is the root tip's function? First, note that "root tip" is an embarrassingly vague designation: roots typically have a root cap produced at their distal ends by a subapical meristem, the latter still more confusingly called an "apical meristem" (see Plates 15 and 16 of Esau 1953). When early (and not so early) physiologists wrote that 1.5 mm of the root tip was cut off, inevitably they were measuring back from the most distal cells on the root cap. Whether such an excision would remove only part of the root cap, or some of the subapical meristem too, or also most of the elongating zone of the root proximal to the meristem, would obviously depend on the size of the root cap. Results from root tip excisions that give no information about the anatomy of the excised piece are therefore essentially meaningless.

Konings (1968) cut different lengths from the tips of pea roots and reported that excising 0.5 mm did not change the total elongation per 6 hours but did completely stop the geotropic response. He asserted that excising 0.5 mm removed the whole root cap, but provided no photograph of longitudinal sections through such root tips to confirm this statement. If only 0.2 mm of the root tip was excised, both the elongation and the geotropic curve progressed as in the intact controls (his Fig. 2 and Table 1). (Two years before, a note by Juniper et al. had asserted, with no data or convincing photographs to back them up, that they could cleanly remove root caps of barley and *Zea*, and that the "capless" roots elongated at normal rates but without geotropic response.) Cerček (1970) found that removing all the root cap of barley completely stopped geotropic curvature, but after 27 hours much of the cap had regenerated and geotropic curva-

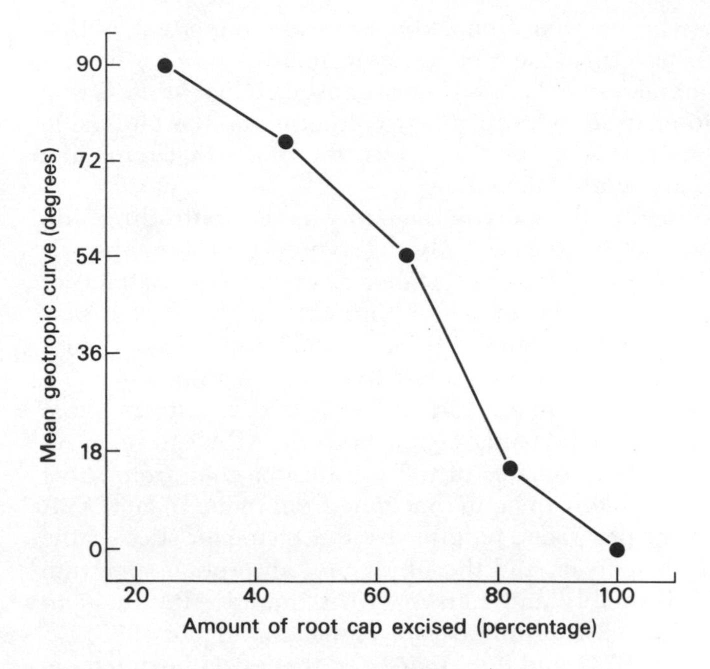

Figure 10-5. The effect of removing increasing percentages of the barley root cap on inhibition of the average positive geotropic curvature of the roots (raw data from Cerček 1970).

ture also reappeared. If less of the root cap was removed, geotropic curvature reappeared more quickly (Figure 10-5). Despite the satisfying sample sizes Cerček used for determining these average egotropic curves ($n = 30$ or more), we still find nothing but his assertion that elongation of decapped roots was normal. Gibbons and Wilkins (1970) looked into the other genus recommended by Juniper et al.; they found that removing the root cap of *Zea* with forceps resulted in loss of geotropic sensitivity that lasted for 24 hours (their Fig. 1), and again we are told that elongation nevertheless continued. Barlow (1974) was no better at providing hard data on elongation rates, but his microphotographs of longisections through root tips convincingly supported his finding that, even though delicately removing root caps did remove the geotropic sensitivity of the root, the restoration of geosensitivity did *not* depend on prior regeneration of part of the root cap. After 24 hours the geotropic sensitivity of both *Zea* and *Triticum* was restored, but no root cap had yet been formed. The only change noticed in the 24-hour roots was that there were more starch grains in the cells of the apical meristem of the roots.

The reason why data on elongation rates are important is that since the 1920s the "tip of the root" (or now, more precisely, the root cap) has been hypothesized to produce a substance that inhibits root elongation and controls geotropism by collecting on the underside of the horizontal root. One might expect, therefore, that removing the cap would *stimulate* elongation.

As faith declined in the view that auxin was the controlling substance from the root tip, the recently discovered inhibitor, abscisic acid (ABA), became an obvious substance to investigate. Extracts of roots revealed a β-inhibitor zone after chromatographing, as do various shoots – a fact known since Kefford's 1955 paper (see Figure 3-3). In addition, it has been shown that illuminating roots decreases their elongation (e.g., Torrey 1952) and results in more β-inhibition in extracts (Masuda 1962). Because ABA had been reported to be present in eluates of the β-inhibitor zone from shoot extracts, it was also likely to be in that zone from roots. In fact, ABA was in extracts of pea roots, judging by the characteristics of the optical rotatory dispersion and the ultraviolet absorption spectrum (Tietz 1971). El-Antably and Larsen (1974) found ABA in *Vicia* roots, also. Added ABA inhibited root elongation (Tietz 1973, El-Antably & Larsen 1974), and light increased the level of extractable ABA (judging by gas–liquid chromatography) as well as decreasing root elongation (Tietz 1974), so the conclusion was drawn that light has an inhibiting effect by increasing production of ABA. Kundu & Audus (1974) concluded that ^{Rf}ABA^{Rf} was mostly in the root cap of *Zea* (using a bioassay that measured the closing of *Commelina* stomates), but they found inhibition at a different zone in chromatographed extracts of root tips that had had their root caps removed. This second inhibitor was not identified. With different bioassays, H. Wilkins & Wain (1974) confirmed the presence of ^{Rf}ABA^{Rf} in *Zea* root caps, although they found detectable amounts only after illumination of the roots with light sufficient to inhibit elongation. Various groups are now in hot pursuit of the hormone presumed to control root geotropism by being redistributed under the influence of gravity. The report by El-Antably and Larsen (1974) that more ABA could be extracted from the lower than from the upper halves of *Vicia* roots (74 as contrasted to 24 ng/g fresh weight in the upper halves) supported ABA's candidacy for the role of geotropic control hormone in roots; but that same paper reported more endogenous gibberellins in the *upper* half (125 vs. 60), so gibberellin (GA) is also a possible controlling hormone.

There is little direct evidence about the longitudinal movement of ABA in roots. One of the few reports that used transport sections found so few counts in receiver blocks that they mostly followed counts extracted from tissue slices after adding ^{14}C-ABA in donor blocks to *Phaseolus* root sections (Hartung & Behl 1974). There was a strong acropetal polarity of movement if longer sections were used (shorter sections showed no significant polarity). Rather surprisingly by comparison with IAA, all the ^{14}C was still with ^{Rf}ABA^{Rf} in these extracts of tissue even after 10 hours of transport. The rate of movement through the bean root sections was 4–5 mm/hour, judging by extrapolated linear regressions. The acropetal movement is restricted almost entirely to the stele (Hartung & Behl 1975a) and is increased more than 100% by illuminating the sections during the transport tests (Hartung & Behl 1975b). Hartung and Behl said that in preliminary experiments they confirmed for IAA the stimulation of acropetal movement by illumination that Scott and Wilkins first reported in 1969 with *Zea* roots. The increased movement of ABA toward the growing zone near the apex of the root obviously might be involved in light's inhibition of root elongation.

Hormones other than auxins and abscisic acid

Thiamine

Thiamine must be added in hormonal concentrations to the medium to obtain continued development in aseptic culture of the excised root tips of most genera tested (Bonner & H. Bonner 1948). In the intact plant thiamine occurs in highest concentration in green leaves. Indications from girdling experiments are that it moves from mature leaves both down the stem toward the roots and up the stem toward the younger leaves (Bonner 1942). Hence, if root tips in excised organ culture are sufficiently similar to roots on the intact plant, thiamine is probably a root growth hormone in plants, the roots being as dependent on other organs (the shoots) for their thiamine supply as we are dependent on other organisms for our supply of thiamine (Vitamin B1, to us). Shoots evolved in land plants before roots (the first vascular plants, the *Psilopsida*, possessing shoots only), and thiamine is a likely candidate as a hormone produced by shoots to regulate root growth. Growth of the intact plant, however, is apparently not limited by thiamine. (Bonner and Greene [1939] did, in fact, report that watering intact plants with thiamine solutions increased plant growth, but neither Arnon [1940] nor Hamner [1940] could duplicate such an effect, and Bonner [1943] admitted he could not get such results again. The guess among Bonner's research associates at the time was that there had been inadequate randomization of plants in the initial experiments.)

As described in Chapter 9, thiamine shows strong basipetal polarity of movement through petiole sections but no polarity through stem sections, whether young or old. Through sections from roots of *Pisum* seedlings thiamine movement is also nonpolar, whether the sections were cut near the root tip or near its base (Jacobs & DeMuth 1977). *Pisum* was selected as one of the genera whose roots need thiamine for development in sterile organ culture. The amount of thiamine moved through the pea roots was also much smaller than that moved through the petiole sections, even after 18 hours averaging only 1.8–1.9% of the thiamine taken up by the section.

Gibberellins

In contrast to IAA and thiamine, gibberellins are apparently produced in root tips. The first fairly specific evidence for production in roots, apart from earlier reports that gibberellic activity was found in extracts of roots, in addition to other plant parts (e.g., Radley 1958; Murakami 1968b), was the report of gibberellin activity in chromatographed extracts of tomato root tips that had been grown for 4–5 weeks in aseptic organ culture (Butcher 1963). It was extremely unlikely that the gibberellin was merely carried out from the shoot, because the cultured tips were from a clone of roots that had been grown in culture for 5 years. More specific evidence for GA production in root tips was soon forthcoming. When 3- to 4-mm long root tips of *Helianthus* were diffused onto agar for 20 hours, and then fresh tips and the prediffused tips as well as the agar were extracted, the extracts chromatographed, and eluates of the zones bioassayed for gibberellin, Jones and Phillips (1966) found statistically significant gibberellin activity at about the same level in all three. That is, the root tips that had diffused out gibberellin for 20 hours contained about as much gibberellin as freshly cut root tips that had not been diffused. In contrast, sections cut 4–8 mm back from the tip, and said to be proximal to the elongation region, showed so little gibberellin activity in extracts of prediffused sections that Jones and Phillips concluded that this more proximal region was not producing gibberellin. These results fitted expectations from earlier experiments that had found gibberellin activity in root exudate (also called xylem exudate or bleeding sap; operationally, this is the liquid collected from the cut surface of a plant that has had the shoot excised near its base) (Phillips & Jones 1964;

Carr et al. 1964; Skene 1967; Kende & Sitton 1967). These reports of GA activity in roots based on bioassays have recently been supplemented by evidence from gas–liquid chromatography that GA-1,2, or 3 was present in 15–20 mm long *Vicia* roots (El-Antably & Larsen 1974) and that GA-3 added to roots of intact *Vicia* stimulated their elongation.

If gibberellin is produced in root tips, one might expect its movement to be polar in the direction *away* from the root tip. (This is by analogy with the polar movement in leaves of auxins, GA, and thiamine, all of which are polar away from the site of synthesis.) Such root-toward-shoot polarity of gibberellin has been shown when GA-3 was added in donors on standard transport sections from near the tip of *Zea* roots (Jacobs & Pruett 1973). A statistically significant difference in movement in the two directions was found whether gibberellin in receiver blocks was estimated directly with the barley endosperm test or GA-3 labeled with ^{14}C was added in the donors with receivers counted directly in a liquid scintillation counter. However, the polarity was slower to develop with GA-3 than it had been for IAA (Figure 10-6). Differences between apical and basal receivers did not become statistically significant until the 14-hour collection. The experiments with the bioassay showed an average of 6.0 ng "GA-3 equivalents" per basal receiver at 18 hours compared to 1.6 ng per apical receiver. The separate experiments with ^{14}C-GA showed identical values for GA in apical receivers and values of the same order of magnitude for basal receivers (calculated on the assumption that all counts in receivers were still with GA-3). Gibberellin activity in the basal receivers showed a decline after reaching its maximum, paralleling the situation for other endogenous hormones whether in shoots (IAA, GA-3, adenine) or roots (IAA) and presumably reflecting metabolism of the hormones. A point of particular interest is that these transport sections from *Zea* roots can move GA and IAA with opposite polarities (Figure 9-1). Hartung and Phillips (1974) confirmed the basipetal polarity at 18 hours of ^{14}C added with GA-3 to *Phaseolus* root sections (although they did not present evidence that the ^{14}C was still with GA or that the receivers showed gibberellin activity). ^{3}H-GA-1 also moved with basipetal polarity, judging solely by counts of tritium, and label from both gibberellins moved with more polarity through stelar than through cortical tissue.

What effects can be attributed to GA, if it is produced in root tips, moving with polarity toward the root base and appearing in "root exudate"? In the hope that the contents of root exudate would give a clue as to what moved up into the stem from the roots of intact

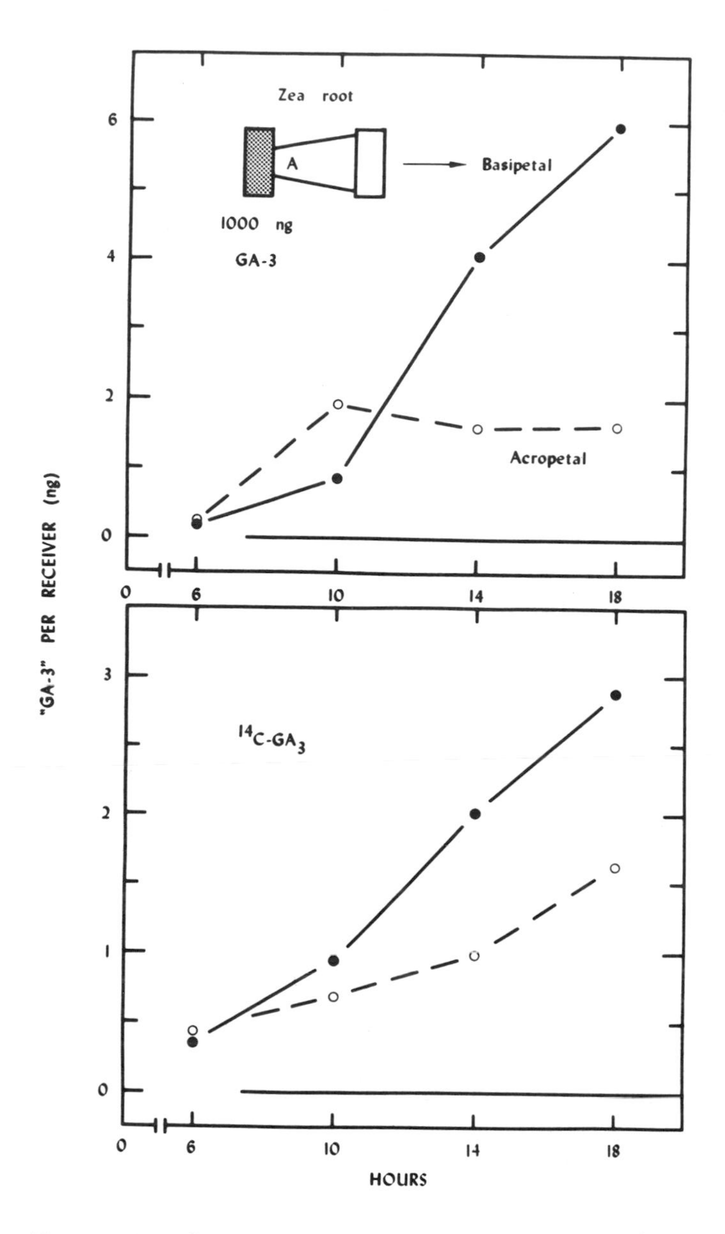

Figure 10-6. The time course and polarity of GA-3 movement through *Zea* root sections and into agar receivers, as estimated by the barley endosperm bioassay (top) and ^{14}C counting after ^{14}C-GA-3 was added in the donors (bottom) (Jacobs & Pruett 1973).

plants, Kende and Sitton (1967) bioassayed for gibberellins in root exudate collected from sunflower plants of various ages. Two of the 10 zones tested showed gibberellin activity. Activity in one of them did not change appreciably as the plants grew from 22–49 days. The other zone, however, increased tenfold in activity at 30 days – when the first signs of flowering were visible as transition stages in dissected shoot tips – and increased at 34 and 49 days. The activity at this zone, calculated from the dwarf-corn bioassay as if it were GA-3, jumped from 0.02 μg/liter of root exudate at 27 days to 0.20 μg/liter at 30 days. Unfortunately, this large rise in gibberellin activity in the root exudate was not accompanied by faster elongation of the stem. The implication was that the increased gibberellin in the root exudate did not seem to be causing more stem elongation. Kende and Sitton also added 1 μM of GA-3 to etiolated pea seedlings with or without their roots and noted that, although the added GA-3 increased stem elongation in both cases, the "relative growth increase in rootless plants is, however, smaller than in the intact ones." Rather puzzlingly, they did not refer to the fact, obvious from their data, that substitution of 1 μM of GA-3 for the roots more than completely replaced the effect of the roots in causing stem elongation for the 7 days of the experiment. (The experimental methods were less than ideal: the cut stems were put into 2% sucrose solutions with no word that conditions were aseptic.) The data of Crozier and Reid (1971) likewise show that when the hormonal dose of 1 μg of GA-3 per seedling was substituted for all the root tips of *Phaseolus* seedlings and elongation was measured 5 days later, GA-3 could replace the effect of root tips on stem elongation. However, like Kende and Sitton, these authors mention only that added GA-3 did not cause as much elongation in "plants minus root tips" as was seen in intact plants when they were treated with the same dose of GA-3. Crozier and Reid were really focused on other problems. They found two major zones of gibberellin activity on chromatograms of "stem and petiole" extracts, ^{Rf}GA-1^{Rf} and ^{Rf}GA-19^{Rf}, with the latter predominant. In roots or "apical buds and leaf blades" ^{Rf}GA-1^{Rf} was predominant, with only a small amount of activity at ^{Rf}GA-19^{Rf}. Seedlings with all root tips excised showed interesting changes in the pattern of endogenous gibberellins. The subapical root segments now had lots of activity at the Rf of GA-19 and none at ^{Rf}GA-1^{Rf}. The same switch occurred in the "apical bud and leaf blades." The "stem and petiole" extracts showed a drastic decrease in ^{Rf}GA-19^{Rf} activity. Crozier and Reid hypothesized that in the intact plant GA-19 was synthesized in the leaves and apical buds, and moved to root tips where it was converted to GA-1, from whence it was exported back to the shoot. (They acknowledged that their data did not preclude the possibility that ^{Rf}GA-19^{Rf} was synthesized in the subapical parts of the roots.) Their hypothesis is attractive because it helps a bit to explain the staggering multiplicity of gibberellins that have been isolated from plants (see Chapter 7).

Cytokinins

Soon after the artificial cytokinin, kinetin, was isolated, Richmond and Lang (1957) made an important discovery: kinetin added to excised leaves of *Xanthium* prevented both the yellowing and the loss of protein-nitrogen that typically accompany leaf senescence. Because rooting of excised leaves had long before been shown to have similar antisenescence effects (e.g., Chibnall 1939), it was natural to hypothesize that on rooted leaf cuttings roots might be synthesizing a natural cytokinin and exporting it to the leaves. Kulaeva (1962) demonstrated that some of kinetin's effects on tobacco leaves did not appear if roots were attached, even if the cells between the roots and the leaf had presumably been killed by "steam girdling." This latter point suggested that the substance(s) from the roots was moving in the transpiration stream through the dead xylem cells. Supporting this interpretation was the kinetinlike effect of sap collected from the plant (although the effect was weak). Her conclusion was that roots provided cytokinins to the xylem sap, which in turn carried it to the leaves. Many later workers have confirmed more directly that "xylem exudate" contains cytokinin activity (e.g., Loeffler & van Overbeek 1964; Kende 1964; Nitsch & Nitsch 1965b; Klämbt 1968). Direct extraction of roots revealed cytokinin activity, as one might expect from the preceding. Weiss and Vaadia (1965) found two zones of activity by the soybean callus assay when they chromatographed extracts of the distal 1 mm of sunflower roots. When they checked extracts of sections only 1–3 mm back from the tip, however, they found little, if any, cytokinin activity. They pointed out that the presence of cytokinins in root tips was in keeping with the well-established fact that root tips do not require added cytokinins for growth in aseptic organ culture. (We see later that this is a bit misleading, ignoring as it does the role of cytokinins in cambial activity.) Two zones of cytokinin activity were also reported by Seth and Wareing (1965), who extracted *Phaseolus* roots. After zeatin was isolated, various people reported ^{Rf}zeatin^{Rf}, or ^{Rf}zeatin glycosides^{Rf}, or both, on their chromatograms of root extracts (Bui-Dang-Ha & Nitsch 1970 from *Cichorium*'s fleshy roots; Radin & Loomis 1971 from radish's fleshy roots;

Yoshida & Oritani 1971, 1972 from *Oryza* roots). Gas–liquid chromatography was used by Babcock and Morris (1970) to identify four cytokinins in extract of pea roots: zeatin and its riboside, and isopentenyl adenine and its riboside. Short and Torrey (1972) confirmed the large amount of cytokinin extractable from the 1-mm distal tip of roots as compared to the next section, using pea roots instead of sunflower. After extensive purification, they found 6.0 mg of "kinetin equivalents" per kg fresh weight in the tip compared to 0.14 mg/kg in sections cut 1–5 mm back; sections cut 5–20 mm from the root tip showed no cytokinin activity. In their assay, which used soybean callus, ^{Rf}zeatin^{Rf} from the tip extract showed most activity, along with its ^{Rf}nucleoside^{Rf} and ^{Rf}nucleotide^{Rf} and another unidentified zone. (Expressing the absolute amounts of cytokinin as "kinetin equivalents" is particularly misleading with zeatin, as Jacobs [1976] pointed out, because zeatin has 100 times the activity of kinetin in the soybean callus bioassay.)

I can find no reference to studies with diffusible cytokinins in roots like van Overbeek's (1939) with auxin or Jones and Phillips's (1966) with gibberellin. (They compared the amount of hormone obtained by "diffusion plus subsequent extraction of the diffused tips" with the amount from direct extraction. If direct extraction gave much less than the sum of the other two, as happened with gibberellin, they concluded that the roots were producing the hormone – not merely acting as a receptacle for hormone that had come down from the shoot.) Hence, we have no evidence, other than the high concentration of cytokinins in root tips, that cytokinins are produced in the root tips.

There are only a few reports on cytokinin movement through transport sections. ^{14}C-kinetin moved but without polarity in 24-hour collections through sections from radish root (Radin & Loomis 1974), thus paralleling results with shoot structures (Figure 9-1). Zeatin was tested for polar transport through root sections of *Pisum* (the genus and organ from which Short and Torrey had isolated ^{Rf}zeatin^{Rf}) to see whether highly active, endogenous cytokinin would move without polarity as adenine and kinetin had been found to do in petiolar sections. Although counts in the receivers were significantly above background when ^{14}C-zeatin was in donors, there was no significant difference between apical and basal receivers by statistical test (Jacobs 1976). Unfortunately, only counts were followed in this root experiment: the tiny amount of labeled zeatin precluded checking with chromatograms. With 1.25 μg of zeatin in donors, there was 1.5 ng of zeatin in 8-hour receivers (if all counts were still with zeatin) and 3.4 ng in 24-hour receivers.

This nonpolar ^{14}C-zeatin movement is at about the same level as the amount of gibberellin moved polarly through *Zea* roots in roughly the same time (2.9 ng ^{14}C-GA-3 in 18 hours, Jacobs & Pruett 1973). Hence, if we can judge from this minute sampling, cytokinins added alone to donor blocks move without polarity through root sections—their apolar movement contrasting with the polarity of IAA and GA-3 in roots but resembling the apolarity of thiamine.

Auxin–cytokinin interactions. An important interaction between cytokinins and auxins has been discovered in roots by Loomis and co-workers. By adding some substances solely through the basal cut of the cultured root tips, Loomis and Torrey found that an auxin and a cytokinin, if applied thus unidirectionally (along with the usual potpourri of minerals and vitamins) would result in sizeable cambial activity. Neither one by itself was sufficient. If *myo*-inositol or certain other cyclitols were added along with the auxin and cytokinin, the number of cambial derivatives approximately doubled (Loomis & Torrey 1964; Torrey & Loomis 1967). Various auxins and various cytokinins were active. The most obvious interpretation was that roots in the intact plant developed an active cambium by virtue of auxin and cytokinin that came down from the shoot. (A similar need for auxins, cytokinins, and *myo*-inositol was demonstrated for cambial activity in turnip roots [Peterson 1973].) When IAA was added along with ^{14}C-kinetin, the nonpolar movement of ^{14}C seen with kinetin alone was changed to acropetally polar movement; that is, IAA increased the movement in the direction from the root base to the root tip so that it was significantly greater than in the basipetal direction (Radin & Loomis 1974). As further evidence that this sort of transport interaction was the basis for the cambial requirement for both auxin and cytokinin, Radin & Loomis reported that radish roots being grown in aseptic culture showed thickening all along the root section only when IAA and the cytokinin (benzyladenine, in this case) were both added at the basal end (i.e., "shoot end") of the root.

But why, one wonders, does a cytokinin have to be added to cultured radish roots to get cambium development, when other investigators have found cytokinins in root exudate and root extracts? Radin and Loomis (1971) capped this lovely series of papers by extracting endogenous kinins from the radish "root" (actually root plus some hypocotyl). Three zones of cytokinin activity were found when they bioassayed chromatograms of the extracts. In developing radish roots, two of these cytokinins increased at 17 days, when freehand sections showed the beginning of cambial activity in roots;

both induced cambial activity when added to the base of radish roots growing in aseptic culture. (They ran to the Rf of zeatin ribonucleotide and "zeatin or zeatin riboside" in the solvents used.) The third cytokinin, by contrast, did not increase "coincident" with the beginning of normal cambial development (its first increase was not until collection of the 32-day seedling), and did not induce cambium when added to the base of roots in culture. The authors' conclusion was that the first two cytokinins were involved in cambial activity in the normal root, probably moving down from the shoot to do so, whereas the third cytokinin was not involved in the control of cambium, might be synthesized in the root, and probably would appear in xylem exudate (since it was present in relatively much higher amounts in xylem than in the outer tissues).

With all this evidence that cytokinins are present in xylem exudate and in extracts of roots, are there situations – other than that of rooted leaf cuttings – where it has been shown that cytokinins when substituted for roots can replace some effect of roots? Kende and Sitton (1967) looked for such effects in rooted and rootless seedlings of pea and sunflower, but found no evidence that any of four added cytokinins could replace the roots in stimulating stem elongation. Mullins (1967) did obtain some indications by such substitution experiments that *Vitis* roots helped maintain the young inflorescence by their cytokinin production: benzyladenine could at least partially replace the roots in preventing atrophy of the inflorescences, but it also resulted in much less shoot growth. Only benzyladenine, of the eight cytokinins tested, gave even this partial replacement. A satisfying case of cytokinins substituting for roots was found in reexamination of van Overbeek's (1937) observation that excising the roots decreased auxin production by coleoptiles. Jordan and Skoog (1971) confirmed the decreased coleoptile growth that excision of roots brought about, but demonstrated that a drop of isopentenyl adenine at 100 μM fully restored growth of the coleoptile to the level of the intact controls. (The same amount added to coleoptiles of intact plants had almost no effect.) The same use of isopentenyl adenine increased the production of diffusible auxin as measured by the *Avena* curve assay. Kinetin and benzylaminopurine were almost as effective as isopentenyl adenine, whereas two chemically related substances known to be lacking cytokinin activity in bioassays had no effect on the coleoptile. Their conclusion was that the roots were normally providing cytokinin to the coleoptile, where the cytokinin maintained coleoptile elongation through its effect on the production of diffusible auxin.

11

Overview

In this last chapter I try to assess our current knowledge of plant hormones from the organismal viewpoint, emphasizing what we have all learned since Went and Thimann covered the known field so well in 1937. I touch briefly on some of the major topics that I have not had space enough to describe in earlier chapters, such as the methods by which hormonal levels are regulated and hypotheses to explain polar movement of hormones. The hormones of higher plants discovered to date are compared with the hormones of higher animals. Finally, I hazard a guess at the major needs and directions of future research.

The most striking change since 1937 is in the number of plant hormones that have been identified. Figure 9-1 gives an indication of this revolution. Instead of "auxin," which might possibly be indole-3-acetic acid (IAA), there are now whole new classes of hormones known: the 50 or more gibberellins, the cytokinins, the inhibitors abscisic acid and the structurally related xanthoxin and phaseic acid, and thiamine (and probably nicotinic acid). Almost as striking a change since 1937 is the extension of knowledge from organs of seedlings growing in the dark on the food reserves in the seed, like the *Avena* coleoptile, to independent green plants growing in the light. The techniques for identification of chemicals have improved to an astonishing degree over the last 40 years and have been quickly – if often uncritically – adopted by the hormone physiologists.

One of the puzzling features in plant hormones research has been the slowness with which the new techniques of statistics have been adopted, particularly in the United States. Bioassays are still needed when isolating a new hormone – and the bioassays need better statistical treatment than they usually get. It is past time for statistical treatment of data to catch up with the chemical instrumentation. It is esthetically jarring to see data collected from ultrasophisticated analytical machines presented as mere averages, instead of being analyzed by equally sophisticated statistical techniques. In addition, the tumbling price of calculators and computers makes such analysis feasible in terms of both time and money.

Control of auxin levels by conjugation or destruction

The mechanisms by which plants control the endogenous levels of hormone have been referred to in passing in several earlier chapters. Two methods have been intensively investigated: IAA destruction and IAA conjugation. Destruction of auxin by crushed tissue was suggested by the observations in the 1920s that no auxin could be obtained from crushed coleoptile tips although it was recovered by diffusion from the excised tip. Thimann (1934) showed that extracts of leaves destroyed auxin, judging by bioassay responses, and he suggested that the destruction was enzymatic. Larsen (1940) provided more specific evidence that enzymatic destruction could occur, and publications from the California Institute of Technology initiated a series of studies on what became known as IAA-oxidase (Tang & Bonner 1947, Galston & Dalberg 1954). (Operationally, IAA-oxidase is not as specific as it sounds. It usually means that the crude or dialyzed juice from ground-up tissue will decrease the IAA level in the surrounding medium, judging by color tests. Purified preparations of the enzyme have been so rare that there is no reason to consider all the IAA-oxidase preparations in the literature to be of the same enzyme.) Galston and Dalberg demonstrated a nice reversed parallel variation between the distribution of elongation in their etiolated pea shoots and the distribution of the IAA-oxidase activity of homogenates of the internodes. They proposed that IAA-oxidase was the endogenous controller of IAA levels and thereby of elongation within the shoot. Later literature has been reviewed by Ray (1958), Galston and Hillman (1961), and Hare (1964). There has been much resistance to considering IAA-oxidase as an endogenous regulator of IAA levels. The resistance comes partly from the discovery of artefacts of technique (the Salkowski color test for IAA is interfered with by various other factors [e.g., Platt & Thimann 1956]), partly from evidence suggesting that such IAA destruction is a function of cut or damaged cells (Briggs et al. 1955), and partly from the later discovery that at least part of the disappearance of IAA was not the result of destruction of the IAA molecule, but rather of IAA conjugation to aspartic acid (Andreae & Good 1955) or glucose (Zenk 1961, Klämbt 1961).

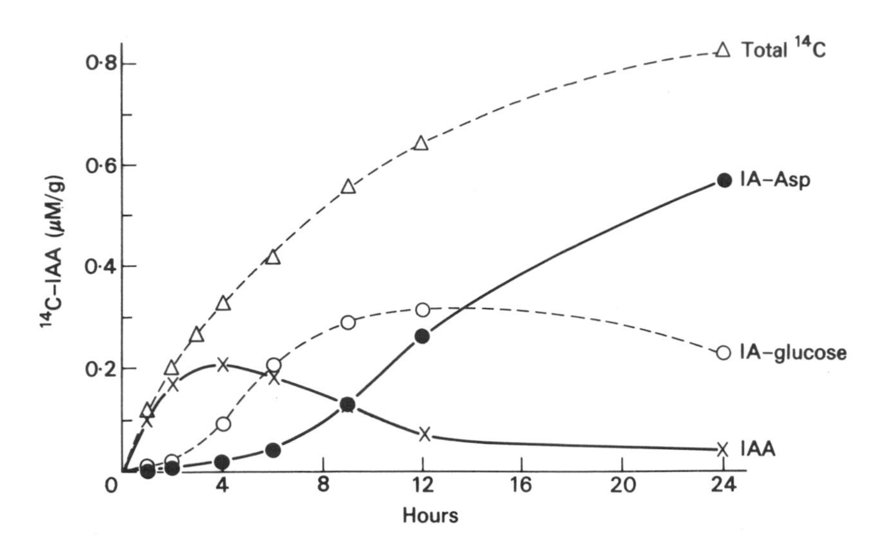

Figure 11-1. The time course of accumulation of IAA, IA-Asp, and indoleacetyl-β-D-glucose in *Hypericum* leaf disks, judging by the Rf of ^{14}C after disks were placed in 5 μM ^{14}C-IAA (after Zenk 1964).

When high concentrations of IAA (100 μM) were added to *Pisum* segments, sizeable amounts of the IAA taken up by the tissue were conjugated into indoleacetylaspartic acid (IA-Asp). No IA-Asp was detected in untreated pea tissue. Because the conjugate was less than one-thousandth as active as IAA in stimulating elongation of pea stems, and because conjugates were considered to be detoxifiers in animal cells, Andreae and Good interpreted IAA's conjugation with aspartic acid as a way for the plant to get rid of excess IAA. Zenk also interpreted IAA conjugation with glucose as a detoxification device.

The time course of conjugation of added ^{14}C-IAA was followed in pea roots (Andreae & van Ysellstein 1960) and *Hypericum* leaf disks (Zenk 1964). Their results were similar with respect to conjugation with aspartic acid, but Zenk showed in addition the time course of IAA's conjugation with glucose (Fig. 11-1). In both pea roots and *Hypericum* leaves the level of free IAA in the treated tissue rose quickly, then started to decline after about 4 hours, as increasing amounts of IAA were conjugated with aspartic acid. In Zenk's tissue, which was selected because it formed both conjugates "in good amounts," conjugation with glucose preceded IA-Asp formation by several hours.

The initial view that IAA conjugation was a detoxification device – and, by implication, was perhaps not functioning in normal metabolism – was weakened by the later discovery of IA-Asp in plants that had not been dosed with 50 or 100 μM of added IAA (Klämbt 1960; Row et al. 1961; Olney 1968). More specific evidence of IA-Asp as a normal constituent was the correlation between the decrease with age in IAA transport through *Coleus* petioles and the concomitant increase in IA-Asp as discussed in Chapter 8 (Veen & Jacobs 1969a, Jacobs 1972) (Figure 8-6). The evidence of Chang and Jacobs (1973) that abscisic acid and senescence factor both speed abscission by decreasing IAA transport to the abscission layer, with the decreased transport the result of conjugating IAA with aspartic acid (see Chapter 8), also supports the view that conjugation is not solely a response to toxic levels of hormone. The regulation of free IAA levels by conjugation with aspartic acid was demonstrated in the opposite direction by Lau and Yang (1973): added kinetin resulted in *more* free ^{14}C-IAA in *Phaseolus* hypocotyl segments and concomitantly less IA-Asp. The increased level of free IAA caused a big increase in ethylene production, and added IA-Asp had no effect on ethylene production. Various other pieces of evidence that IA-Asp is hormonally inactive came from studies of cotton and bean in which IA-Asp added to debladed petioles showed much less inhibition of abscission that did IAA (Robinson et al. 1968; Craker et al. 1970). Coleoptiles, however, seem to be an exception: IA-Asp is just as effective as IAA in stimulating coleoptile elongation (Andreae & Good 1955), and IAA added to coleoptile sections is not converted to ^{Rf}IA-Asp^{Rf} unless the primary leaf is included inside the coleoptile (Winter & Thimann 1966). These observations suggest that coleoptiles can convert IA-Asp back into free IAA. The lack of conjugation in the coleoptile probably explains the unusually high percentage of applied IAA that is transported through coleoptiles compared to green shoots (see Chapter 9).

The relative importance of conjugation and destruction as methods of regulating IAA levels still needs to be worked out. Andreae and Ysellstein pointed out that the earlier studies on IAA-oxidase, which assumed that IAA gone from the medium had been destroyed, were made suspect by Andreae's discovery that much of the IAA was not destroyed but conjugated. However, some destruction of IAA by decarboxylation does occur. One of the few quantitative balance sheets in the literature reported that after 24 hours' incubation with ^{14}C-IAA, pea root tips showed only ^{14}C-IA-Asp in the tissue (17% of the IAA originally added in the medium),

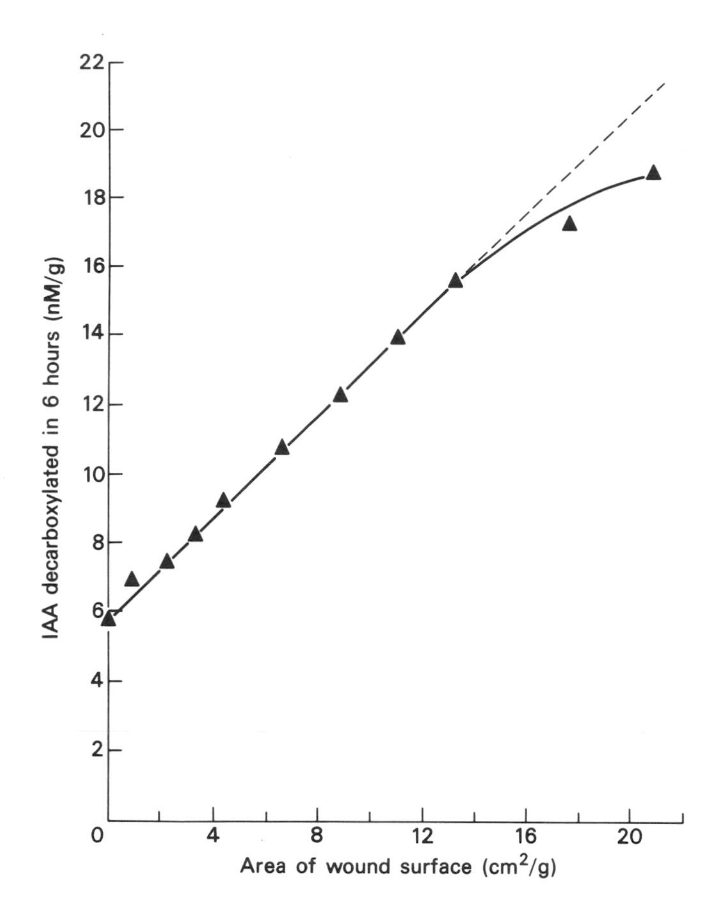

Figure 11-2. The influence of wound surface area on the amount of $^{14}CO_2$ given off by *Helianthus* hypocotyl segments placed in 10 μM of carboxy-labeled IAA (after Zenk & Müller 1964).

whereas the medium contained 19% of the added IAA that had been decarboxylated, about 6% degraded without decarboxylation and 53% unchanged IAA (Andreae et al. 1961). We very much need such balance sheets for other tissues, particularly for tissues that – unlike roots – clearly use IAA as a correlative hormone. A particularly elegant paper by Zenk and Müller (1964) demonstrated a strong wounding effect on the decarboxylation of IAA in *Helianthus* hypocotyls, with the amount of decarboxylation in 6 hours increasing linearly with total wound area (Figure 11-2). Extrapolation of the straight line back to zero wound area indicated decarboxylation still occurring at the rate of 5.8 nM IAA/g/6 hour, a value exactly confirmed by coating each cut end of a long section with collodion to make it impermeable to water and thus an approximation to "unwounded tissue." Hence, although all the earlier guesses about the destructive effect of wound on IAA were confirmed quantitatively by Zenk and Müller, the "unwounded" tissue still destroyed substantial amounts of IAA by decarboxylation.

The roles of destruction and conjugation in controlling the levels of other hormones is just beginning to be worked on, although the existence and importance of gibberellin conjugates is well established by the research of the Halle group (Sembdner et al. 1974).

Interactions and sequential action of hormones

Interactions of the various hormones have been mentioned in various earlier chapters and they have been the object of many reports that I have not had space to discuss. A surprising number of the physiological effects of added gibberellins (GA), cytokinins, or abscisic acid are explainable by their effects on IAA metabolism or transport. The increased levels of free IAA that result from GA treatment have been attributed to decreased conjugation of IAA with aspartic acid (Fang et al. 1960) or lowered IAA-oxidase activity (e.g., Watanabe & Stutz 1960). A major need in the future is to describe the interactions of the hormones with specific reference to the levels of each hormone that occur in specific plant organs. At present this is essentially an unexplored subject.

The endogenous sequence of hormone action has been elucidated even less. It has been frequently observed that the action of added gibberellic acid is more restricted to young tissue than is the action of IAA; and a comparison of the sparse data on endogenous gibberellin production with the more extensive data on IAA production suggests that maximal production of GA may occur in younger leaves than those producing most IAA. An intriguing paper by Wright (1966) described the elongation response of sections cut from various ages of wheat coleoptiles to added GA, IAA, and kinetin. He reported that very young stages were most sensitive to GA, somewhat older coleoptiles gave most response to kinetin, and still older coleoptiles (54 hours and older) showed the classical large response to added IAA (Figure 11-3). This sequence is not dependent on concomitant cell divisions, as Rose and Adamson demonstrated with gamma-irradiated wheat (1969). It is reasonable to expect that such sequential action of hormones would also occur normally, but there is a dearth of specific information on this point.

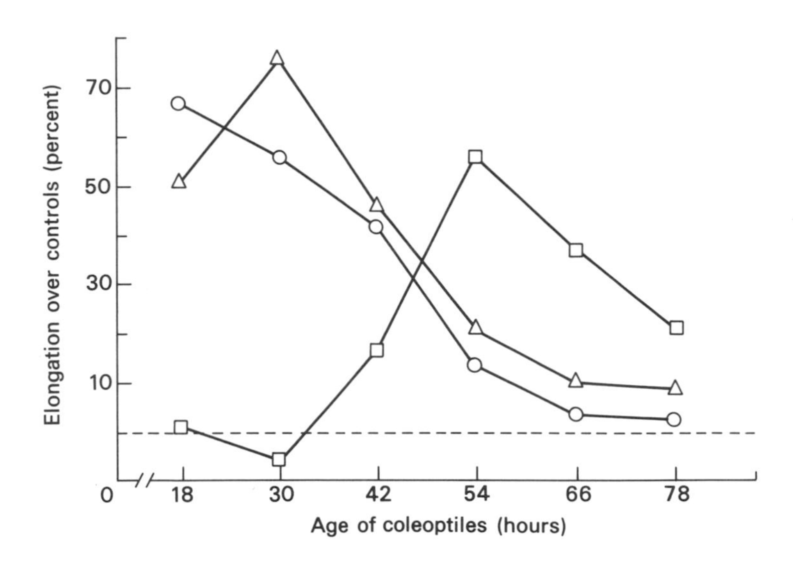

Figure 11-3. Sequential elongation response of wheat coleoptiles of various ages to GA-3 (○), kinetin (△), and IAA (□) (after Wright 1966).

Explanations of polar transport

The polarity of auxin movement is so striking and has been known for such a long time that various hypotheses have been suggested to explain it. The first was the electrical polarity hypothesis: plants show an electrical polarity and it was suggested that polar IAA movement resulted from the negatively charged ion of IAA moving toward the positively charged base of the coleoptile (Went 1932). Clark (1937a) confirmed the "apical negativity" of intact shoots and added that cut sections also showed apical negativity as would be expected if the electrical polarity is the cause of the polarity of auxin transport. However, Clark (1937b, 1938) concluded that the electrical polarity was not the cause of the auxin polarity, because reversing the electrical polarity with an externally imposed electrical potential had no effect on auxin transport. Nevertheless, parallel variation to a striking quantitative degree was demonstrated between the rate and polarity of basipetal movement of an electric wave after IAA had been added to decapitated *Avena* coleoptile and the rate and polarity of IAA movement (Newman 1963). The two phenomena are obviously very closely associated.

Protoplasmic streaming was ruled out as a major source of auxin movement by Clark's experiments (1938) demonstrating that when streaming was stopped by treating sections with saponin, the full amount of IAA transport occurred. Clark also showed that all IAA transport could be stopped by pretreating coleoptile sections with sodium glycocholate; here streaming continued, the normal electrical polarity was maintained, oxygen uptake was unchanged, and the sections still gave the normal elongation response to IAA. Polar movement of IAA through sections in which protoplasmic streaming had been stopped was confirmed recently by Cande et al. (1973), using cytochalasin B to stop the streaming.

Ingenious experiments, involving centrifuging coleoptile cell contents to either the apical or basal ends and then using cytochalasin to prevent their redistribution, led Goldsmith and Ray (1973) to the conclusion that polar transport of IAA is due to secretion localized at the basal plasma membrane.

The currently most popular way of explaining polar auxin transport is the chemiosmotic hypothesis of Rubery and Sheldrake (1974), taken up by Raven (1975) and discussed extensively by Goldsmith in her recent review (1977). The basic idea of this hypothesis is that cells accumulate IAA when the cytoplasmic pH is higher than the wall pH; cell polarity could then result from one end of a cell having a higher ratio of permeability of the IAA in anionic form to the permeability of undissociated IAA than does the other end of the cell.

Comparison of plant hormones with animal hormones

All the hormones so far known from higher plants are relatively small molecules, with molecular weights of only a few hundred. This is in striking contrast to the situation in vertebrates, where some hormones are proteins (e.g., the growth hormone from the pituitary gland, with a molecular weight of perhaps as much as 50,000), many are big polypeptides (e.g., melanocyte-stimulating hormone, with 22 amino acids in its polypeptide, or follicle-stimulating hormone, with a molecular weight of 23,000 or so), and even a small polypeptide such as insulin has a molecular weight of 6000. Of course, some vertebrate hormones are as small as those known from plants. Thyroxine, the steroid sex hormones, and some of the hormones involved in transmission of the nerve impulse across neural synapses are all of the order of magnitude of known angiosperm hormones. As mentioned earlier, serotonin, one of the hormones involved in polar transmission across synapses, is closely related in chemical structure to indoleacetic acid, the major plant hormone with polar transport through tissues. So far no steroid hormones have been found operative in angiosperms, although their importance in vertebrate reproduction has encouraged the hypothesis that the elusive "flowering hormone" might be a steroid. In fact, the angiosperm hormones with the closest relation to flowering – namely, the gibberellins – share with the animal steroid sex hormones a common early pathway in their synthesis from mevalonic acid.

Of course, the absence of polypeptides and proteins among the currently known angiosperm hormones could easily be a mere accident of the time of discovery. The "flowering hormone" might turn out to be a polypeptide or protein.

Major research areas for the future

Several of the most important areas for future research have already been covered, such as the need for knowledge of interactions and of the sequence of action of the hormones, and the need for knowledge of the methods of regulating hormonal levels. However, I should like to emphasize what seem to me to be the six areas of greatest importance in terms of our current ignorance. At the organ and cell level these are the internal factors that control flowering, the development of roots, the senescence of shoots, and the differentiation of cells. For cell differentiation we have extensive information about the hormonal control of only two cell types, tracheary cells and sieve-tube elements. Within the cell the sites and the mechanisms of hormone activity are the topics of major interest and major importance.

Final word to students

Remember that this book represents only one person's view. You cannot understand a field as broad and as replete with literature as plant hormones and plant development by reading only one author. Other botanists emphasize different points and consider different topics to be more important than those selected here. And for every advantage one author has there will usually be a matching disadvantage. For instance, I have the advantage of having carried out research on all the hormones, but that carries with it the disadvantage that I cannot possibly know the literature on just one of those hormones as thoroughly as someone who has restricted his life's research to a single type of hormone. So read widely. Check the books and the yearly volumes entitled the *Annual Review of Plant Physiology*, and read the appropriate volumes in the original *Ency-*

clopedia of Plant Physiology and in its New Series, currently being published.

And, as final advice, try to do better research than that you find in the literature, better than your professor does, better than has been done before. Experimental science is exhilarating work in which *your* performance is the major factor determining how closely you can approach perfection.

References

Aberg, B. 1957. Auxin relations in roots. *Ann. Rev. Plant Physiol. 8,* 153–180.

Abrol, B. K., & Audus, L. J. 1973. The lateral transport of 2,4-dichlorophenoxyacetic acid in horizontal hypocotyl segments of *Helianthus annuus. J. Exp. Bot. 24,* 1209–1223.

Addicott, F. T., Carns, H. R., Lyon, J. L., Smith, O. E., & McMeans, J. L. 1964. On the physiology of abscisins. In *Régulateurs naturels de la croissance végétale,* J. P. Nitsch (ed.), pp. 687–703. Paris: C.N.R.S.

– & Lynch, R. S. 1951. Acceleration and retardation of abscission by indoleacetic acid. *Science 114,* 688–689.

– & Lyon, J. L. 1969. Physiology of abscisic acid and related substances. *Ann. Rev. Plant Physiol. 20,* 139–164.

Aloni, R. 1976. Polarity of induction and pattern of primary phloem fiber differentiation in *Coleus. Amer. J. Bot. 63,* 877–889.

– 1979. The role of auxin and gibberellin in differentiation of primary phloem fibers. *Plant Physiol.* (in press).

– & Jacobs, W. P. 1977a. Polarity of tracheary regeneration in young internodes of *Coleus* (Labiatae). *Amer. J. Bot. 64,* 395–403.

– & Jacobs, W. P. 1977b. The time course of sieve tube and vessel regeneration and their relation to phloem anastomoses in mature internodes of *Coleus. Amer. J. Bot. 64,* 615–621.

Andreae, W. A., & Good, N. E. 1955. The formation of indoleacetylaspartic acid in pea seedlings. *Plant Physiol. 30,* 380–382.

Andreae, W. A., Robinson, J. R., & van Ysselstein, M. W. H. 1961. Studies on 3-indoleacetic acid metabolism. VII. Metabolism of radioactive 3-indoleacetic acid by pea roots. *Plant Physiol. 36,* 783–787.

– & van Ysselstein, M. W. 1960. Studies on 3-indoleacetic acid metabolism. VI. 3-indoleacetic acid uptake and metabolism by pea roots and epicotyls. *Plant Physiol. 35,* 225–232.

Anker, L. 1967. Geotropism and tip regeneration of the *Avena* coleoptile in the presence of gibberellic acid. *Acta Bot. Neerl. 16,* 205–210.

Arnon, D. 1940. Vitamin B_1 in relation to the growth of green plants. *Science 92,* 264–266.

Audus, L. J. 1953. *Plant Growth Substances.* New York: Interscience.

Avery, G. S., Jr. 1935. Differential distribution of a phytohormone in the developing leaf of *Nicotiana,* and its relation to polarized growth. *Bull. Torrey Bot. Club 62,* 313–330.

– & Burkholder, P. R. 1936. Polarized growth and cell studies on the *Avena* coleoptile, phytohormone test object. *Bull. Torrey Bot. Club 63*, 1–15.

– Creighton, H. B., & Hock, C. W. 1939. A low cost chamber for phytohormone tests. *Amer. J. Bot. 26*, 360–365.

Babcock, D. F., & Morris, R. O. 1970. Quantitative measurement of isoprenoid nucleosides in transfer ribonucleic acid. *Biochemistry* 9, 3701–3705.

Bailey, I. W., & Smith, A. C. 1942. Degeneriaceae, a new family of flowering plants from Fiji. *J. Arnold Arbor. 23*, 356–365.

– & Swamy, B. G. L. 1951. The conduplicate carpel of dicotyledons and its initial trends of specialization. *Amer. J. Bot. 38*, 373–379.

Bailiss, K. W., & Hill, T. A. 1971. Biological assays for gibberellins. *Bot. Rev. 37*, 437–479.

Bandurski, R. S., & Schulze, A. S. 1974. Concentrations of indole-3-acetic acid and its esters in *Avena* and *Zea*. *Plant Physiol. 54*, 257–262.

Bara, M. 1957. La quantité fonctionelle relative de l'hétéroauxine (IAA) dans les réactions phototropiques et géotropiques, et l'action de la gravitation sur la production hormonale. *Rev. Fac. Sci. Univ. Istanbul* (*B*) 22, 209–236.

Barlow, P. W. 1974. Recovery of geotropism after removal of the root cap. *J. Exp. Bot. 25*, 1137–1146.

Batra, M. W., Edwards, K. L., & Scott, T. K. 1975. Auxin transport in roots: its characteristics and relationship to growth. In *The Development and Function of Roots*, J. G. Torrey & D. T. Clarkson (eds.), pp. 299–325. New York: Academic Press.

Batt, S., & Venis, M. A. 1976. Separation and localization of two classes of auxin binding sites in corn coleoptile membranes. *Planta 130*, 15–21.

– Wilkins, M. B., & Venis, M. A. 1976. Auxin binding to corn coleoptile membranes: kinetics and specificity. *Planta 130*, 7–13.

Bayer, M. H. 1972. Transport and accumulation of IAA-^{14}C in tumour-forming *Nicotiana* hybrids. *J. Exp. Bot. 23*, 801–812.

– 1973. Effect of kinetin on auxin uptake and distribution in normal and tumor-prone *Nicotiana* shoots. *Plant Cell Physiol. 14*, 293–298.

Beevers, L. 1966. Effect of gibberellic acid on the senescence of leaf discs of nasturtium (*Tropaeolum majus*). *Plant Physiol. 41*, 1074–1076.

Bellandi, D. M., & Dörffling, K. 1974. Transport of abscisic acid-2-C-14 in intact peas seedlings. *Physiol. Plant. 32*, 365–368.

Benayoun, J., Aloni, R., & Sachs, T. 1975. Regeneration around wounds and the control of vascular differentiation. *Ann. Bot. 39*, 447–454.

Bendana, F. E., Galston, A. W., Kaur-Sawhney, R., & Penny, P. J. 1965. Recovery of labeled ribonucleic acid following administration of labeled-auxin to green pea stem sections. *Plant Physiol. 40*, 977–983.

Bennet-Clark, T. A., Tambiah, M. S., & Kefford, N. P. 1952. Estimation of plant growth substances by partition chromatography. *Nature 169*, 452–453.

Bennett, P. A., & Chrispeels, M. J. 1972. *De novo* synthesis of ribonuclease and β-1,3-glucanase by aleurone cells of barley. *Plant Physiol. 49*, 445–447.

Bentley, J. A. 1958. The naturally occurring auxins and inhibitors. *Ann. Rev. Plant Physiol.* 9, 47–80.

Berger, J., & Avery, G. S., Jr. 1944. Isolation of an auxin precursor and an auxin (indoleacetic acid) from maize. *Amer. J. Bot. 31*, 199–203.
Bernier, G. 1969. *Sinapis alba L.* In *The Induction of Flowering*, L. T. Evans (ed.), pp. 305–327. Ithaca: Cornell University Press.
– 1976. La nature complexe du stimulus floral et des facteurs de floraison. In *Etudes de biologie végétale*, Hommage au Professor P. Chouard, R. Jacques (ed.), pp. 243–264. Paris: Imprimerie Louis-Jean.
Beslow, D. T., & Rier, J. P. 1969. Sucrose concentration and xylem regeneration in *Coleus* internodes *in vitro*. *Plant Cell Physiol. 10*, 69–77.
Beyer, A. 1925. Untersuchungen über den Traumatotropismus der Pflanzen. *Biol. Zentralbl. 45*, 683–702, 746–768.
Biale, J. B., & Halma, F. F. 1937. The use of heteroauxin in rooting of subtropicals. *Proc. Amer. Soc. Hort. Sci. 35*, 443–447.
Blaauw, A. H. 1909. Die Perzeption des Lichtes. *Rec. Trav. Bot. Neerl. 5*, 209–372.
– 1918. Licht und Wachstum. *Meded. Landbouwhogesch. Wageningen 15*, 89–204.
Blaauw, O. H., & Blaauw-Jansen, G. 1970. The phototropic responses of *Avena* coleoptiles. *Acta Bot. Neerl. 19*, 755–763.
Black, M. K., & Osborne, D. J. 1965. Polarity of transport of benzyladenine, adenine and indole-3-acetic acid in petiole segments of *Phaseolus vulgaris*. *Plant Physiol. 40*, 676–680.
Blake, J. 1972. A specific bioassay for the inhibition of flowering. *Planta 103*, 126–128.
Bonde, E. K. 1953. Growth inhibitors and auxin in leaves of cocklebur. *Physiol. Plant. 6*, 234–239.
Bonnemain, J. L. 1971. Transport et distribution des traceurs après application de AIA-2-^{14}C sur les feuilles de *Vicia faba*. *C. R. Acad. Sci. Paris 273*, 1699–1702.
Bonner, D. M. 1940. Leaf growth factors. Ph.D. thesis, California Institute of Technology, Pasadena.
– & Haagen-Smit, A. J. 1939. Leaf growth factors. II. The activity of pure substances in leaf growth. *Proc. Nat. Acad. Sci. U.S.A. 25*, 184–188.
Bonner, D. M., Haagen-Smit, A. J., & Went, F. W. 1939. Leaf growth hormones. I. A bio-assay and source for leaf growth hormones. *Bot. Gaz. 101*, 128–144.
Bonner, J. 1933. The action of the plant growth hormone. *J. Gen. Physiol. 17*, 63–76.
– 1937. Vitamin B_1: a growth factor for higher plants. *Science 85*, 183–184.
– 1940. On the growth-factor requirements of isolated roots. *Amer. J. Bot. 27*, 692–701.
– 1942. Transport of thiamine in the tomato plant. *Amer. J. Bot. 29*, 136–142.
– 1943. Effects of application of thiamine to *Cosmos*. *Bot. Gaz. 104*, 475–479.
– 1949. Limiting factors and growth inhibitors in the growth of the *Avena* coleoptile. *Amer. J. Bot. 36*, 323–332.
– Bandurski, R. S., & Millerd, A. 1953. Linkage of respiration to auxin-induced water uptake. *Physiol. Plant. 6*, 511–522.
– & Bonner, D. 1948. Note on induction of flowering in *Xanthium*. *Bot. Gaz. 110*, 154–156.

– & Bonner, H. 1948. The B vitamins as plant hormones. In *Vitamins and Hormones*, Vol. 6, R. Harris & K. V. Thimann (eds.), pp. 225–275. New York: Academic Press.
– & Greene, J. 1939. Further experiments on the relation of vitamin B_1 to the growth of green plants. *Bot. Gaz. 101*, 491–500.
– & Thurlow, J. 1949. Inhibition of photoperiodic induction in *Xanthium* by applied auxin. *Bot. Gaz. 110*, 613–624.
– & Wildman, S. G. 1946. Contributions to the study of auxin physiology. *Growth Symp. 6*, 51–68.
Bonner, J. T. 1952. *Morphogenesis.* Princeton, N.J.: Princeton University Press.
Bonnett, H. T., Jr., & Torrey, J. G. 1965a. Auxin transport in *Convolvulus* roots cultured *in vitro. Plant Physiol. 40*, 813–818.
– & Torrey, J. G. 1965b. Chemical control of organ formation in root segments of *Convolvulus* cultured *in vitro. Plant Physiol. 40*, 1228–1236.
– & Torrey, J. G. 1966. Comparative anatomy of endogenous bud and lateral root formation in *Convolvulus arvensis* roots cultured *in vitro. Amer. J. Bot. 53*, 496–507.
Booth, A. 1958. Nonhormonal growth promotion shown by aqueous extracts. *J. Exp. Bot. 9*, 306–310.
– & Wareing, P. F. 1958. Growth-promoting activity of potato tuber extracts in the coleoptile straight-growth test: quantitative aspects of indoleacetic acid detection. *Nature 182*, 406.
Boroughs, H. 1954. Separation of phosphatase activity from the bulk protein of leaves. *Arch. Biochem. Biophys. 53*, 94–98.
Borthwick, A. W. 1905. The production of adventitious roots and their relation to bird's eye formation (*Maserholz*) in the wood of various trees. *Notes Roy. Bot. Gard. Edinb. 4*, 15–36.
Borthwick, H. A., & Parker, M. W. 1938. Photoperiodic perception in Biloxi soybeans. *Bot. Gaz. 100*, 374–387.
– Parker, M. W., & Hendricks, S. B. 1950. Recent developments in the control of flowering by photoperiod. *Amer. Nat. 84*, 117–134.
Bottelier, H. P. 1959. On the relation betwen auxin concentration and effect in different tests. *Proc. Kon. Akad. Wetensch. Amst. 62*, 493–504.
Böttger, M. 1970a. Die hormonale Regulation des Blattfalls bei *Coleus rehneltianus* Berger. I. Die Wechselwirkung von Indole-3-essigsäure, Gibberellin- und Abscisinsäure auf Explantate. *Planta 93*, 190–204.
– 1970b. Die hormonale Regulation des Blattfalls bei *Coleus rehneltianus* Berger. II. Die näturliche Rolle von Abscisinsäure im Blattfallprozess. *Planta 93*, 205–213.
Bottomley, W., Kefford, N. P., Zwar, J. A., & Goldacre, P. L. 1963. Kinin activity from plant extracts. I. Biological assay and sources of activity. *Aust. J. Bio. Sci. 16*, 395–406.
Bouillenne, R., & Went, F. 1933. Recherches expérimentales sur la néoformation des racines dans les plantules et les boutures des plantes supérieures. (Substances formatrices de racines.) *Ann. Jard. Bot. Buitenzorg 43*, 25–202.
Boysen-Jensen, P. 1910. Ueber die Leitung des phototropischen Reizes in *Avena*-keimpflanzen. *Ber. Deut. Bot. Ges. 28*, 118–120.
– 1911. La transmission de l'irritation phototropique dans l'*Avena. Bull. Acad. Roy. Den. 1*, 3–24.

– 1913. Ueber die Leitung des phototropischen Reizes in der *Avena*-koleoptile. *Ber. Deut. Bot. Ges. 31,* 559–566.
– 1933a. Ueber die durch einseitige Lichtwirkung hervorgerufene transversale Leitung des Wuchsstoffes in der *Avena* Koleoptile. *Planta 19,* 335–344.
– 1933b. Die Bedeutung des Wuchsstoffes fur das Wachstum und die geotropische Krümmung der Wurzeln von *Vicia faba. Planta 20,* 688–698.
Brandes, H., & Kende, H. 1968. Studies on cytokinin-controlled bud formation in moss protonemata. *Plant Physiol. 43,* 827–837.
Brauner, L. 1955. Ueber die Funktion der Spitzenzone beim Phototropismus der *Avena*-koleoptile. *Z. Bot. 43,* 467–498.
Bremer, A. H. 1931. Einfluss der Tageslänge auf die Wachstumsphasen des Salats. Genetische Untersuchungen. *Gartenbauwiss. 4,* 469–483.
Brian, P. W., & Hemming, H. G. 1955. The effect of gibberellic acid on shoot growth of pea seedlings. *Physiol. Plant. 8,* 669–681.
Bridges, I. G., Hillman, J. R., & Wilkins, M. B. 1973. Identification and localisation of auxin in primary roots of *Zea mays* by mass spectrometry. *Planta 115,* 189–192.
Briggs, D. E. 1964. Origin and distribution of amylase in malt. *J. Inst. Brewing 70,* 14–24.
– 1966. Residues from organic solvents showing gibberellin-like biological activity. *Nature 210,* 419–421.
– 1973. Hormones and carbohydrate metabolism in germinating cereal grains. In *Biosynthesis and Its Control in Plants,* B. V. Milborrow (ed.), pp. 219–277. London: Academic Press.
Briggs, W. R. 1963a. Red light, auxin relationships, and the phototropic responses of corn and oat coleoptiles. *Amer. J. Bot. 50,* 196–207.
– 1963b. Mediation of phototropic responses of corn coleoptiles by lateral transport of auxin. *Plant Physiol. 38,* 237–247.
– Steeves, T. A., Sussex, I. M., & Wetmore, R. H. 1955. A comparison of auxin destruction by tissue extracts and intact tissues of the fern *Osmunda cinnamomea* L. *Plant Physiol. 30,* 148–155.
Brown, A. B. 1935. Cambial activity, root habit and sucker shoot development in two species of poplar. *New Phytol. 34,* 163–179.
Brown, H. T., & Morris, G. H. 1890. Researches on the germination of some of the *Gramineae. J. Chem. Soc. 57,* 458–528.
Bruce, M. I., & Zwar, J. A. 1966. Cytokinin activity of some substituted ureas and thioureas. *Proc. Roy. Soc. Lond. (B) 165,* 245–265.
– Zwar, J. A., & Kefford, N. P. 1965. Chemical structure and plant kinin activity – the activity of urea and thiourea derivatives. *Life Sci. 4,* 461–466.
Bui, D. H. D., & Nitsch, J. P. 1970. Isolation of zeatin riboside from the chicory root. *Planta 95,* 119–126.
Bukovac, M. J., & Wittwer, S. H. 1961. Biological evaluation of gibberellins A_1, A_2, A_3 and A_4 and some of their derivatives. In *Plant Growth Regulation,* R. M. Klein, (ed.), pp. 505–520. Ames: Iowa State University Press.
Bünning, E. 1937. Phototropismus and Carotinoide. III. Weitere Untersuchungen an Pilzen und höheren Pflanzen. *Planta 27,* 583–610.
– 1955. Weitere Untersuchungen über die Funktion gelber Pigmente beim Phototropismus der *Avena*-koleoptile. *Z. Bot. 43,* 167–174.

Bünsow, R., & Harder, R. 1956. Blütenbildung von *Bryophyllum* durch Gibberellin. *Naturwissenschaften 43*, 479–480.
Butcher, D. N. 1963. The presence of gibberellins in excised tomato roots. *J. Exp. Bot. 14*, 272–280.
Buy, H. G. du, & Nuernbergk, E. 1930. Ueber das Wachstum der Koleoptile und des Mesokotyls von *Avena sativa* unter verschiedenen Bedingungen (III). *Proc. Kon. Akad. Wetensch. Amst. 33*, 542–556.
– & Nuernbergk, E. 1934. Phototropismus und Wachstum der Pflanzen. II. *Ergebn. Biol. 10*, 207–322.
Camus, G. 1949. Recherches sur le role des bourgeons dans les phénomènes de morphogénèse. *Rev. Cytol. Biol. Veg. 11*, 1–199.
Cande, W. Z., Goldsmith, M. H. M., & Ray, P. M. 1973. Polar auxin transport and auxin-induced elongation in the absence of cytoplasmic streaming. *Planta 111*, 279–296.
Cane, A. R., & Wilkins, M. B. 1970. Auxin transport in roots. VI. Movement of IAA through different zones of *Zea* roots. *J. Exp. Bot. 21*, 212–218.
Canny, M. J. 1973. *Phloem Translocation.* London: Cambridge University Press.
Caplin, S. M., & Steward, F. C. 1948. Effect of coconut milk on the growth of explants from carrot root. *Science 108*, 655–657.
Carlson, M. C. 1938. The formation of nodal adventitious roots in *Salix cordata*. *Amer. J. Bot. 25*, 721–725.
– 1950. Nodal adventitious roots in willow stems of different ages. *Amer. J. Bot. 37*, 555–561.
Carns, H. R., Addicott, F. T., Baker, K. C., & Wilson, R. K. 1961. Acceleration and retardation of abscission by gibberellic acid. In *Plant Growth Regulation*, R. M. Klein (ed.), pp. 559–565. Ames: Iowa State University Press.
Carr, D. J. 1953. On the nature of photoperiodic induction. I. Photoperiodic treatments applied to detached leaves. *Physiol. Plant. 6*, 672–679.
– 1967. The relationship between florigen and the flower hormones. *Ann. N.Y. Acad. Sci. 144*, 305–312.
– & Melchers, G. 1954. Auslösung von Blütenbildung bei der Kurztagpflanze *Kalanchoë blossfeldiana* in Langtagbedingungen durch Pfropfpartner. *Z. Naturforsch. 9b*, 215–218.
– Reid, D. M., & Skene, K. G. M. 1964. The supply of gibberellins from the root to the shoot. *Planta 63*, 382–392.
Cathey, H. M. 1968. Response of *Dianthus caryophyllus* L. (carnation) to synthetic abscisic acid. *Proc. Amer. Soc. Hort. Sci. 93*, 560–568.
Cerček, L. 1970. Effect of x-ray irradiation on regeneration and geotropic function of barley root caps. *Int. J. Radiat. Biol. 17*, 187–194.
Chailakhyan, M. K. 1936. New facts in support of the hormonal theory of plant development. *C. R. (Dokl.) Acad. Sci. URSS 13*, 79–83.
– 1937. Concerning the hormonal nature of plant development processes. *C. R. (Dokl.) Acad. Sci. URSS 16*, 227–230.
– 1947. On the nature of the inhibitory effect of leaves upon flowering. *C. R. (Dokl.) Acad. Sci. URSS 55*, 69–72.
– 1957. Effect on gibberellin on growth and flowering in plants. (In Russian.) *C. R. (Dokl.) Acad. Sci. URSS 117*, 1077–1080.
– 1958. Hormonale Faktoren des Pflanzenblühens. *Biol. Zentralbl. 77*, 641–662.

– & Butenko, R. G. 1957. Translocation of assimilants from leaves to shoots during different photoperiodic regimes of plants. *Sov. Plant Physiol. 4,* 426–438.
– & Lozhnikova, V. N. 1960. Gibberellin-like substances in higher plants and their effect on growth and flowering. (In Russian.) *Sov. Plant Physiol 7,* 521–530.
– & Lozhnikova, V. N. 1966. Effect of interruption of darkness by light on plant gibberellins. *Sov. Plant Physiol. 13,* 734–741.
Champagnat, P. 1965. Physiologie de la croissance et de l'inhibition des bourgeons: dominance apicale et phénomènes analogues. *Encycl. Plant Physiol. 15* (1), 1106–1164.
Chang, Y. P. 1971. The movement of indoleacetic acid in *Coleus* petioles as affected by abscisic acid, senescence factor and kinetin. Ph.D. thesis, Princeton University, Princeton, N.J.
– & Jacobs, W. P. 1972. The contrast between active transport and diffusion of indole-3-acetic acid in *Coleus* petioles. *Plant Physiol. 50,* 635–639.
– & Jacobs, W. P. 1973. The regulation of abscission and IAA by senescence factor and abscisic acid. *Amer. J. Bot. 60,* 10–16.
Chatterjee, S. K., and Leopold, A. C. 1964. Kinetin and gibberellin actions on abscission processes. *Plant Physiol. 39,* 334–337.
Chibnall, A. C. 1939. *Protein Metabolism in the Plant.* New Haven: Yale University Press.
Cholodny, N. 1924. Ueber die hormonale Wirkung der Organspitze bei der geotropischen Krümmung. *Ber. Deut. Bot. Ges. 42,* 356–362.
– 1926. Beiträge zur Analyse der geotropischen Reaktion. *Jahrb. Wiss. Bot. 65,* 447–459.
– 1934. Ueber die Bildung und Leitung des Wuchshormons bei der Wurzeln. *Planta 21,* 517–530.
Chouard, P. 1957. La journée courte ou l'acide gibbérellique comme succedanées du froid pour la vernalisation d'une plante vivace en rosette, le *Scabiosa succisa* L. *C. R. Acad. Sci. Paris 245,* 2520–2522.
Chrispeels, M. J., & Varner, J. E. 1967. Gibberellic acid-enhanced synthesis and release of α-amylase and ribonuclease by isolated barley aleurone layers. *Plant Physiol. 42,* 398–406.
Clark, J., & Bonga, J. M. 1963. Evidence for indole-3-acetic acid in balsam fir, *Abies balsamea* (L.) Mill. *Can. J. Bot. 41,* 165–173.
Clark, W. G. 1937a. Electrical polarity and auxin transport. *Plant Physiol. 12,* 409–440.
– 1937b. Polar transport of auxin and electrical polarity in coleoptile of *Avena. Plant Physiol. 12,* 737–754.
– 1938. Electrical polarity and auxin transport. *Plant Physiol. 13,* 529–552.
Cleland, C. F. 1972. The use of aphids in the search for hormonal factors controlling flowering. In *Plant Growth Substances 1970,* D. J. Carr, (ed.), pp. 753–757. New York: Springer-Verlag.
– 1974a. Isolation of flower-inducing and flower-inhibitory factors from aphid honeydew. *Plant Physiol. 54,* 899–903.
– 1974b. The influence of salicylic acid on flowering and growth in the long-day plant *Lemna gibba* G3. *Roy. Soc. N.Z. Bull. 12,* 553–557.
– & Ajami, A. 1974. Identification of the flower-inducing factor isolated from aphid honeydew as being salicylic acid. *Plant Physiol. 54,* 904–906.
– & Briggs, W. S. 1969. Gibberellin and CCC effects on flowering and

growth in the long-day plant *Lemna gibba* G3. *Plant Physiol. 44*, 503–507.
– & Zeevaart, J. A. D. 1970. Gibberellins in relation to flowering and stem elongation in the long day plant *Silene armeria. Plant Physiol. 46*, 392–400.
Cleland, R. 1971. Cell wall extension. *Ann. Rev. Plant Physiol. 22*, 197–222.
– 1976. Kinetics of hormone-induced H^+ excretion. *Plant Physiol. 58*, 210–213.
Clutter, M. E. 1960. Hormonal induction of vascular tissue in tobacco pith *in vitro. Science 132*, 548–549.
Cohen, D., Robinson, J. B., & Paleg, L. G. 1966. Decapitated peas and diffusible gibberellins. *Aust. J. Biol. Sci. 19*, 535–543.
Collins, G. G., Jenner, C. F., & Paleg, L. G. 1972. The metabolism of soluble nucleotides in wheat aleurone layers treated with gibberellic acid. *Plant Physiol. 49*, 404–410.
Comer, A. E. 1972. The role of cell division in wound vessel differentiation in *Coleus* pith explants. Ph.D. thesis, University of Minnesota, Minneapolis.
Commoner, B., Fogel, S., & Muller, W. H. 1943. The mechanism of auxin action. The effect of auxin on water-absorption by potato-tuber tissue. *Amer. J. Bot. 30*, 23–28.
– & Thimann, K. V. 1941. On the relation between growth and respiration in the *Avena* coleoptile. *J. Gen. Physiol. 24*, 279–296.
Coombe, B. G., Cohen, D., & Paleg, L. G. 1967a. Barley endosperm bioassay for gibberellins. I. Parameters of the response system. *Plant Physiol. 42*, 105–112.
– Cohen, D., & Paleg, L. G. 1967b. Barley endosperm bioassay for gibberellins. II. Application of the method. *Plant Physiol 42*, 113–119.
Cooper, J. P. 1954. Studies on growth and development in *Lolium*. IV. Genetic control of heading responses in local populations. *J. Ecol. 42*, 521–556.
Cornforth, J. W., Milborrow, B. V., & Ryback, G. 1965. Synthesis of (±) abscisin II. *Nature 206*, 715.
– Milborrow, B. V., Ryback, G., Rothwell, K., & Wain, R. L. 1966. Identification of the yellow lupin (from *Lupinus luteus*) growth inhibitor as (+)-abscisin II ([+]-dormin). *Nature 211*, 742–743.
– Milborrow, B. V., Ryback, G., & Wareing, P. F. 1965. Identity of sycamore 'dormin' with abscisin II. *Nature 205*, 1269–1270.
Couillerot, J. P., & Bonnemain, J. L. 1975. Transport et devenir des molécules marquées après l'application d'acide gibbérellique-^{14}C sur les jeunes feuilles de tomate. *C. R. Acad. Sci. Paris 280*, 1453–1456.
Craker, L. E., Chadwick, A. V., & Leather, G. R. 1970. Abscission. Movement and conjugation of auxin. *Plant Physiol. 46*, 790–793.
Crozier, A., Kuo, C. C., Durley, R. C., & Pharis, R. P. 1970. The biological activities of 26 gibberellins in nine plant bioassays. *Can. J. Bot. 48*, 867–877.
– & Reid, D. M. 1971. Do roots synthesize gibberellins? *Can. J. Bot. 49*, 967–975.
Curtis, O. F., & Clark, D. G. 1950. *An Introduction to Plant Physiology*. New York: McGraw-Hill.
Cutler, H. G., & Vlitos, A. J. 1962. The natural auxins of the sugar cane. II.

Acidic, basic, and neutral growth substances in roots and shoots from twelve days after germination of vegetative buds to maturity. *Physiol. Plant.* *15*, 27–42.

Czaja, A. T. 1935. Wurzelwachstum, Wuchsstoff and die Theorie der Wuchsstoffwirkung. *Ber. Deut. Bot. Ges.* *53*, 221–245.

Danckwardt-Lillieström, C. 1957. Kinetin induced shoot formation from isolated roots of *Isatis tinctoria*. *Physiol. Plant.* *10*, 794–797.

Darwin, C. 1880. *The Power of Movement in Plants*. London: Murray.

Das, N. K., Patau, K., & Skoog, F. 1956. Initiation of mitosis and cell division by kinetin and indoleacetic acid in excised tobacco pith tissue. *Physiol. Plant.* *9*, 640–651.

Davies, C. R., Seth, A. K., & Wareing, P. F. 1966. Auxin and kinetin interaction in apical dominance. *Science* *151*, 468–469.

Davies, P. J. 1973. Current theories on the mode of action of auxin. *Bot. Rev.* *39*, 139–169.

– & Galston, A. W. 1971. Labeled indole-macromolecular conjugates from growing stems supplied with labeled indoleacetic acid. *Plant Physiol.* *47*, 435–441.

– & Mitchell, E. K. 1972. Transport of indoleacetic acid in intact roots of *Phaseolus coccineus*. *Planta* *105*, 139–154.

Davis, L. A., & Addicott, F. T. 1972. Abscisic acid: correlations with abscission and with development in the cotton fruit. *Plant Physiol.* *49*, 644–648.

dela Fuente, R. K., & Leopold, A. C. 1966. Kinetics of polar auxin transport. *Plant Physiol.* *41*, 1481–1484.

– & Leopold, A. C. 1968. Lateral movement of auxin in phototropism. *Plant Physiol.* *43*, 1031–1036.

– & Leopold, A. C. 1970. Time course of auxin stimulations of growth. *Plant Physiol.* *46*, 186–189.

Delisle, A. L. 1938. Morphogenetical studies in the development of successive leaves in *Aster*, with respect to relative growth, cellular differentiation, and auxin relationships. *Amer. J. Bot.* *25*, 420–430.

DeMaggio, A. E. 1972. Induced vascular tissue differentiation in fern gametophytes. *Bot. Gaz.* *133*, 311–317.

Denffer, D., von. 1950. Blühhormon oder Blühhemmung? Neue Gesichtspunkte zur Physiologie der Blütenbildung. *Naturwissenschaften* *37*, 296–301, 317–321.

DeRopp, R. S. 1945. Studies in the physiology of leaf growth. I. The effect of various accessory growth factors on the growth of the first leaf of isolated stem tips of rye. *Ann. Bot.* *9*, 369–381.

Dijkman, M. J. 1934. Wuchsstoff und geotropische Krümmung bei *Lupinus*. *Rec. Trav. Bot. Neerl.* *31*, 391–450.

Doershug, M. R., & Miller, C. O. 1967. Chemical control of adventitious organ formation in *Lactuca sativa* explants. *Amer. J. Bot.* *54*, 410–413.

Dolk, H. E. 1930. Geotropie en groeistof. *Rec. Trav. Bot. Neerl.* *33*, 509–585 (English translation, 1936).

Dore, J. 1965. Physiology of regeneration in cormophytes. *Encycl. Plant Physiol.* *15*(2), 1–91.

Dörffling, K., Bellandi, D. M., Böttger, M., Lückel, H., & Menzer, U. 1973. Abscisic acid: properties of transport and effect on distribution of potassium and phosphorus. *Proc. Res. Inst. Pomol. Skierniewice* *3*, 259–272.

– & Böttger, M. 1968. Transport von Abszisinsäure in Explantaten, Blattstiel- und Internodealsegmenten von *Coleus rheneltianus*. *Planta 80*, 299–308.
– Böttger, M., Martin, D., Schmidt, V. & Borowski, D. 1978. Physiology and chemistry of substances accelerating abscission in senescent petioles and fruit stalks. *Physiol. Plant. 43*, 292–296.
Dostál, R. 1950. Morphogenetic experiments with *Bryophyllum verticillatum*. *Acta Acad. Sci. Nat. Moravo-Silesia (Brno) 22*, 57–98.
– & Hosek, M. 1937. Ueber den Einfluss von Heteroauxin auf die Morphogenese bei *Circaea* (das Sachssche Phänomen). *Flora (Jena) 131*, 263–286.
Dowler, M. J., Rayle, D. L., Cande, W. Z., Ray, P. M., Durand, H., & Zenk, M. H. 1974. Auxin does not alter the permeability of pea segments to tritium-labeled water. *Plant Physiol. 53*, 229–232.
Earle, E. D. 1968. Induction of xylem elements in isolated *Coleus* pith. *Amer. J. Bot. 55*, 302–305.
El-Antably, H. M. M., & Larsen, P. 1974. Distribution of gibberellins and abscisic acid in geotropically stimulated *Vicia faba* roots. *Physiol. Plant. 32*, 322–329.
– Wareing, P. F., & Hillman, J. 1967. Some physiological responses to D, L abscisin (dormin). *Planta 73*, 74–90.
Eliasson, L. 1961. Responses of pea roots to growth substances. *Physiol. Plant. 14*, 803–812.
Elliot, M. C., & Greenwood, M. S. 1974. Indol-3yl-acetic acid in roots of *Zea mays*. *Phytochemistry 13*, 239–241.
Engvild, K. C. 1975. Natural chlorinated auxins labelled with radioactive chloride in immature seeds. *Physiol. Plant. 34*, 286–287.
Esau, K. 1953. *Plant Anatomy*. New York: Wiley.
Eschrich, W. 1968. Translokation radioaktiv markierter Indolyl-3-essigsäure in Siebröhren von *Vicia faba*. *Planta 78*, 144–157.
Evans, L. T. 1962. Day-length control of inflorescence initiation in the grass *Rottboellia exaltata* L.f. *Aust. J. Biol. Sci. 15*, 291–303.
– 1966. Abscisin II: Inhibitory effect on flower induction in a long-day plant. *Science 151*, 107–108.
Evans, M. L. 1974. Rapid responses to plant hormones. *Ann. Rev. Plant Physiol. 25*, 195–223.
– & Ray, P. M. 1969. Timing of the auxin response in coleoptiles and its implications regarding auxin action. *J. Gen. Physiol. 53*, 1–20.
Faber, E.-R. 1936. Wuchsstoffversuche an Keimwurzeln. *Jahrb. Wiss. Bot. 83*, 439–469.
Fang, S. C., Bourke, J. B., Stevens, V. L., & Butts, J. S. 1960. Influences of gibberellic acid on metabolism of indoleacetic acid, acetate, and glucose in roots of higher plants. *Plant Physiol. 35*, 251–255.
Farrar, K. R. 1962. The zeanins: a group of growth-promoting compounds from immature maize kernels. *Biochem. J. 83*, 220–224.
Ferguson, J. L. 1971. Analysis of factors influencing the curvature responses of coleoptiles to plant growth hormone. Ph.D. thesis, University of Iowa, Ames.
Filner, P., & Varner, J. E. 1967. A test for *de novo* synthesis of enzymes: density labelling with H_2O^{18} of barley α-amylase induced by gibberellic acid. *Proc. Nat. Acad. Sci. U.S.A. 58*, 1520–1526.

Finney, D. J. 1964. *Statistical Method in Biological Assay.* 2nd ed. New York: Hafner.
Fletcher, R. A., & Osborne, D. J. 1965. Regulation of protein and nucleic acid synthesis by gibberellin during leaf senescence. *Nature 207,* 1176–1177.
– & Osborne, D. J. 1966. Gibberellin as a regulator of protein and ribonucleic acid synthesis during senescence in leaf cells of *Taraxacum officinale. Can. J. Bot. 44,* 739–745.
Fosket, D. E., & Roberts, L. W. 1964. Induction of wound-vessel differentiation in isolated *Coleus* stem segments in vitro. *Amer. J. Bot. 51,* 19–25.
Fox, J. E. 1966. Incorporation of a kinin, *N*,6-benzyladenine into soluble RNA. *Plant Physiol. 41,* 75–82.
– 1969. The cytokinins. In *Physiology of Plant Growth and Development,* M. B. Wilkins, (ed.), pp. 85–123. New York: McGraw-Hill.
– & Miller, C. O. 1959. Factors in corn steep water promoting growth of plant tissues. *Plant Physiol. 34,* 577–579.
– & Weis, J. S. 1965. Transport of the kinin, N_6-benzyladenine: non-polar or polar? *Nature 206,* 678–679.
Francis, G. E., Mulligan, W., & Wormall, A. 1959. *Isotopic Tracers.* 2nd ed. New York: Oxford University Press.
Freundlich, H. F. 1908. Entwicklung und Regeneration von Gefässbündeln in Blattgebilden. *Jahrb. Wiss. Bot. 46,* 137–206.
Frydman, V. M., & Wareing, P. F. 1973. Phase change in *Hedera helix* L. *J. Exp. Bot. 24,* 1131–1148.
Fukui, H. N., De Vries, J. E., Wittwer, S. H., & Sell, H. M. 1957. Ethyl-3-indoleacetate: an artifact in extracts of immature corn kernels. *Nature 180,* 1205.
Galston, A. W. 1949a. Transmission of the floral stimulus in soybean. *Bot. Gaz. 110,* 495–501.
– 1949b. Riboflavin-sensitized photo-oxidation of indoleacetic acid and related compounds. *Proc. Nat. Acad. Sci. U.S.A. 35,* 10–17.
– & Baker, R. S. 1949. Studies on the physiology of light action. II. The photodynamic action of riboflavin. *Amer. J. Bot. 36,* 773–780.
– & Dalberg, L. Y. 1954. The adaptive formation and physiological significance of indoleacetic acid oxidase. *Amer. J. Bot. 41,* 373–380.
– & Hillman, W. S. 1961. The degradation of auxin. *Encycl. Plant Physiol. 14,* 647–670.
– & Purves, W. K. 1960. The mechanism of action of auxin. *Ann. Rev. Plant Physiol. 11,* 239–276.
Gandar, J. C., & Nitsch, C. 1967. Isolement de l'ester méthylique d'un acide chloro-3-indolylacétique à partir de graines immatures de Pois, *Pisum sativum* L. C. R. Acad. Sci. Paris 265, 1795–1798.
Gardner, F. E., & Cooper, W. C. 1943. Effectiveness of growth substances in delaying abscission of *Coleus* petioles. *Bot. Gaz. 105,* 80–89.
Garner, W. W., & Allard, H. A. 1920. Effect of the relative length of day and night and other factors of the environment on growth and reproduction in plants. *J. Agric. Res. 18,* 553–606.
– & Allard, H. A. 1923. Further studies in photoperiodism, the response of the plant to relative length of day and night. *J. Agric. Res. 23,* 871–920.
– & Allard, H. A. 1931. Effect of abnormally long and short alternations of

light and darkness on growth and development of plants. *J. Agric. Res. 42,* 629–651.

Gaur, B. K., & Leopold, A. C. 1955. The promotion of abscission by auxin. *Plant Physiol. 30,* 487–490.

Gautheret, R. J. 1959. *La Culture des tissus végétaux.* Paris: Masson.

Gee, H. 1972. Localization and uptake of ^{14}C-IAA in relation to xylem regeneration in *Coleus* internodes. *Planta 108,* 1–9.

Gentcheff, G., & Gustafsson, A. 1940. The cultivation of plant species from seed to flower and flower to seed in different agar solutions. *Hereditas 25,* 250–255.

Gibbons, G. S. B., & Wilkins, M. B. 1970. Growth inhibitor production by root caps in relation to geotropic responses. *Nature 226,* 558–559.

Gmelin, R. 1964. Occurrence, isolation, and properties of glucobrassicin and neoglucobrassicin. In *Régulateurs naturels de la croissance végétale,* J. P. Nitsch (ed.), pp. 159–167. Paris: C.N.R.S.

– & Virtanen, A. I. 1961. Glucobrassicin, the precursor of 3-indolylacetonitrile, ascorbigen and SCN in *Brassica oleracea* species. *Suom. Kem. B 34,* 15–18.

Goldschmidt, E. E., & Monselise, S. P. 1968. Native growth inhibitors from citrus shoots. Partition, bioassay, and characterization. *Plant Physiol. 43,* 113–116.

– & Monselise, S. P. 1972. Hormonal control of flowering in citrus and some other woody perennials. In *Plant Growth Substances 1970,* D. J. Carr (ed.), pp. 758–766. New York: Springer-Verlag.

Goldsmith, M. H. M. 1959. Characteristics of the translocation of indoleacetic acid in the coleoptile of *Avena.* Ph.D. thesis, Harvard University, Cambridge, Mass.

– 1966. Movement of indoleacetic acid in coleoptiles of *Avena sativa* L. II. Suspension of polarity by total inhibition of the basipetal transport. *Plant Physiol. 41,* 15–27.

– 1967. Movement of pulses of labeled auxin in corn coleoptiles. *Plant Physiol. 42,* 258–263.

– 1977. The polar transport of auxin. *Ann. Rev. Plant Physiol. 28,* 439–478.

– Cataldo, D. A., Karn, J., Brenneman, T., & Trip, P. 1974. The rapid nonpolar transport of auxin in the phloem of intact *Coleus* plants. *Planta 116,* 301–317.

– & Ray, P. M. 1973. Intracellular localization of the active process in polar transport of auxin. *Planta 111,* 297–314.

– & Thimann, K. V. 1962. Some characteristics of movement of indoleacetic acid in coleoptiles of *Avena.* I. Uptake, destruction, immobilization, and distribution of IAA during basipetal translocation. *Plant Physiol. 37,* 492–505.

Goldthwaite, J. J., & Laetsch, W. M. 1968. Control of senescence in *Rumex* leaf discs by gibberellic acid. *Plant Physiol. 43,* 1855–1858.

Good, N. E., & Andreae, W. A. 1957. Malonyltryptophan in higher plants. *Plant Physiol. 32,* 561–566.

Goodwin, R. H. 1937. The role of auxin in leaf development in *Solidago* species. *Amer. J. Bot. 24,* 43–51.

Gordon, S. A., Cameron, E., & Shen-Miller, J. 1973. Polarity and rate of transport of cyclic adenosine 3′,5′-monophosphate in the coleoptile. *Plant Physiol. 52,* 105–110.

– & Sanchez Nieva, F. 1949. The biosynthesis of auxin in the vegetative pineapple. I. Nature of the active auxin. *Arch. Biochem. 20,* 356–366.
– & Wildman, S. G. 1943. The conversion of tryptophane to a plant growth substance by conditions of mild alkalinity. *J. Biol. Chem. 147,* 389–398.
Gorter, C. J. 1932. Groeistofproblemen bij Wortels. *Diss. Utrecht.*
– 1957. Abscission as a bio-assay for the determination of plant growth regulators. *Physiol. Plant. 10,* 858–868.
– & Veen, H. 1966. Auxin transport in explants of *Coleus. Plant Physiol.* 41, 83–86.
Gorton, B. S., & Eakin, R. E. 1957. Development of the gametophyte in the moss *Tortella caespitosa. Bot. Gaz. 119,* 31–38.
Greenwood, M. S., & Goldsmith, M. H. M. 1970. Polar transport and accumulation of indole-3-acetic acid during root regeneration by *Pinus lambertiana* embryos. *Planta 95,* 297–313.
– & Hillman, J. R., Shaw, S., & Wilkins, M. B. 1973. Localization and identification of auxin in roots of *Zea mays. Planta 109,* 369–374.
– Shaw, S., Hillman, J. R., Ritchie, A., & Wilkins, M. B. 1972. Identification of auxin from *Zea* coleoptile tips by mass spectrometry. *Planta 108,* 179–183.
Gregory, F. G., & Hancock, C. R. 1955. The rate of transport of natural auxin in woody shoots. *Ann. Bot. 76,* 450–465.
Groat, J. I., & Briggs, D. E. 1969. Gibberellins and α-amylase formation in germinating barley. *Phytochemistry 8,* 1615–1627.
Guern, J., Doree, M., & Sadorge, P. 1968. Transport, metabolism and biological activity of some cytokinins. In *Biochemistry and Physiology of Plant Growth Substances,* F. Wightman & G. Setterfield (eds.), pp. 1155–1167. Ottawa: Runge Press.
– & Hugon, É. 1966. Remarques à propos du transport de la 6-benzylaminopurine marquée au ^{14}C dans les jeunes plantes étiolées de *Cicer arietinum* L. *C. R. Acad. Sci. Paris 262,* 2226–2229.
Guttenberg, H. von. 1959. Ueber die Perzeption des phototropen Reizes. *Planta 53,* 412–433.
György, P. 1954. Chemistry [of biotin]. In *The Vitamins,* Vol. I, W. H. Sebrell, Jr., & R. S. Harris (eds.), pp. 527–571. New York: Academic Press.
Haagen-Smit, A. J., Dandliker, W. B., Wittwer, S. H., & Murneek, A. E. 1946. Isolation of 3-indoleacetic acid from immature corn kernels. *Amer. J. Bot. 33,* 118–120.
– Leech, W. D., & Bergren, W. R. 1941. Estimation, isolation and identification of auxins in plant material. *Science 93,* 624–625.
– Leech, W. D., & Bergren, W. R. 1942. The estimation, isolation and identification of auxins in plant materials. *Amer. J. Bot. 29,* 500–506.
Haberlandt, G. 1913. Zur Physiologie der Zellteilung. *Sitzungsber. Preuss. Akad. Wiss. 16,* 318–345.
Hager, A., & Schmidt, R. 1968. Auxintransport und Phototropismus. I. Die lichtbedingte Bildung eines Hemmstoffes für den Transport von Wuchsstoffen in Koleoptilen. *Planta 83,* 347–371.
Halaban, R. 1968. The flowering response of *Coleus* in relation to photoperiod and the circadian rhythm of leaf movement. *Plant Physiol. 43,* 1894–1898.
Halevy, A. H., & Cathey, H. M. 1960. Effects of structure and concentration

of gibberellins on the growth of cucumber seedlings. *Bot. Gaz. 122*, 63–67.
Hall, R. H., Csonka, L., David, H., & McLennan, B. 1967. Cytokinins in the soluble RNA of plant tissues. *Science 156*, 69–71.
Halliday, M. B. W., & Wangermann, E. 1972. Leaf abscission in *Coleus*. I. Abscission zone formation and the effect of auxin on abscission. *New Phytol. 17*, 649–663.
Hamilton, R. H., Bandurski, R. S., & Grigsby, B. H. 1961. Isolation of indole-3-acetic acid from corn kernels and etiolated corn seedlings. *Plant Physiol. 36*, 354–359.
Hamner, C. L. 1940. Effects of vitamin B_1 upon the development of some flowering plants. *Bot. Gaz. 102*, 156–168.
Hamner, K. C., & Bonner, J. 1938. Photoperiodism in relation to hormones as factors in floral initiation and development. *Bot. Gaz. 100*, 388–431.
Hancock, C. R., & Barlow, H. W. B. 1953. The assay of growth substances by a modified straight growth method. *Annual Report East Malling Research Station for 1952*, pp. 88–94.
Harada, H. 1960. Extraction de deux substances de floraison. *Ann. Physiol. Veg. 2*, 249–254.
– 1962. Etude des substances naturelles de croissance en relation avec la floraison. Isolement d'une substance de montaison. *Rev. Gen. Bot. 69*, 201–297.
– & Nitsch, J. P. 1959a. Changes in endogenous growth substances during flower development. *Plant Physiol. 34*, 409–415.
– & Nitsch, J. P. 1959b. Extraction d'une substance provoquant la floraison chez *Rudbeckia speciosa* Wend. *Bull. Soc. Bot. Fr. 106*, 451–454.
– & Nitsch, J. P. 1961. Isolement et propriétés physiologiques d'une substance de montaison. *Ann. Physiol. Veg. 3*, 193–208.
– & Nitsch, J. P. 1967. Isolation of gibberellins A_1, A_3, A_9 and of a fourth growth substance from *Althaea rosea* Cav. *Phytochemistry 6*, 1695–1703.
– & Yokota, T. 1970. Isolation of gibberellin A_8-glucoside from shoot apices of *Althaea rosea*. *Planta 92*, 100–104.
Harder, R., & Bünsow, R. 1958. Ueber die Wirkung von Gibberellin auf Entwicklung und Blutenbildung der Kurztagpflanze *Kalanchoe Blossfeldiana*. *Planta 51*, 201–222.
– & van Senden, H. 1949. Antagonistische Wirkung von Wuchsstoff und "Blühhormon." *Naturwissenschaften 36*, 348–349.
Hare, R. C. 1964. Indoleacetic acid oxidase. *Bot. Rev. 30*, 129–165.
Harel, S. 1969. Modification of 2,4-dichlorophenoxyacetic acid movement in bean petioles by light. *Plant Physiol. 44*, 615–617.
Harris, R. M. 1953. Auxin relations in a dwarf-1 allele of *Zea mays* L. Ph.D. thesis, University of California, Los Angeles.
Hartung, W., & Behl, R. 1974. Transport und Stoffwechsel von 2-(^{14}C) Abscisinsäure in Wurzelsegmenten von *Phaseolus coccineus* L. *Planta 120*, 299–305.
– & Behl, R. 1975a. Lokalisation des akropetalen Transports von 2-(^{14}C) Abscisinsäure in Wurzeln von *Phaseolus coccineus* L. und Hinweise für einen Radialtransport von ABA zwischen Zentralzylinder und Rindenzylinder. *Planta 122*, 53–59.
– & Behl, R. 1975b. Die Wirkung von Licht auf den Transport von 2-(^{14}C)Abscisinsäure in Bohnenwurzelsegmenten. *Planta 122*, 61–65.

– & Phillips, I. D. J. 1974. Basipetally polarized transport of (^{3}H)gibberellin A_1 and (^{14}C)gibberellin A_3 and acropetal polarity of (^{14}C) indole-3-acetic acid in stelar tissues of *Phaseolus coccineus* roots. *Planta 118,* 311–322.
Hashimoto, Y., & Yamaki, T. 1960. Comparative effectiveness of gibberellins A_1, A_2, A_3, and A_4, with special reference to that of A_4. *Bot. Mag. (Tokyo) 173,* 64–68.
Haupt, W. 1956. Gibt es Beziehungen zwischen Polarität und Blütenbildung? *Ber. Deut. Bot. Ges. 69,* 61–66.
Haxo, F. T., & Blinks, L. R. 1950. Photosynthetic action spectra of marine algae. *J. Gen. Physiol. 33,* 389–422.
Hayashi, T. 1940. Biochemical studies on "Bakanae" fungus of rice. VI. Effect of gibberellin on the activity of amylase in germinated cereal grains. *Bull. Agric. Chem. Soc. Japan 16,* 531–538.
Heide, O. M. 1965. Effects of 6-benzylamino-purine and 1-naphthaleneacetic acid on the epiphyllous bud formation in *Bryophyllum. Planta 67,* 281–296.
– 1967. The auxin level of *Begonia* leaves in relation to their regeneration ability. *Physiol. Plant. 20,* 886–902.
– 1968. Auxin level and regeneration of *Begonia* leaves. *Planta 81,* 153–159.
– 1972. The role of cytokinin in regeneration processes. In *Hormonal Regulation in Plant Growth and Development,* H. Kaldewey & Y. Vardar (eds.), pp. 207–219. Weinheim, West Germany: Verlag Chemie.
– & Skoog, F. 1967. Cytokinin activity in *Begonia* and *Bryophyllum. Physiol. Plant. 20,* 771–780.
Hejnowicz, A., & Tomaszewski, M. 1969. Growth regulators and wood formation in *Pinus silvestris. Physiol. Plant. 22,* 984–992.
Hemberg, T. 1949. Growth-inhibiting substances in terminal buds of *Fraxinus. Physiol. Plant. 2,* 37–44.
– 1972. The effect of kinetin on the occurrence of acid auxin in *Coleus blumei. Physiol. Plant. 26,* 98–103.
– & Larsson, U. 1972. Interaction of kinetin and indoleacetic acid in the *Avena* straight-growth test. *Physiol. Plant. 26,* 104–107.
Henbest, H. B., Jones, E. R. H., & Smith, G. F. 1953. Isolation of a new plant-growth hormone, 3-indolylacetonitrile. *J. Chem. Soc.* 3796–3801.
Hertel, R. 1962. Der Auxintransport in der Koleoptile von *Zea mays* L. Diss. Ludwig-Maximilians University, Munich.
– Evans, M. L., Leopold, A. C., & Sell, H. M. 1969. The specificity of the auxin transport system. *Planta 85,* 238–249.
– & Flory, R. 1968. Auxin movement in corn coleoptiles. *Planta 82,* 123–144.
– & Leopold, A. C. 1963. Versuche zur Analyse des Auxintransports in der Koleoptile von *Zea mays* L. *Planta 59,* 535–562.
– Thomson, K. S., & Russo, V. E. A. 1972. *In-vitro* auxin binding to particulate fractions from corn coleoptiles. *Planta 107,* 325–340.
Heyn, A. N. J. 1931. Der Mechanismus der Zellstreckung. *Rec. Trav. Bot. Neerl. 28,* 113–244.
– 1935. The chemical nature of some growth hormones as determined by the diffusion method. *Proc. Kon. Akad. Wetensch. Amst. 38,* 1074–1081.
– 1940. The physiology of cell elongation. *Bot. Rev. 6,* 515–574.

Hicks, P. A. 1928. The carbon/nitrogen ratio in the wheat plant. *New Phytol. 27*, 1–46.
Hillman, S. K., & Phillips, I. D. J. 1970. Transport and metabolism of indol-3yl-(acetic acid-2-^{14}C) in pea roots. *J. Exp. Bot. 21*, 959–967.
Hoad, G. V. 1967. (+)-Abscisin II, ((+)-dormin) in phloem exudate of willow. *Life Sci. 6*, 1113–1118.
Hocking, T. J., Hillman, J. R., & Wilkins, M. B. 1972. Movement of abscisic acid in *Phaseolus vulgaris* plants. *Nat. New Biol. 235*, 124–125.
Hodson, H. K., & Hamner, K. C. 1970. Floral inducing extract from *Xanthium*. *Science 167*, 384–385.
Hofinger, M., Gaspar, T., & Darimont, E. 1970. Occurrence, titration and enzymatic degradation of 3- (3-indolyl)-acrylic acid in *Lens culinaris* Med. extracts. *Phytochemistry 9*, 1757–1761.
Holst, U.-B. 1971. Some properties of inhibitor β from *Solanum tuberosum* compared to abscisic acid. *Physiol. Plant. 24*, 392–396.
Horton, R. F., & Fletcher, R. A. 1968. Transport of the auxin, picloram, through petioles of bean and *Coleus* and stem sections of pea. *Plant Physiol. 43*, 2045–2048.
Houck, D. F., & LaMotte, C. E. 1977. Primary phloem regeneration without concomitant xylem regeneration: its hormone control in *Coleus*. *Amer. J. Bot. 64*, 799–809.
Housley, S., & Griffiths, N. M. 1962. The occurrence of 3-indolylacetic acid in seeds and shoots of pea. *Phyton 19*, 85–93.
– & Taylor, W. C. 1958. Studies on plant growth hormones. VI. The nature of inhibitor-β in potato. *J. Exp. Bot. 9*, 458–471.
Hull, H. M., Went, F. W., & Yamada, N. 1954. Fluctuations in sensitivity of the *Avena* test due to air pollutants. *Plant Physiol. 29*, 182–187.
Igoshi, M., Yamaguchi, I., Takahashi, N., & Hirose, K. 1971. Plant growth substances in the young fruit of *Citrus unshiu*. *Agric. Biol. Chem. 35*, 629–631.
Ingersoll, R. B., & Smith, O. E. 1970. Movement of (RS)-abscisic acid in the cotton explant. *Plant Physiol 45*, 576–578.
– & Smith, O. E. 1971. Transport of abscisic acid. *Plant Cell Physiol. 12*, 301–309.
Iversen, T.-H., & Aasheim, T. 1970. Decarboxylation and transport of auxin in segments of sunflower and cabbage roots. *Planta 93*, 354–362.
– Aasheim, T., & Pedersen, K. 1971. Transport and degradation of auxin in relation to geotropism in roots of *Phaseolus vulgaris*. *Physiol. Plant. 25*, 417–424.
Jablonski, J. R., & Skoog, F. 1954. Cell enlargement and cell division in excised tobacco pith tissue. *Physiol. Plant. 7*, 16–24.
Jacobs, M., & Hertel, R. 1978. Auxin binding to subcellular fractions from *Cucurbita* hypocotyls: in vitro evidence for an auxin transport carrier. *Planta 142*, 1–10.
– & Ray, P. M. 1976. Rapid auxin-induced decrease in free space pH and its relationship to auxin-induced growth in maize and pea. *Plant Physiol. 58*, 203–209.
Jacobs, W. P. 1950a. Auxin-transport in the hypocotyl of *Phaseolus vulgaris* L. *Amer. J. Bot. 37*, 248–254.
– 1950b. Control of elongation in the bean hypocotyl by the ability of the hypocotyl tip to transport auxin. *Amer. J. Bot. 37*, 551–555.

– 1951. Auxin relationships in an intercalary meristem: further studies on the gynophore of *Arachis hypogaea* L. *Amer. J. Bot. 38,* 307–310.
– 1952. The role of auxin in differentiation of xylem around a wound. *Amer. J. Bot. 39,* 301–309.
– 1954. Acropetal auxin transport and xylem regeneration – a quantitative study. *Amer. Nat. 88,* 327–337.
– 1955. Studies on abscission: the physiological basis of the abscission-speeding effect of intact leaves. *Amer. J. Bot. 42,* 594–604.
– 1956. Internal factors controlling cell differentiation in the flowering plants. *Amer. Nat. 90,* 163–169.
– 1958. Further studies on the relation between auxin and abscission of *Coleus* leaves. *Amer. J. Bot. 45,* 673–675.
– 1959. What substance normally controls a given biological process? I. Formulation of some rules. *Dev. Biol. 1,* 527–533.
– 1961. The polar movement of auxin in the shoots of higher plants: its occurrence and physiological significance. In *Plant Growth Regulation,* R. M. Klein (ed.), pp., 397–409. Ames: Iowa State University Press.
– 1962. Longevity of plant organs: internal factors controlling abscission. *Ann. Rev. Plant Physiol. 13,* 403–436.
– 1964. The role of native growth substances in controlling the shedding of organs and abscission. *Proceedings of the 16th International Horticultural Congress* (*of 1962*), Vol. 5, pp. 619–625.
– 1967. Comparison of the movement and vascular differentiation effects of the endogenous auxin and of phenoxyacetic weed-killers in stems and petioles of *Coleus* and *Phaseolus. Ann. N.Y. Acad. Sci. 144,* 102–117.
– 1968. Hormonal regulation of leaf abscission. *Plant Physiol. 43,* 1480–1495.
– 1970. Regeneration and differentiation of sieve tube elements. *Int. Rev. Cytol. 28,* 239–273.
– 1972. The movement of plant hormones: auxins, gibberellins, and cytokinins. In *Plant Growth Substances 1970,* D. J. Carr (ed.), pp. 701–709. New York: Springer-Verlag.
– 1976. Apolar movement of zeatin through *Coleus* petioles and *Pisum* roots as estimated by bioassay and radioactive labelling. *Amer. J. Bot. 63,* 571–577.
– 1977. Polarity of indoleacetic acid in young *Coleus* stems. *Plant Physiol. 60,* 95–97.
– 1978a. Does the induction of flowering by photoperiod change the polarity or other characteristics of IAA transport in petioles for the short-day plant *Xanthium? Plant Physiol. 61,* 307–310.
– 1978b. Regulation of development by the differential polarity of various hormones as well as by effects of one hormone on the polarity of another. In *Regulation of Developmental Processes in Plants,* H. R. Schütte & D. Gross (eds.), pp. 361–380. Jena: G. Fischer Verlag.
– & Aloni, R. 1978. Evidence that IAA does not move basipetally against a concentration gradient in *Coleus blumei* internodes. *Ann. Bot. 42,* 989–991.
– & Case, D. B. 1965. Auxin transport, gibberellin, and apical dominance. *Science 148,* 1729–1731.
– Danielson, J., Hurst, V., & Adams, P. 1959. What substance normally controls a given biological process? II. The relation of auxin to apical dominance. *Dev. Biol. 1,* 534–554.

– & DeMuth, P. J. 1977. Non-polar movement of thiamine in stems of tomato and roots of pea. *Bot. Gaz. 138*, 266–269.
– & Kaldewey, H. 1970. Polar movement of gibberellic acid through young *Coleus* petioles. *Plant Physiol. 45*, 539–541.
– Kaushik, M. P., & Rochmis, P. G. 1964. Does auxin inhibit the abscission of *Coleus* leaves by acting as a growth hormone? *Amer. J. Bot. 51*, 893–897.
– & Kirk, S. C. 1966. Effect of gibberellic acid on elongation and longevity of *Coleus* petioles. *Plant Physiol. 41*, 487–490.
– & McCready, C. C. 1967. Polar transport of growth regulators in pith and vascular tissues of *Coleus* stems. *Amer. J. Bot. 54*, 1035–1040.
– McCready, C. C., & Osborne, D. J. 1966. Transport of the auxin 2,4-dichlorophenoxyacetic acid through abscission zones, pulvini, and petioles of *Phaseolus vulgaris*. *Plant Physiol. 41*, 725–730.
– & Morrow, I. B. 1957. A quantitative study of xylem development in the vegetative shoot apex of *Coleus*. *Amer. J. Bot. 44*, 823–842.
– & Pruett, P. 1972. The polar movement of gibberellin through *Coleus* petioles. In *Hormonal Regulation in Plant Growth and Development*. H. Kaldewey & Y. Vardar (eds.), pp. 45–55. Weinheim, West Germany: Verlag Chemie.
– & Pruett, P. E. 1973. The time-course of polar movement of gibberellin through *Zea* roots. *Amer. J. Bot. 60*, 896–900.
– & Raghavan, V. 1962. Studies on the floral histogenesis and physiology of *Perilla*. I. Quantitative analysis of flowering in *P. frutescens* (L.) Britt. *Phytomorphology 12*, 144–167.
– Raghavan, V., & Kaushik, M. P. 1965. New applications of tissue culture: floral induction in *Perilla* and petiolar abscission in *Coleus*. In *Proceedings of an International Conference on Plant Tissue Culture*, P. R. White & A. R. Grove (eds.), pp. 225–241. Berkeley, Calif.: McCutchan Publishing Corporation.
– Shield, J. A., Jr., & Osborne, D. J. 1962. Senescence factor and abscission of *Coleus* leaves. *Plant Physiol. 37*, 104–106.
Jacobsen, J. V., & Varner, J. E. 1967. Gibberellic acid-induced synthesis of protease by isolated aleurone layers of barley. *Plant Physiol. 42*, 1596–1600.
Jeffs, R. A., & Northcote, D. H. 1966. Experimental induction of vascular tissue in an undifferentiated plant callus. *Biochem. J. 101*, 146–152.
Jelsema, C., Ruddat, M., Morré, D. J., & Williamson, F. A. 1977. Specific binding of gibberellin A_1 to aleurone grain fractions from wheat endosperm. *Plant Cell Physiol. 18*, 1009–1019.
Jepson, J. B. 1958. Indolylacetamide – a chromatographic artifact from the natural indoles, indolylacetylglucosiduronic acid and indolylpyruvic acid. *Biochem. J. 69*, 22P.
Johnston, E. S. 1934. Phototropic sensitivity in relation to wave length. *Smithson. Misc. Coll.* 92, (11), 1–17.
Jones, D. F., MacMillan, J., & Radley, M. 1963. Plant hormones. III. Identification of gibberellic acid in immature barley and immature grass. *Phytochemistry 2*, 307–314.
Jones, R. L. 1971. Gibberellic acid-enhanced release of β-1,3-glucanase from barley aleurone cells. *Plant Physiol. 47*, 412–416.
– & Lang, A. 1968. Extractable and diffusible gibberellins from light- and dark-grown pea seedlings. *Plant Physiol. 43*, 629–634.

– & Phillips, I. D. J. 1964. Agar-diffusion technique for estimating gibberellin production by plant organs. *Nature 204,* 497–499.
– & Phillips, I. D. J. 1966. Organs of gibberellin synthesis in light-grown sunflower plants. *Plant Physiol. 41,* 1381–1386.
– & Varner, J. E. 1967. The bioassay of gibberellins. *Planta 72,* 155–161.
Jordan, W. R., & Skoog, F. 1971. Effects of cytokinins on growth and auxin in coleoptiles of derooted *Avena* seedlings. *Plant Physiol. 48,* 97–99.
Jost, L. 1893. Ueber Beziehungen zwischen der Blattentwickelung und der Gefässbildung in den Pflanze. *Bot. Ztg. 51,* 89–138.
– 1940. Zur Physiologie der Gefässbildung. *Z. Bot. 35,* 114–149.
– 1942. Ueber Gefässbrücken. *Z. Bot. 38,* 161–215.
– & Reiss, E. 1936. Zur Physiologie der Wuchsstoffe. II. Einfluss des Heteroauxins auf Längen- und Dickenwachstum. *Z. Bot. 30,* 335–376.
Kaan Albest, A. von 1934. Anatomische und physiologische Untersuchungen über die Entstehung von Siebröhrenverbindungen. *Z. Bot. 27,* 1–94.
Kaldewey, H. 1964. Papier- und dünnschichtchromatographische Trennung, Farbreaktionen und biologischer Test einiger Wachstumsregulatoren aus Fruchtstielen von *Fritillaria meleagris* L. In *Régulateurs naturels de la croissance végétale,* J. P. Nitsch (ed.), pp. 421–443. Paris: C.N.R.S.
– 1965. Wuchsstofftransport, Temperatur und Pflanzenalter. *Ber. Deut. Bot. Ges. 78* (128)–(143).
– & Jacobs, W. P. 1975. The influence of humidity and gibberellic acid on movement and immobilization of radiocarbon within petioles of *Coleus blumei* Benth. after application of indole-3-(acetic acid-2-^{14}C) to leaves of intact plants. *Biochem. Physiol. Pflanz. 168,* 401–410.
Kato, J. 1958. Nonpolar transport of gibberellin through pea stem and a method for its determination. *Science 128,* 1008–1009.
Katunskij, V. M. 1936. Short periodical illumination as a method of controlling the development of plant organisms. *C. R. (Dokl.) Acad. Sci. URSS 3,* 303–304.
Kaufman, P. B., Ghosheh, N. S., Nakosteen, L., Pharis, R. P., Durley, R. C., & Morf, W. 1976. Analysis of native gibberellins in the internode, nodes, leaves, and inflorescence of developing *Avena* plants. *Plant Physiol. 58,* 131–134.
Kaushik, M. P. 1965. Exact substitution of leaf blade by beta-indolylacetic acid with respect to the retardation of abscission of debladed petioles of *Coleus blumei* Benth. *Indian J. Plant Physiol. 8,* 23–35.
Kefford, N. P. 1955. The growth substances separated from plant extracts by chromatography. *J. Exp. Bot. 6,* 129–151.
– 1962. The inactivity of 1-docosanol in some plant growth tests in relation to the auxin of Maryland Mammoth tobacco. *Aust. J. Biol. Sci. 15,* 304–311.
– & Caso, O. H. 1966. A potent auxin with unique chemical structure – 4-amino-3,5,6-trichloropicolinic acid. *Bot. Gaz. 127,* 159–163.
Kendall, F. H., Park, C. K., & Mer, C. L. 1971. Indole-3-acetic acid metabolism in pea seedlings. A comparative study using carboxyl- and ring-labelled isomers. *Ann. Bot. 35,* 565–579.
Kende, H. 1964. Preservation of chlorophyll in leaf sections by substances obtained from root exudate. *Science 145,* 1066–1067.

– 1971. The cytokinins. *Int. Rev. Cytol. 31*, 301–338.
– & Gardner, G. 1976. Hormone binding in plants. *Ann. Rev. Plant Physiol.* 27, 267–290.
– & Sitton, D. 1967. The physiological significance of kinetin- and gibberellin-like root hormones. *Ann. N.Y. Acad. Sci. 144*, 235–243.
Kentzer, T., & Libbert, E. 1961. Blockade des Gibberellinsäure-transports in Hypocotylsegmenten durch Trijodbenzoesäure. Zugleich ein neuer Agarblocktest auf Gibberellin. *Planta 56*, 23–27.
Ketellapper, H. J., & Barbaro, A. 1966. The role of photoperiod, vernalization and gibberellic acid in floral induction in *Coreopsis grandiflora* Nutt. *Phyton 23*, 33–41.
Khalifah, R. A., Lewis, L. N., & Coggins, C. W., Jr. 1963. New natural growth promoting substances in young citrus fruit. *Science 142*, 399–400.
– Lewis, L. N., & Coggins, C. W., Jr. 1966. Differentiation between indoleacetic acid and the citrus auxin by column chromatography. *Plant Physiol. 41*, 208–210.
Kinet, J. M., Bodson, M., Jacqmard, A., & Bernier, G. 1975. The inhibition of flowering by abscisic acid in *Sinapis alba* L. *Z. Pflanzenphysiol. 77*, 70–74.
Kirk, S. C., & Jacobs. W. P. 1968. The movement of 3-indoleacetic acid-^{14}C in roots of *Lens* and *Phaseolus*. In *Biochemistry and Physiology of Plant Growth Substances*, F. Wightman & G. Setterfield (eds.), pp. 1077–1094. Ottawa: Runge Press.
– Morrow, I. B., & Jacobs, W. P. 1967. Developmental morphology of the *Xanthium* shoot apex after photo-induction. *Phytomorphology 17*, 410–419.
Klämbt, H. D. 1960. Indol-3-acetylasparaginsäure, ein natürlich vorkommendes Indolderivat. *Naturwissenschaften 47*, 398.
– 1961. Wachstumsinduktion und Wuchsstoffmetabolismus im Weizenkoleoptilzylinder . II. Stoffwechselprodukte der Indol-3-essigsäure und der Benzoesäure. *Planta 56*, 618–631.
– 1967. Nachweis eines Cytokinins aus *Agrobacterium tumefaciens* und sein Vergleich mit dem Cytokinin aus *Corynebacterium fascians. Wiss. Z. Univ. Rostock, Math.-Naturwiss. 16*, 623–625.
– 1968. Cytokinine aus *Helianthus annuus. Planta 82*, 170–178.
– Thies, G., & Skoog, F. 1966. Isolation of cytokinins from *Corynebacterium fascians. Proc. Nat. Acad. Sci. U.S.A. 56*, 52–59.
Klein, R. M., & Weisel, B. W. 1964. Determinant growth in the morphogenesis of bean hypocotyls. *Bull. Torrey Bot. Club 91*, 217–224.
Knegt, E., & Bruinsma, J. 1973. A rapid, sensitive and accurate determination of indolyl-3-acetic acid. *Phytochemistry 12*, 753–756.
Knott, J. E. 1934. Effect of a localized photoperiod on spinach. *Proc. Amer. Soc. Hort. Sci. 31*, 152–154.
Koevenig, J. L. 1973. Nonpolar movement of N_6-genzyladenine-^{14}C in coleoptile, stem, petiole and floral organ sections. *Can. J. Bot. 51*, 2079–2083.
– & Jacobs, W. P. 1972. Effect of light on basipetal movement of indoleacetic acid in green stem sections of *Coleus. Plant Physiol. 49*, 866–867.
– & Sillix, D. 1973. Movement of IAA in sections from spider flower (*Cleome hassleriana*) stamen filaments. *Amer. J. Bot. 60*, 231–235.
Kögl, F., Erxleben, H., & Haagen-Smit, A. J. 1933. Ueber ein Phytohormon

der Zellstreckung. Zur Chemie des krystallisierten Auxins. V. Mitteilung. *Z. Physiol. Chem. 216,* 31–44.
– Haagen-Smit, A. J., & Erxleben, H. 1934. Ueber ein neues Auxin ("Heteroauxin") aus Harn. XI. Mitteilung. *Z. Physiol. Chem. 228,* 90–103.
– & Kostermans, D. G. F. R. 1934. Hetero-auxin als Stoffwechselprodukt niederer pflanzlicher Organismen. Isolierung aus Hefe. XIII. Mitteilung. *Z. Physiol. Chem. 228,* 113–121.
Koning, R., Tkaczyk, A., Kaufman, P. B., Pharis, R. P., & Morf, W. 1977. Regulation of internodal extension in *Avena* shoots by the inflorescence, nodes, leaves, and intercalary meristem *Physiol. Plant. 40,* 119–124.
Konings, H. 1968. The significance of the root cap for geotropism. *Acta Bot. Neerl. 17,* 203–211.
Koshimizu, K., Fukui, H., Mitsui, T., & Ogawa, Y. 1966. Identity of lupin inhibitor with abscisin II and its biological activity on growth of rice seedlings. *Agric. Biol. Chem. 30,* 941–943.
– Kusaki, T., Mitsui, T., & Matsubara, S. 1967. Isolation of cytokinin, (−)-dihydrozeatin, from immature seeds of *Lupinus luteus. Tetrahedron Lett. 14,* 1317–1320.
Köves, E. 1957. Papierchromatographische Untersuchungen der ätherlöslichen keimung- und wachstums-hemmenden Stoffe der Haferspelze. *Acta Biol. Szeged 3,* 179–187.
Kramer, M., & Went, F. W. 1949. The nature of the auxin in tomato stem tips. *Plant Physiol. 24,* 207–221.
Krishnamoorthy, H. N. 1974. *Gibberellins and Plant Growth.* New York: Wiley.
Kruszewski, S. P., & Jacobs, W. P. 1974. Polarity of thiamine movement through tomato petioles. *Plant Physiol. 54,* 310–311.
Kruyt, W., & Veldstra, H. 1947. Researches on plant growth regulators. XIII. Leaf growth factors. I. II. *Proc. Kon. Akad. Wetensch. Amst. 50,* 1142–1149, 1317–1323.
Kulaeva, O. N. 1962. The effects of roots on leaf metabolism in relation to the action of kinetin on leaves. *Sov. Plant Physiol. 9,* 182–188.
Kundu, K. K., & Audus, L. J. 1974. Root-growth inhibitors from root tips of *Zea mays* L. *Planta 117,* 183–186.
Kuraishi, S. 1959. Effect of kinetin analogs on leaf growth. *Sci. Pap. Coll. Gen. Educa., Univ. Tokyo. 9,* 67–104.
– & Muir, R. M. 1964a. The relationship of gibberellin and auxin in plant growth. *Plant Cell Physiol. 5,* 61–69.
– & Muir, R. M. 1964b. The mechanisms of gibberellin action in the dwarf pea. *Plant Cell Physiol. 5,* 259–271.
Kursanov, A. L., Kulaeva, O. N., Sveshnikova, I. N., Popova, E. A., Bolyakina, Y. P., Klyachko, N. L., & Vorobeva, I. P. 1964. Restoration of cellular structures and metabolism in yellow leaves due to the action of 6-benzylaminopurine. *Sov. Plant Physiol. 11,* 710–719.
Küster, E. 1916. Beiträge zur Kenntniss des Laubfalles. *Ber. Deut. Bot. Ges. 34,* 184–193.
Kutáček, M. & Procházka, Z. 1964. Méthodes de détermination et d'isolement des composés indoliques chez les crucifères. In *Régulateurs naturels de la croissance végétale,* J. P. Nitsch (ed.), pp. 445–456. Paris: C.N.R.S.

Kuyper, J., & Wiersum, L. K. 1936. Occurrence and transport of a substance causing flowering in the soya bean (*Glycine max* L.), *Proc. Kon. Akad. Wetensch. Amst. 39*, 1114–1122.
Lahiri, A. N., & Audus, L. J. 1960. Growth substances in the roots of *Vicia faba*. *J. Exp. Bot. 11*, 341–350.
Laibach, F. 1932. Pollenhormon und Wuchsstoff. *Ber. Deut. Bot. Ges. 50*, 383–390.
– 1933. Wuchsstoffversuche mit lebenden Orchideenpollinien. *Ber. Deut. Bot. Ges. 51*, 336–340.
– 1947. Crossbreedings between *Coleus*-species of a long and short day character. *Office of Military Govt. for Germany (U.S.) FIAT Report* No. 1135, pp. 5–20.
– & Kornmann, P. 1933. Zur Frage des Wuchsstofftransportes in der Haferkoleoptile. *Planta 21*, 396–418.
LaMotte, C. E., & Jacobs, W. P. 1962. Quantitative estimation of phloem regeneration in *Coleus* internodes. *Stain Technol. 37*, 63–73.
– & Jacobs, W. P. 1963. A role of auxin in phloem regeneration in *Coleus* internodes. *Dev. Biol. 8*, 80–98.
Lang, A. 1942. Beiträge zur Genetik des Photoperiodismus. I. Faktorenanalyse des Kurztagcharakters von *Nicotiana tabacum* "Maryland-Mammut." *Z. Indukt. Abstamm. Vererbungsl. 80*, 210–219.
– 1956a. Stem elongation in a rosette plant, induced by gibberellic acid. *Naturwissenschaften 43*, 257–258.
– 1956b. Induction of flower formation in biennial *Hyoscyamus* by treatment with gibberellin. *Naturwissenschaften 43*, 284–285.
– 1956c. Gibberellin and flower formation. *Naturwissenschaften 43*, 544.
– 1957. The effect of gibberellin upon flower formation. *Proc. Nat. Acad. Sci. U.S.A. 43*, 709–717.
– 1960. Gibberellin-like substances in photoinduced and vegetative *Hyoscyamus* plants. *Planta 54*, 498–504.
– 1961. Auxins in flowering. *Encycl. Plant Physiol. 14*, 909–950.
– 1965. Physiology of flower initiation. *Encycl. Plant Physiol. 15* (Part 1), 1380–1536.
– 1970. Gibberellins: structure and metabolism. *Ann. Rev. Plant Physiol. 21*, 537–570.
– & Melchers, G. 1943. Die photoperiodische Reaktion von *Hyoscyamus niger*. *Planta 33*, 653–702.
Lange, S. 1927. Die Verteilung der Lichtempfindlichkeit in der Spitze der Haferkoleoptile. *Jahrb. Wiss. Bot. 67*, 1–51.
Larsen, P. 1940. Untersuchungen über den thermolabilen wuchsstoffoxydierenden Stoff in *Phaseolus*-Keimpflanzen. *Planta 30*, 673–682.
– 1955. Growth substances in higher plants. In *Modern Methods of Plant Analysis*, Vol. 3, K. Paech and M. V. Tracey (eds.), pp. 565–625. Berlin: Springer-Verlag.
Lau, O. L., & Yang, S. F. 1973. Mechanism of a synergistic effect of kinetin on auxin-induced ethylene production. Suppression of auxin conjugation. *Plant Physiol. 51*, 1011–1014.
Laudi, G. 1956. Studi sulla fisiologia dell'abscissione: Influenza della presenza di rami normali, foglie e gemme ascellari sull'abscissione di piccioli privati del lembo. *Nuovo Giorn. Bot. Ital. 63*, 204–211.
– & Gerola F. 1956. Studi sulla fisiologia dell' abscissione: nuove indagini

sul suo meccanismo mediante modificazioni sperimentali dei normali fenomeni correlativi. *Nuovo Giorn. Bot. Ital. 63,* 336–344.

Lazer, L., Baumgartner, W. E., & Dahlstrom, R. V. 1961. Determination of endogenous gibberellins in green malt by isotopic, derivative dilution procedures. *Agric. Food Chem.* 9(1), 24–26.

Leike, H. 1967. Beeinflussung des Auxintransports durch Kinetin. *Wiss. Z. Univ. Rostock, Math.-Naturwiss. 16,* 501–502.

Lek, H. A. A. van der. 1924. Over de wortelvorming van houtige stekken. *Meded. Landbouwhogesch. Wageningen 28,* 1–230.

Lembi, C. A., Morré, D. J., Thomson, K. S., & Hertel, R. 1971. *1-N*-naphthylphthalamic-acid (NPA)-binding activity of a plasma membrane-rich fraction from maize coleoptiles. *Planta 99,* 37–45.

Lenton, J. R., Perry, V. M., & Saunders, P. F. 1972. Endogenous abscisic acid in relation to photoperiodically induced bud dormancy. *Planta 106,* 13–22.

Leopold, A. C. 1949. Flower initiation in total darkness. *Plant Physiol. 24,* 530–533.

– 1955. *Auxins and Plant Growth.* Berkeley: University of California Press.

– 1964. The polarity of auxin transport. *Brookhaven Symp. Biol. 16,* 218–233.

– & dela Fuente, R. K. 1967. The polarity of auxin transport. *Ann. N.Y. Acad. Sci. 144,* 94–101.

– & dela Fuente, R. K. 1968. A view of polar auxin transport. In *The Transport of Plant Hormones,* Y. Vardar (ed.), pp. 24–47. Amsterdam: North-Holland.

– & Hall, O. F. 1966. Mathematical model of polar auxin transport. *Plant Physiol. 41,* 1476–1480.

– & Lam, S. L. 1961. Polar transport of three auxins. In *Plant Growth Regulation,* R. M. Klein (ed.), pp. 411–418. Ames: Iowa State University Press.

– & Lam, S. L. 1962. The auxin transport gradient. *Physiol. Plant. 15,* 631–638.

Leshem, Y. 1973. *The Molecular and Hormonal Basis of Plant-Growth Regulation.* New York: Pergamon Press.

Letham, D. S. 1963a. Purification of factors inducing cell division extracted from plum fruitlets. *Life Sci. 2,* 152–157.

– 1963b. Zeatin, a factor inducing cell division from *Zea mays. Life Sci. 2,* 569–573.

– 1966. Purification and probable identity of a new cytokinin in sweet corn extracts. *Life Sci. 5,* 551–554.

– 1967a. Regulators of cell division in plant tissues. V. A comparison of the activities of zeatin and other cytokinins in five bioassays. *Planta 74,* 228–242.

– 1967b. Chemistry and physiology of kinetin-like compounds. *Ann. Rev. Plant Physiol. 18,* 349–364.

– 1968. A new cytokinin bioassay and the naturally occuring cytokinin complex. In *Biochemistry and Physiology of Plant Growth Substances,* F. Wightman and G. Setterfield (eds.), pp. 19–31. Ottawa: Runge Press.

– & Miller, C. O. 1965. Identity of kinetin-like factors from *Zea mays. Plant Cell Physiol. 6,* 355–359.

– Shannon, J. S., & McDonald, I. R. 1964. The structure of zeatin, a factor inducing cell division. *Proc. Chem. Soc.* (*Lond.*), pp. 230–231.

Libbert, E. 1964. Kontrolliert Auxin die Apikaldominanz? *Physiol. Plant. 17*, 371–378.

– & Kanter, B. 1966. Auxineinflüsse auf polare Erscheinungen bei Aufnahme, Transport und Metabolismus von Adenin in Hypocotylsegmenten von *Helianthus annuus*. *Flora* (*A*) *157*, 51–67.

– Kentzer, T., & Steyer, B. 1961. Die Wirkungen einiger Adenin-Verbindungen einschliesslich ATP auf Keimung, Keimlingsentwicklung und Auxintransport. *Flora 151*, 663–669.

Lincoln, R. G., Mayfield, D. L., & Cunningham, A. 1961. Preparation of a floral initiating extract from *Xanthium*. *Science 133*, 756.

Lindner, R. C. 1940. Factors affecting regeneration of the horseradish root. *Plant Physiol. 15*, 161–181.

Linser, H., Mayr, H., & Maschek, F. 1954. Papierchromatographie von zellstreckend wirksamen Indolkörpern aus *Brassica*-Arten. *Planta 44*, 103–120.

Linsmaier, E. M., & Skoog, F. 1965. Organic growth factor requirements of tobacco tissue cultures. *Physiol. Plant. 18*, 100–127.

Little, C. H. A., & Goldsmith, M. H. M. 1967. Effect of inversion on growth and movement of indole-3-acetic acid in coleoptiles. *Plant Physiol. 42*, 1239–1245.

Little, E. C. S., & Blackman, G. E. 1963. The movement of growth regulators in plants. III. Comparative studies of transport in *Phaseolus vulgaris*. *New Phytol. 62*, 173–197.

Little, T. M., & Kantor, J. H. 1941. Inheritance of earliness of flowering in the sweet pea. *J. Hered. 32*, 379–383.

Liu, W. C. N., & Carns, H. R. 1961. Isolation of abscisin, an abscission accelerating substance. *Science 134*, 384–385.

Lockhart, J. A., & Deal, P. H. 1960. Prevention of red light inhibition of stem growth in the Cucurbitaceae by gibberellin A_4. *Naturwissenschaften 6*, 141–142.

Loeffler, J. E., & Overbeek, J. van. 1964. Kinin activity in coconut milk. In *Régulateurs naturels de la croissance végétale*, J. P. Nitsch (ed.), pp. 77–82. Paris: C.N.R.S.

Lona, F. 1946. Sui fenomeni di induzione, post-effetto e localizzazione fotoperiodica. L'induzione antogena indiretta della foglie primordiali di *Xanthium italicum* Moretti. *Nuovo Giorn. Bot. Ital. 53*, 548–575.

– 1948. La fioritura della brevidiurna *Chenopodium amaranticolor* Coste et Reyn. coltivata in soluzione nutritizia con saccarosio, in assenza di stimolo fotoperiodico euflorigeno. *Nuovo Giorn. Bot. Ital. 55*, 559–562.

– 1949a. L'induzione fotoperiodica di foglie staccate. *Boll. Soc. Ital. Biol. Sper. 25*, 761–763.

– 1949b. La fioritura delle brevidiurne a notte continua. *Nuovo Giorn. Bot. Ital. 56*, 479–515.

– 1956a. Observazioni orientative cuia l'effeto dell'acido gibberellico sullo svilluppo reproduttivo di alcune longidiurne e brevidiurne. *L'ateneo Parmense 27*, 865–875.

– 1956b. L'azione dell'acido-gibberellico sull'accrescimento-caulinare di talune piante erbacee in condizioni esterne controllate. *Nuovo Giorn. Bot. Ital. 63*, 61–76.

– & Bocchi, A. 1956. Sviluppo vegetativo e riproduttivo di alcune longidiurne in rapporto all'azione dell'acido gibberellico. *Nuovo Giorn. Bot. Ital. 63,* 469–486.
– & Bocchi, A. 1957. Effetti morfogenetici ed organogenetici provocati dalla cinetina (kinetin) su piante erbacee in condizioni esterne controllate. *Nuovo Giorn. Bot. Ital. 64,* 236–246.
Lonberg-Holm, K. K. 1967. Nucleic acid synthesis in seedlings. *Nature 213,* 454–457.
Long, E. M. 1939. Photoperiodic induction as influenced by environmental factors. *Bot. Gaz. 101,* 168–188.
Loomis, R. S., & Torrey, J. G. 1964. Chemical control of vascular cambium initiation in isolated radish roots. *Proc. Nat. Acad. Sci. U.S.A. 52,* 3–11.
Luckwill, L. C. 1956. Two methods for the bioassay of auxins in the presence of growth inhibitors. *J. Hort. Sci. Lond. 31,* 89–98.
– 1957. Studies of fruit development in relation to plant hormones. IV. Acidic auxins and growth inhibitors in leaves and fruits of the apple. *J. Hort. Sci. Lond. 32,* 18–33.
MacLeod, A. M., & Millar, A. S. 1962. The effects of gibberellic acid on barley endosperm. *J. Inst. Brewing 68,* 322–332.
MacMillan, J. 1972. Application of GLC-MS to hormone studies. In *Plant Growth Substances 1970,* D. J. Carr (ed.), pp. 790–797. New York: Springer-Verlag.
– & Pryce, R. J. 1968. Recent studies of endogenous plant growth substances using combined gas chromatography-mass spectrometry. In *Plant Growth Regulators,* pp. 36–50. London: Society of Chemical Industry Monograph No. 31.
Mai, G. 1934. Korrelationsuntersuchungen an entspreiteten Blattstielen mittels lebender Orchideenpollinien als Wuchsstoffquelle. *Jahrb. Wiss. Bot. 79,* 681–713.
Marcovitch, S. 1924. The migration of *Aphididae* and the appearance of sexual forms as affected by the relative length of the daily light exposure. *J. Agric. Res. 27,* 513–522.
Marumo, S., Abe, H., Hattori, H., & Munakata, K. 1968. Isolation of a novel auxin, methyl 4-chloro-indoleacetate from immature seeds of *Pisum sativum. Agric. Biol. Chem. 32,* 117–118.
Masuda, Y. 1962. Effect of light on a growth inhibitor in wheat roots. *Physiol. Plant. 15,* 780–790.
Maurer, H. R., & Chalkley, G. R. 1967. Some properties of a nuclear binding site of estradiol. *J. Mol. Biol. 27,* 431–441.
Mayfield, D. L. 1964. Floral-inducing extracts of *Helianthus* and of *Xanthium:* some chemical and physiological properties. In *Régulateurs naturels de la croissance végétale,* J. P. Nitsch (ed.), pp. 621–633. Paris: C.N.R.S.
McCalla, D. R., Morré, D. J., & Osborne, D. J. 1962. The metabolism of a kinin, benzyladenine. *Biochim. Biophys. Acta 55,* 522–528.
McCready, C. C. 1958. A direct-plating method for the precise assay of carbon-14 in small liquid samples. *Nature 181,* 1406.
– 1963. Movement of growth regulators in plants. I. Polar transport of 2,4-dichlorophenoxyacetic acid in segments from the petioles of *Phaseolus vulgaris. New Phytol. 62,* 3–18.
– & Jacobs, W. P. 1963a. Movement of growth regulators in plants. II. Polar

transport of radioactivity from indoleacetic acid-(^{14}C) and 2,4-dichlorophenoxyacetic acid-(^{14}C) in petioles of *Phaseolus vulgaris*. *New Phytol.* *62*, 19–34.
– & Jacobs, W. P. 1963b. Movement of growth regulators in plants. IV. Relationships between age, growth and polar transport in petioles of *Phaseolus vulgaris*. *New Phytol.* *62*, 360–366.
– & Jacobs, W. P. 1967. Movement of growth regulators in plants. V. A further note on the relationship between polar transport and growth. *New Phytol.* *66*, 485–488.
– Osborne, D. J., & Black, M. K. 1965. Promotion by kinetin of the polar transport of two auxins. *Nature* *208*, 1065–1067.
McKinney, H. H., & Sando, W. J. 1935. Earliness of sexual reproduction as influenced by temperature and light in relation to growth phases. *J. Agric. Res.* *51*, 621–641.
Melchers, G., & Lang, A. 1941. Weitere Untersuchungen zur Frage der Blühhormone. *Biol. Zentralbl.* *61*, 16–39.
Menschick, R., Hild, V., & Hager, A. 1977. Decarboxylierung von Indolylessigsäure im Zusammenhang mit den Phototropismus in *Avena*-Koleoptilen. *Planta* *133*, 223–228.
Michniewicz, M., & Lang, A. 1962. Effect of nine different gibberellins on stem elongation and flower formation in cold-requiring and photoperiodic plants grown under non-inductive conditions. *Planta* *58*, 549–563.
Milborrow, B. V. 1968. Identification and measurement of (+)-abscisic acid in plants. In *Biochemistry and Physiology of Plant Growth Substances*, F. Wightman and G. Setterfield (eds.), pp. 1531–1545. Ottawa: Runge Press.
Miller, C. O. 1956. Similarity of some kinetin and red light effects. *Plant Physiol.* *31*, 318–319.
– 1961. A kinetin-like compound in maize. *Proc. Nat. Acad. Sci. U.S.A.* *47*, 170–174.
– 1963. Kinetin and kinetin-like compounds. In *Modern Methods of Plant Analysis*, Vol. 6, H. F. Linskens and M. V. Tracey (eds.), pp. 194–202. Berlin: Springer-Verlag.
– 1965. Evidence for the natural occurrence of zeatin and derivatives: compounds from maize which promote cell division. *Proc. Nat. Acad. Sci. U.S.A.* *54*, 1052–1058.
– 1967. Cytokinins in *Zea mays*. *Ann. N.Y. Acad. Sci.* *144*, 251–257.
– 1968. Naturally-occurring cytokinins. In *Biochemistry and Physiology of Plant Growth Substances*, F. Wightman and G. Setterfield (eds.), pp. 33–45. Ottawa: Runge Press.
– 1975. Cell-division factors from *Vinca rosea* L. crown gall tumor tissue. *Proc. Nat. Acad. Sci., U.S.A.* *72*, 1883–1886.
– Skoog, F., Okumura, F. S., Von Saltza, M. H., & Strong, F. M. 1955a. Structure and synthesis of kinetin. *J. Amer. Chem. Soc.* *77*, 2662.
– Skoog, F., Von Saltza, M. H., & Strong, F. M. 1955b. Kinetin, a cell division factor from deoxribonucleic acid. *J. Amer. Chem. Soc.* *77*, 1392.
– & Witham, F. H. 1964. A kinetin-like factor from maize and other sources. In supplement to *Régulateurs naturels de la croissance végétale*, pp. I–VI. Paris: C.N.R.S.
Morgan, D. G., & Söding, H. 1958. Ueber die Wirkungsweise von

Phthalsäuremono-α-naphthylamid (PNA) auf das Wachstum der Haferkoleoptile. *Planta 52,* 235–249.

Morris, D. A., Briant, R. E., & Thomson, P. G. 1969. The transport and metabolism of ^{14}C-labelled indoleacetic acid in intact pea seedlings. *Planta 89,* 178–197.

– & Kadir, G. O. 1972. Pathways of auxin transport in the intact pea seedling (*Pisum sativum* L.). *Planta 107,* 171–182.

– Kadir, G. O., & Barry, A. J. 1973. Auxin transport in intact pea seedlings (*Pisum sativum* L.): the inhibition of transport by 2,3,5-triiodobenzoic acid. *Planta 110,* 173–182.

Mothes, K. 1960. Ueber das Altern der Blätter und die Möglichkeit ihrer Wiederverjüngung. *Naturwissenschaften 15,* 337–351.

– & Engelbrecht, L. 1961. Kinetin-induced directed transport of substances in excised leaves in the dark. *Phytochemistry 1,* 58–62.

Mullins, M. G. 1967. Morphogenetic effects of roots and of some synthetic cytokinins in *Vitis vinifera* L. *J. Exp. Bot. 18,* 206–214.

– 1972. Auxin and ethylene in adventitious root formation in *Phaseolus aureus* (Roxb.). In *Plant Growth Substances 1970,* D. J. Carr (ed.), pp. 526–533. New York: Springer-Verlag.

Murakami, Y. 1968a. A new rice seedling test for gibberellins, "microdrop method," and its use for testing extracts of rice and morning glory. *Bot. Mag. (Tokyo) 81,* 33–43.

– 1968b. Gibberellin-like substances in roots of *Oryza sativa, Pharbitis nil* and *Ipomoea batatas* and the site of their synthesis in the plant. *Bot. Mag. (Tokyo) 81,* 334–343.

Myrayama, K., & Ueda, K. 1973. Short-term response of *Pisum* stem segments to indole-3-acetic acid. *Plant Cell Physiol. 14,* 973–979.

Myers, R. M. 1939. Factors affecting the abscission of the leaves of *Coleus blumei. Ill. State Acad. Sci. Trans. 32,* 80–81.

– 1940. Effect of growth substances on the absciss layer in leaves of *Coleus. Bot. Gaz. 102,* 323–338.

Nanda, K. K., Krishnamoorthy, H. N., Anuradha, T. A., & Lal, K. 1967. Floral induction by gibberellic acid in *Impatiens balsamina,* a qualitative short-day plant. *Planta 76,* 367–370.

Naqvi, S. M. 1963. Transport studies with C^{14}-indoleacetic acid and C^{14}-2,4-dichlorophenoxyacetic acid in *Coleus* stems. Ph.D. thesis, Princeton University, Princeton, N.J.

– 1972. Possible role of abscisic acid in phototropism. *Z. Pflanzenphys. 67,* 454–456.

– & Engvild, K. C. 1974. Action of abscisic acid on auxin transport and its relation to phototropism. *Physiol. Plant. 30,* 283–287.

– & Gordon, S. A. 1965. Auxin transport in flowering and vegetative shoots of *Coleus blumei* Benth. *Plant Physiol. 40,* 116–118.

– & Gordon, S. A. 1966. Auxin transport in *Zea mays* L. coleoptiles. I. Influence of gravity on the transport of indoleacetic acid-2-^{14}C. *Plant Physiol. 41,* 1113–1118.

– & Gordon, S. A. 1967. Auxin transport in *Zea mays* coleoptiles. II. Influence of light on the transport of indoleacetic acid-2-^{14}C. *Plant Physiol. 42,* 138–143.

Naylor, A. W. 1941. Effects of some environmental factors on photoperiodic induction of beet and dill. *Bot. Gaz. 102,* 557–575.

Naylor, J. M. 1966. Dormancy studies in seed of *Avena fatua*. 5. On the response of aleurone cells to gibberellic acid. *Can. J. Bot. 44*, 19–32.
Neame, K. D., & Homewood, C. A. 1974. *Liquid Scintillation Counting*. New York: Wiley.
Newman, I. A. 1963. Electric potentials and auxin translocation in *Avena*. *Aust. J. Biol. Sci. 16*, 629–646.
– 1970. Auxin transport in *Avena*. I. Indoleacetic acid-^{14}C distributions and speeds. *Plant Physiol. 46*, 263–272.
Niedergang-Kamien, E., & Skoog, F. 1956. Studies on polarity and auxin transport in plants. I. Modification of polarity and auxin transport by triiodobenzoic acid. *Physiol. Plant. 9*, 60–73.
Nielsen, N. 1924. Studies on the transmission of stimuli in the coleoptile of *Avena*. *Dan. Bot. Ark. 4*(8).
Nissl, D., & Zenk, M. H. 1969. Evidence against induction of protein synthesis during auxin-induced initial elongation of *Avena* coleoptiles. *Planta 89*, 323–341.
Nitsch, C., & Nitsch, J. P. 1960. An artifact in chromatography of indolic auxins. *Plant Physiol. 35*, 450–454.
Nitsch, J. P. 1956. Methods for the investigation of natural auxins and growth inhibitors. In *Chemistry and Mode of Action of Plant Growth Substances*, R. L. Wain and F. Wightman (eds.), pp. 3–31. London: Butterworths.
– 1957. Growth responses of woody plants to photoperiodic stimuli. *Proc. Amer. Soc. Hort. Sci. 70*, 512–525.
– 1965. Physiology of flower and fruit development. *Encycl. Plant Physiol. 15*(1), 1537–1647.
– & Nitsch, C. 1959. Photoperiodic effects in woody plants: evidence for the interplay of growth-regulating substances. In *Photoperiodism*, R. B. Withrow (ed.), pp. 225–242. Washington: AAAS.
– & Nitsch, C. 1965a. Présence d'une phytokinine dans le cambium. *Bull. Soc. Bot. Fr. 112*, 1–10.
– & Nitsch, C. 1965b. Présence de phytokinines et autres substances de croissance dans la sève d'*Acer saccharum* et de *Vitis vinifera*. *Bull. Soc. Bot. Fr. 112*, 11–18.
Ohkuma, K., Addicott, F. T., Smith, O. E., & Thiessen, W. E. 1965. The structure of abscisin II. *Tetrahedron Lett. 29*, 2529–2535.
– Lyon, J. L., Addicott, F. T., & Smith, O. E. 1963. Abscisin II, an abscission-accelerating substance from young cotton fruit. *Science 142*, 1592–1593.
Ohwaki, Y. 1970. Thin-layer chromatography of diffusible auxin. *Sci. Rep. Tohoku Univ. 4 Ser. Biol. 35*, 69–89.
Okuda, M. 1954. Flower formation of *Xanthium canadense* under long day conditions induced by grafting with long day plants. *Bot. Mag.* (*Tokyo*) *66*, 247–255.
Olney, H. O. 1968. Growth substances from *Veratrum tenuipetalum*. *Plant Physiol. 43*, 293–302.
Oota, Y. 1975. Short-day flowering of *Lemna gibba* G3 induced by salicylic acid. *Plant Cell Physiol. 16*, 1131–1135.
Ordin, L., Cleland, R., & Bonner, J. 1957. Methyl esterification of cell wall constituents under the influence of auxin. *Plant Physiol. 32*, 216–220.

Osborne, D. J. 1955. Acceleration of abscission by a factor produced in senescent leaves. *Nature 176*, 1161–1163.
– 1959. Control of leaf senescence by auxins. *Nature 183*, 1459–1460.
– & Black, M. K. 1964. Polar transport of a kinin, benzyladenine. *Nature 201*, 97.
– Horton, R. F., & Black, M. K. 1968. Senescence in excised petiole segments: the relevance to auxin and kinin transport. In *The Transport of Plant Hormones*, Y. Vardar (ed.), pp. 79–96. Amsterdam: North-Holland.
– Jackson, M. B., & Milborrow, B. V. 1972. Physiological properties of abscission accelerator from senescent leaves. *Nature New Biol. 240*, 98–101.
– & McCalla, D. R. 1961. Rapid bioassay for kinetin and kinins using senescing leaf tissue. *Plant Physiol. 36*, 219–221.
– & McCready, C. C. 1965. Transport of the kinin, N_6-benzyladenine: Non-polar or polar? *Nature 206*, 679–680.
Overbeek, J. van. 1933. Wuchsstoff, Lichtwachstumsreaktion und Phototropismus bei *Raphanus*. *Rec. Trav. Bot. Neerl. 30*, 537–626.
– 1935. The growth hormone and the dwarf type of growth in corn. *Proc. Nat. Acad. Sci. U.S.A. 21*, 292–299.
– 1937. Effects of roots on the production of auxin by the coleoptile. *Proc. Nat. Acad. Sci. U.S.A. 23*, 272–276.
– 1938. Auxin production in seedlings of dwarf maize. *Plant Physiol. 13*, 587–598.
– 1939. Is auxin produced in roots? *Proc. Nat. Acad. Sci. U.S.A. 25*, 245–248.
– 1941. A quantitative study of auxin and its precursors in coleoptiles. *Amer. J. Bot. 28*, 1–10.
– Conklin, M., & Blakeslee, A. 1941. Factors in coconut milk essential for growth and development of *Datura* embryos. *Science 94*, 350–351.
– Conklin, M., & Blakeslee, A. F. 1942. Cultivation *in vitro* of small *Datura* embryos. *Amer. J. Bot. 29*, 472–477.
– Gordon, S. A., & Gregory, L. E. 1946. An analysis of the function of the leaf in the process of root formation in cuttings. *Amer. J. Bot. 33*, 100–107.
– & Gregory, L. E. 1945. A physiological separation of two factors necessary for the formation of roots on cuttings. *Amer. J. Bot. 32*, 336–341.
– Siu, R., & Haagen-Smit, A. J. 1944. Factors affecting the growth of *Datura* embryos *in vitro*. *Amer. J. Bot. 31*, 219–224.
Paál, A. 1919. Ueber phototropische Reizleitung. *Jahrb. Wiss. Bot. 58*, 406–458.
Paleg, L. G. 1960a. Physiological effects of gibberellic acid: I. On carbohydrate metabolism and amylase activity of barley endosperm. *Plant Physiol. 35*, 293–299.
– 1960b. Physiological effects of gibberellic acid. II. On starch hydrolyzing enzymes of barley endosperm. *Plant Physiol. 35*, 902–906.
– 1964. Cellular localization of the gibberellin-induced response of barley endosperm. In *Régulateurs naturels de la croissance végétale*, J. P. Nitsch (ed.), pp. 303–317. Paris: C.N.R.S.
– Aspinall, D., Coombe, B., & Nicholls, P. 1964. Physiological effects of gibberellic acid. VI. Other gibberellins in three test systems. *Plant Physiol. 39*, 286–290.

Palmer, G. H. 1972. Transport of [^{14}C] gibberellic acid in the barley embryo. *J. Inst. Brewing 78*, 470–471.
Palmer, J. H., & Halsall, D. M. 1969. Effect of transverse stimulation, gibberellin and indoleacetic acid upon polar transport of IAA-C^{14} in the stem of *Helianthus annuus*. *Physiol. Plant. 22*, 59–67.
Pavillard, J., & Beauchamp, C. 1957. La constitution auxinique de tabacs sains ou atteints de maladies à virus: présence et rôle de la scopolétine. *C. R. Acad. Sci. Paris 244*, 1240–1243.
Pegg, G. F., & Selman, I. W. 1959. An analysis of the growth response of young tomato plants to infection by *Verticillium albo-atrum*. II. The production of growth substances. *Ann. Appl. Biol. 47*, 222–231.
Pengelly, W., & Meins, F., Jr. 1977. A specific radioimmunoassay for nanogram quantities of the auxin, indole-3-acetic acid. *Planta 136*, 173–180.
Penner, J. 1960. Ueber den Einfluss von Gibberellin auf die photoperiodisch bedingten Blühvorgänge bei *Bryophyllum*. *Planta 55*, 542–572.
Person, C., Samborski, D. J., & Forsyth, F. R. 1957. Effects of benzimidazole on detached wheat leaves. *Nature 180*, 1294–1295.
Peterson, R. L. 1973. Control of cambial activity in roots of turnip (*Brassica rapa*). *Can. J. Bot. 51*, 475–480.
Phillips, I. D. J. 1971a. Effect of relative hormone concentration on auxin–gibberellin interaction in correlative inhibition of axillary buds. *Planta 96*, 27–34.
– 1971b. *Introduction to the Biochemistry and Physiology of Plant Growth Hormones*. New York: McGraw-Hill.
– & Hartung, W. 1974. Basipetal and acropetal transport of [3,4-^{3}H] gibberellin A_1 in short and long segments of *Phaseolus coccineus* second internode. *Planta 116*, 109–121.
– & Jones, R. L. 1964. Gibberellin-like activity in bleeding sap of root systems of *Helianthus annuus* detected by a new dwarf pea epicotyl assay and other methods. *Planta 63*, 269–278.
– Vlitos, A. J., & Cutler, H. 1959. The influence of gibberellic acid upon the endogenous growth substances of the Alaska pea. *Contrib. Boyce Thompson Inst. 20*, 111–120.
Phinney, B. O. 1956. Growth response of single-gene dwarf mutants in maize to gibberellic acid. *Proc. Nat. Acad. Sci. U.S.A. 42*, 185–189.
– 1961. Dwarfing genes in *Zea mays* and their relation to the gibberellins. In *Plant Growth Regulation*, R. M. Klein (ed.), pp. 489–501. Ames: Iowa State University Press.
– & West, C. A. 1961. Gibberellins and plant growth. *Encycl. Plant Physiol. 14*, 1185–1227.
– West, C. A., Ritzel, M., & Neely, P. M. 1957. Evidence for "gibberellin-like" substances from flowering plants. *Proc. Nat. Acad. Sci.* U.S.A. *43*, 398–404.
Pickard, B. G., & Thimann, K. V. 1964. Transport and distribution of auxin during tropistic response. II. The lateral migration of auxin in phototropism of coleoptiles. *Plant Physiol. 39*, 341–350.
Platt, R. S., Jr., & Thimann, K. V. 1956. Interference in Salkowski assay of indoleacetic acid. *Science 123*, 105–106.
Plett, W. 1921. Untersuchungen über die Regenerationserscheinungen an Internodien. *Diss. Hamburg.*

Plummer, T. H., & Leopold, A. C. 1957. Chemical treatment for bud formation in *Saintpaulia. Proc. Amer. Soc. Hort. Sci. 70*, 442–444.
Pohl, R. 1961. Die Wirkung der Wuchsstoffe auf die Zellwand. *Encycl. Plant Physiol. 14*, 703–742.
Pol, P. A. van de. 1972. Floral induction, floral hormones and flowering. *Meded. Landbouwhogesch. Wageningen 72*, 1–89.
Polevoy, V. V. 1967. The action of auxin on electropotential, growth, respiration and synthesis of RNA in sections of corn seedlings. Order of reactions. *Wiss. Z. Univ. Rostock, Math.-Naturwiss. 16*, 477–478.
Pollard, C. J. 1969. A survey of the sequence of some effects of gibberellic acid in the metabolism of cereal grains. *Plant Physiol. 44*, 1227–1232.
Poovaiah, B. W., & Leopold, A. C. 1976. Effects of inorganic solutes on the binding of auxin. *Plant Physiol. 58*, 783–785.
Prakash, G. 1976. A senescence factor and foliar abscission in *Catharanthus roseus. Ann. Bot. 40*, 537–541.
Pryce, R. J. 1972. Gallic acid as a natural inhibitor of flowering in *Kalanchoe blossfeldiana. Phytochemistry 11*, 1911–1918.
Raadts, E., & Söding, H. 1957. Chromatographische Untersuchungen über die Wuchsstoffe der Haferkoleptile. *Planta 49*, 47–60.
Raalte, M. H. van. 1937. On factors determining the auxin content of the root tip. *Rec. Trav. Bot. Neerl. 34*, 279–332.
Radin, J. W., & Loomis, R. S. 1971. Changes in the cytokinins of radish roots during maturation. *Physiol. Plant. 25*, 240–244.
– & Loomis, R. S. 1974. Polar transport of kinetin in tissues of radish. *Plant Physiol. 53*, 348–351.
Radley, M. 1956. Occurrence of substances similar to gibberellic acid in higher plants. *Nature 178*, 1070–1071.
– 1958. The distribution of substances similar to gibberellic acid in higher plants. *Ann. Bot. 22*, 297–307.
– 1959. The occurrence of gibberellin-like substances in barley and malt. *Chem. Ind.* (Lond.) 877-878.
– 1963. Gibberellin content of spinach in relation to photoperiod. *Ann. Bot. 27*, 373–377.
– 1970. The effects of changing photoperiod on the gibberellin content of spinach. In *Cellular and Molecular Aspects of Floral Induction*, G. Bernier (ed.), pp. 280–283. London: Longman.
– & Dear, E. 1958. Occurrence of gibberellin-like substances in the coconut. *Nature 182*, 1098.
Raghavan, V. 1961. Studies on the floral histogenesis and physiology of *Perilla*. III. Effects of indoleacetic acid on the flowering of apical buds and explants in culture. *Amer. J. Bot. 48*, 870–876.
– & Jacobs, W. P. 1961. Studies on the floral histogenesis and physiology of *Perilla*. II. Floral induction in cultured apical buds of *P. frutescens. Amer. J. Bot. 48*, 751–760.
Rasmussen, H. P., & Bukovac, M. J. 1966. Naphthaleneacetic acid: localization in the abscission zone of the bean. *Science 152*, 217–218.
Rathbone, M. P., & Hall, R. H. 1972. Concerning the presence of the cytokinin, N^6- (Δ^2-isopentenyl)adenine, in cultures of *Corynebacterium fascians. Planta 108*, 93–102.
Raven, J. A. 1975. Transport of indoleacetic acid in plant cells in relation to

pH and electrical potential gradients, and its significance for polar IAA transport. *New Phytol. 74*, 163–172.

Ray, P. M. 1958. Destruction and auxin. *Ann. Rev. Plant Physiol. 9*, 81–118.

– 1974. The biochemistry of the action of indoleacetic acid on plant growth. *Recent Advances in Phytochemistry 7*, 93–122.

– 1977. Auxin-binding sites of maize coleoptiles are localized on membranes of the endoplasmic reticulum. *Plant Physiol. 59*, 594–599.

– Dohrmann, U., & Hertel, R. 1977. Characterization of naphthaleneacetic acid binding to receptor sites on cellular membranes of maize coleoptile tissue. *Plant Physiol. 59*, 357–364.

– & Ruesink, A. W. 1962. Kinetic experiments on the nature of the growth mechanism in oat coleoptile cells. *Dev. Biol. 4*, 377–397.

Rayle, D. L., Ouitrakul, R., & Hertel, R. 1969. Effect of auxins on the auxin transport system in coleoptiles. *Planta 87*, 49–53.

Redemann, C. T., Wittwer, S. H., & Sell, H. M. 1951. The fruit-setting factor from the ethanol extracts of immature corn kernels. *Arch. Biochem. Biophys. 32*, 80–84.

Reeve, D. R., & Crozier, A. 1974. Gibberellin bioassays. In *Gibberellins and Plant Growth*, H. N. Krishnamoorthy (ed.), pp. 35–64. New York: Wiley.

Reiff, B., & Guttenberg, H. von. 1961. Der polare Wuchsstofftransport von *Helianthus annuus* in seiner Abhängigkeit von Alter, Quellungszustand und Kohlenhydratversorgung des Gewebes. *Flora 151*, 44–72.

Reinders, D. E. 1934. The sensibility for light of the base of normal and decapitated coleoptiles of *Avena. Proc. Kon. Akad. Wetensch. Amst. 37*, 308–314.

Resende, F. 1952. "Long-short" day plants. *Port. Acta Biol.* (A) *3*, 318–321.

Richmond, A. E., & Lang, A. 1957. Effect of kinetin on protein content and survival of detached *Xanthium* leaves. *Science 125*, 650–651.

Robbins, W. J., & Bartley, M. A. 1937. Thiazole and the growth of excised tomato roots. *Proc. Nat. Acad. Sci. U.S.A. 23*, 385–388.

Roberts, L. W., & Fosket, D. E. 1966. Interaction of gibberellic acid and indoleacetic acid in the differentiation of wound vessel members. *New Phytol. 65*, 5–8.

Roberts, R. H. 1951. The induction of flowering with a plant extract. In *Plant Growth Substances*, F. Skoog (ed.), pp. 347–350. Madison : University of Wisconsin Press.

Robinson, B. J., Forman, M., & Addicott, F. T. 1968. Auxin transport and conjugation in cotton explants. *Plant Physiol. 43*, 1321–1323.

Roland, J. C. 1969. Mise en évidence sur coupes ultrafines de formations polysaccharidiques directement associées au plasmalemme. *C. R. Acad. Sci. Paris 269*, 939–942.

Rose, R. J., & Adamson, D. 1969. A sequential response to growth substances in coleoptiles from γ-irradiated wheat. *Planta 88*, 274–281.

Rossetter, F. N., & Jacobs, W. P. 1953. Studies on abscission: the stimulating role of nearby leaves. *Amer. J. Bot. 40*, 276–280.

Rothert, W. 1894. Ueber Heliotropismus. *Beitr. Biol. Pflanz. 7*, 1–212.

Row, V. V., Sanford, W. W., & Hitchcock, A. E. 1961. Indole-3-acetyl-*d,l*-aspartic acid as a naturally-occurring indole compound in tomato seedlings. *Contrib. Boyce Thompson Inst. 21*, 1–10.

Rowan, W. 1926. On photoperiodism, reproductive periodicity, and the annual migration of birds and certain fishes. *Proc. Boston Soc. Nat. Hist. 38*, 147–189.
Rubery, P. H., & Sheldrake, A. R. 1974. Carrier-mediated auxin transport. *Planta 118*, 101–121.
Rubinstein, B., & Leopold, A. C. 1963. Analysis of the auxin control of bean leaf abscission. *Plant Physiol. 38*, 262–267.
Ruddat, M., & Pharis, R. P. 1966. Participation of gibberellin in the control of apical dominance in soybean and redwood. *Planta 71*, 222–228.
Russell, B. 1945. *A History of Western Philosophy*. New York: Simon & Schuster.
Sabnis, D. D., Hirshberg, G., & Jacobs, W. P. 1969. Radioautographic analysis of the distribution of label from ^{3}H-indoleacetic acid supplied to isolated *Coleus* internodes. *Plant Physiol. 44*, 27–36.
Sachs, J. 1880. 1882. Stoff und Form der Pflanzenorgane. I and II. *Arb. Bot. Inst. Wurzburg 2*, 452–488; 689–718.
Sachs, R. M. 1956. Floral initiation in *Cestrum nocturnum*. I. A long-short day plant. *Plant Physiol. 31*, 185–192.
Sachs, T., & Thimann, K. V. 1967. The role of auxins and cytokinins in the release of buds from dominance. *Amer. J. Bot. 54*, 136–144.
Samuels, R. M. 1961. Bacterial-induced fasciation in *Pisum sativum* var. Alaska. Ph.D. thesis, Indiana University, Bloomington.
Sargent, J. A. 1968. The role of growth regulators in determining the penetration of 2,4-D into the primary leaves of *Phaseolus vulgaris*. In *The Transport of Plant Hormones*, Y. Vardar (ed.), pp. 345–362. Amsterdam: North-Holland.
Scharfetter, E., Rottenburg, T., & Kandeler, R. 1978. The effect of EDDHA and salicylic acid on flowering and vegetative development in *Spirodela punctata*. *Z. Pflanzenphys. 87*, 445–454.
Schmitz, R. Y., Skoog, F., Playtis, A. J., & Leonard, N. J. 1972. Cytokinins: synthesis and biological activity of geometric and position isomers of zeatin. *Plant Physiol. 50*, 702–705.
Schocken, V. 1949. The genesis of auxin during the decomposition of proteins. *Arch. Biochem. 23*, 198–204.
Schrank, A. R., & Murrie, D. G. 1962. The absorption and utilization of tryptophan by *Avena* coleoptiles. *Physiol. Plant. 15*, 683–692.
Schraudolf, H., & Reinert, J. 1959. Interaction of plant growth regulators in regeneration processes. *Nature 184*, 465–466.
Schwabe, W. W. 1956. Evidence for a flowering inhibitor produced in long days in *Kalanchoe blossfeldiana*. *Ann. Bot. 20*, 1–14.
– 1959. Studies of long-day inhibition in short-day plants. *J. Exp. Bot. 10*, 317–329.
– 1972. Flower inhibition in *Kalanchoe blossfeldiana*. Bioassay of an endogenous long-day inhibitor and inhibition by (±) abscisic acid and xanthoxin. *Planta 103*, 18–23.
– & Wimble, R. H. 1976. Control of flower initiation in long- and short-day plants–a common model approach. In *Perspectives in Experimental Biology*, Vol. 2, N. Sunderland (ed.), pp. 41–57. Oxford: Pergamon Press.
Scott, T. K., & Briggs, W. R. 1960. Auxin relationships in the Alaska pea (*Pisum sativum*). *Amer. J. Bot. 47*, 492–499.

– & Briggs, W. R. 1962. Recovery of native and applied auxin from the light-grown "Alaska" pea seedling. *Amer. J. Bot. 49,* 1056–1063.
– Case, D. B., & Jacobs, W. P. 1967. Auxin–gibberellin interaction in apical dominance. *Plant Physiol. 42,* 1329–1333.
– & Jacobs, W. P. 1963. Auxin in *Coleus* stems: limitation of transport at higher concentrations. *Science 139,* 589–590.
– & Jacobs, W. P. 1964. Critical assessment of techniques for identifying the physiologically significant auxins in plants. In *Régulateurs naturels de la croissance végétale,* J. P. Nitsch (ed.), pp. 457–474. Paris: C.N.R.S.
– & Wilkins, M. B. 1968. Auxin transport in roots. II. Polar flux of IAA in *Zea* roots. *Planta 83,* 323–334.
– & Wilkins, M. B. 1969. Auxin transport in roots. IV. Effects of light on IAA movement and geotropic responsiveness in *Zea* roots. *Planta 87,* 249–258.
Sembdner, G., Adam, G., Lischewski, M., Sych, F. J., Schulze, C., Knöfel, D., Müller, P., Schneider, G., Liebisch, H. W., & Schreiber, K. 1974. Biological activity and metabolism of conjugated gibberellins. In *Plant Growth Substances 1973,* pp. 349–355. Tokyo: Hirokawa Publishing Co.
Seth, A. K., Davies, C. R., & Wareing, P. F. 1966. Auxin effects on the mobility of kinetin in the plant. *Science 151,* 587–588.
– & Wareing, P. F. 1965. Isolation of a kinin-like root-factor in *Phaseolus vulgaris. Life Sci. 4,* 2275–2280.
Setterfield, G., & Bayley, S. T. 1961. Structure and physiology of cell walls. *Ann. Rev. Plant Physiol. 12,* 35–62.
Seubert, E. 1925. Ueber Wachstumsregulatoren in der Koleoptile von *Avena. Z. Bot. 17,* 49–88.
Shantz, E. M., & Steward, F. C. 1952. Coconut milk factor: the growth-promoting substances in coconut milk. *J. Amer. Chem. Soc. 74,* 6133.
– & Steward, F. C. 1955. The identification of compound A from coconut milk as 1,3-diphenylurea. *J. Amer. Chem. Soc. 77,* 6351–6353.
Shaw, G., & Wilson, D. V. 1964. A synthesis of zeatin. *Proc. Chem. Soc. (Lond.),* 231.
Shaw, S., & Wilkins, M. B. 1974. Auxin transport in roots. X. Relative movement of radioactivity from IAA in the stele and cortex of *Zea* root segments. *J. Exp. Bot. 25,* 199–207.
Sheldrake, A. R. 1973. Auxin transport in secondary tissues. *J. Exp. Bot. 24,* 87–96.
Shen-Miller, J., Cooper, P., & Gordon, S. A. 1969. Phototropism and photoinhibition of basipolar transport of auxin in oat coleoptiles. *Plant Physiol. 44,* 491–496.
– & Gordon, S. A. 1966. Hormonal relations in the phototropic response. III. The movement of C^{14}-labeled and endogenous indoleacetic acid in phototropically stimulated *Zea* coleoptiles. *Plant Physiol. 41,* 59–65.
Sherwin, J. E., & Gordon, S. A. 1974. Linear velocity of cyclic adenosine 3′,5-monophosphate transport in corn coleoptiles. *Plant Physiol. 53,* 416–418.
Shibaoka, H. 1961. Studies on the mechanism of growth inhibiting effect of light. *Plant Cell Physiol. 2,* 175–197.
– & Thimann, K. V. 1970. Antagonisms between kinetin and amino acids. Experiments on the mode of action of cytokinins. *Plant Physiol. 46,* 212–220.

– & Yamaki, T. 1959. A sensitized *Avena* curvature test and identification of the diffusible auxin in *Avena* coleoptile. *Bot. Mag. (Tokyo)*, *72*, 152–158.

Shindy, W. W., Asmundson, C. M., Smith, O. E., & Kumamoto, J. 1973. Absorption and distribution of high specific radioactivity 2-^{14}C-abscisic acid in cotton seedlings. *Plant Physiol.* *52*, 443–447.

– & Smith, O. E. 1975. Identification of plant hormones from cotton ovules. *Plant Physiol.* *55*, 550–554.

Short, K. C., & Torrey, J. G. 1972. Cytokinins in seedling roots of pea. *Plant Physiol.* *49*, 155–160.

Shropshire, W., Jr., & Withrow, R. B. 1958. Action spectrum of phototropic tip-curvature of *Avena*. *Plant Physiol.* *33*, 360–365.

Simon, S. 1908. Experimentelle Untersuchungen über die Entstehung von Gefässverbindungen. *Ber. Deut. Bot. Ges.* *26*, 364–396.

Sinnott, E. W., & Bloch, R. 1945. The cytoplasmic basis of intercellular patterns in vascular differentiation. *Amer. J. Bot.* *32*, 151–156.

Skene, K. G. M. 1967. Gibberellin-like substances in root exudate of *Vitis vinifera*. *Planta* *74*, 250–262.

Skoog, F. 1938. Absorption and translocation of auxin. *Amer. J. Bot.* *25*, 361–372.

– & Armstrong, D. J. 1970. Cytokinins. *Ann. Rev. Plant Physiol.* *21*, 359–384.

– & Leonard, N. J. 1968. Sources and structure: activity relationships of cytokinins. In *Biochemistry and Physiology of Plant Growth Substances*, F. Wightman & G. Setterfield (eds.), pp. 1–18. Ottawa: Runge Press.

– & Miller, C. O. 1957. Chemical regulation of growth and organ formation in plant tissues cultured *in vitro*. *Symp. Soc. Exp. Biol.* *11*, 118–131.

– & Tsui, C. 1948. Chemical control of growth and bud formation in tobacco stem segments and callus cultured *in vitro*. *Amer. J. Bot.* *35*, 782–787.

– & Tsui, C. 1951. Growth substances and the formation of buds in plant tissues. In *Plant Growth Substances*, F. Skoog (ed.), pp. 263–285. Madison: University of Wisconsin Press.

Smith, C. W., & Jacobs, W. P. 1968. The movement of IAA-^{14}C in the hypocotyl of *Phaseolus vulgaris*. In *Transport of Plant Hormones*, Y. Vardar (ed.), pp. 48–59. Amsterdam: North-Holland.

Smith, I. 1960. *Chromatographic and Electrophoretic Techniques*. 2nd ed. Vol. 1. New York: Interscience.

Smith, O. E., Lyon, J. L., Addicott, F. T., & Johnson, R. E. 1968. Abscission physiology of abscisic acid. In *The Biochemistry and Physiology of Plant Growth Substances*, F. Wightman and G. Setterfield (eds.), pp. 1547–1560. Ottawa: Runge Press.

Smith, P. F. 1945. Auxin in leaves and its inhibitory effect on bud growth in guayule. *Amer. J. Bot.* *32*, 270–276.

Snedecor, G. W. 1956. *Statistical Methods*. 5th ed. Ames: Iowa State College Press.

Snow, R. 1924. Conduction of excitation in stem and leaf of *Mimosa pudica*. *Proc. Roy. Soc. Lond. (B)* *98*, 188–201.

Snyder, W. E. 1948. Mechanism of the photoperiodic response of *Plantago lanceolata* L., a long-day plant. *Amer. J. Bot.* *35*, 520–525.

Söding, H. 1925. Zur Kenntnis der Wuchshormone in der Haferkoleoptile. *Jahrb. Wiss. Bot.* *64*, 587–603.

– 1952. *Die Wuchsstofflehre.* Stuttgart: Georg Thieme Verlag.

Stark, P. 1921. Studien über traumatotrope und haptotrope Reizleitungsvorgänge mit besonderer Berücksichtigung der Reizübertragung auf fremde Arten und Gattungen. *Jahrb. Wiss. Bot. 60,* 67–134.

– & Drechsel, O. 1922. Phototropische Reizleitungsvorgänge bei Unterbrechung des organischen Zusammenhangs. *Jahrb. Wiss. Bot. 61,* 339–371.

Steeves, T. A., & Briggs, W. R. 1960. Morphogenetic studies on *Osmunda cinnamomea* L. The auxin relationships of expanding fronds. *J. Exp. Bot. 11,* 45–67.

Steyer, B. 1967. Phototropismus dicotyler Keimpflanzen. *Wiss. Z. Univ. Rostock Math.-Naturwiss. 16,* 559–560.

Stoddart, J. L. 1962. The effect of gibberellin on a nonflowering genotype of red clover. *Nature 194,* 1064–1065.

– & Lang, A. 1968. The effect of daylength on gibberellin synthesis in leaves of red clover (*Trifolium pratense* L.). In *Biochemistry and Physiology of Plant Growth Substances,* F. Wightman and G. Setterfield (eds.), pp. 1371–1383. Ottawa: Runge Press.

Strong, F. M. 1958. Kinetin and kinins. In *Topics in Microbial Chemistry,* F. M. Strong (ed.), Chap. 3. New York: Wiley.

Sussex, I. M., Clutter, M. E., & Goldsmith, M. H. M. 1972. Wound recovery by pith cell redifferentiation: structural changes. *Amer. J. Bot.* 59, 797–804.

Takimoto, A. 1955. Flowering response to various combinations of light and dark periods in *Silene armeria. Bot. Mag. (Tokyo) 68,* 308–314.

Tang, Y. W., & Bonner, J. 1947. The enzymatic inactivation of indoleacetic acid. I. Some characteristics of the enzyme contained in pea seedlings. *Arch. Biochem. 13,* 11–25.

Tautvydas, K. J. 1971. Mass isolation of pea nuclei. *Plant Physiol. 47,* 499–503.

– & Galston, A. W. 1972. Binding of indoleacetic acid to isolated pea nuclei. In *Plant Growth Substances 1970,* D. J. Carr (ed.), pp. 256–264. New York: Springer-Verlag.

Taylor, A. O. 1965. Some effects of photoperiod on the biosynthesis of phenylpropane derivatives in *Xanthium. Plant Physiol. 40,* 273–280.

Terpstra, W. 1953. Extraction and identification of growth substances. *Meded. Bot. Lab. Rijksuniv. 4* (Thesis).

– 1956. Some factors influencing the abscission of debladed leaf petioles. *Acta Bot. Neerl. 5,* 157–170.

– Konings, H., Veen, H., & Blaauw-Jansen, G. 1962. On von Guttenberg's conception of IAA as an activator of auxin-precursors. *Proc. Kon. Akad. Wetensch. Amst. (C) 65,* 160–163.

Thimann, K. V. 1934. Studies on the growth hormone of plants. VI. The distribution of the growth substance in plant tissues. *J. Gen. Physiol. 18,* 23–34.

– 1935a. On the plant growth hormone produced by *Rhizopus suinus. J. Biol. Chem. 109,* 279–291.

– 1935b. On an analysis of the activity of two growth-promoting substances on plant tissues. *Proc. Kon. Akad. Wetensch. Amst. 38,* 896–912.

– 1951. The synthetic auxins: relation between structure and activity. In *Plant Growth Substances,* F. Skoog, (ed.), pp. 21–36. Madison: University of Wisconsin Press.

– & Curry, G. M. 1960. Phototropism and phototaxis. *Comp. Biochem. 1,* 243–309.
– & Sachs, T. 1966. The role of cytokinins in the "fasciation" disease caused by *Corynebacterium fascians. Amer. J. Bot. 53,* 731–739.
– & Skoog, F. 1933. Studies on the growth hormone of plants. III. The inhibiting action of the growth substance on bud development. *Proc. Nat. Acad. Sci. U.S.A. 19,* 714–716.
– & Skoog, F. 1934. On the inhibition of bud development and other functions of growth substance in *Vicia Faba. Proc. Roy. Soc. Lond. (B) 114,* 317–339.
Thompson, N. P. 1965. The influence of auxin on regeneration of xylem and sieve tubes around a stem wound. Ph.D. thesis, Princeton University, Princeton, N.J.
– 1967. The time course of sieve tube and xylem cell regeneration and their anatomical orientation in *Coleus* stems. *Amer. J. Bot. 54,* 588–595.
– 1968. Polarity of IAA-^{14}C and 2,4-D-^{14}C transport and vascular regeneration in isolated internodes of peanut. In *Biochemistry and Physiology of Plant Growth Substances,* F. Wightman & G. Setterfield (eds.), pp. 1205–1213. Ottawa: Runge Press.
– 1970. The transport of auxin and regeneration of xylem in okra and pea stems. *Amer. J. Bot. 57,* 390–393.
– & Jacobs, W. P. 1966. Polarity of IAA effect on sieve-tube and xylem regeneration in *Coleus* and tomato stems. *Plant Physiol. 41,* 673–682.
Thompson, P. A., & Guttridge, C. G. 1960. The role of leaves as inhibitors of flower induction in strawberry. *Ann. Bot. 24,* 482–490.
Thomson, K. S. 1972. The binding of naphthylphthalamic acid (NPA), an inhibitor of auxin transport, to particulate fractions of corn coleoptiles. In *Hormonal Regulation in Plant Growth and Development,* H. Kaldewey & Y. Vardar (eds.), pp. 83–88. Weinheim, West Germany: Verlag Chemie.
Thornton, R. M., & Thimann, K. V. 1967. Transient effects of light on auxin transport in the *Avena* coleoptile. *Plant Physiol. 42,* 247–257.
Thurlow, J. 1948. Certain aspects of photoperiodism. Ph.D. thesis, California Institute of Technology, Pasadena.
Tietz, A. 1971. Nachweis von Abscisinsäure in Wurzeln. *Planta 96,* 93–96.
– 1973. Abscisinsäure und Keimlingswachstum. *Z. Planzenphysiol. 68,* 382–384.
– 1974. Der Einfluss von Licht auf Wachstum und Abscisinsäuregehalt der Erbsenwurzel. *Biochem. Physiol. Pflanz. 165,* 387–392.
Torrey, J. 1952. Effects of light on elongation and branching in pea roots. *Plant Physiol. 27,* 591–602.
– 1953. The effect of certain metabolic inhibitors on vascular tissue differentiation in isolated pea roots. *Amer. J. Bot. 40,* 525–533.
– 1958a. Endogenous bud and root formation by isolated roots of *Convolvulus* grown *in vitro. Plant Physiol. 33,* 258–263.
– 1958b. Differential mitotic response of diploid and polyploid nuclei to auxin and kinetin treatment. *Science 128,* 1148.
– & Loomis, R. S. 1967. Auxin–cytokinin control of secondary vascular tissue formation in isolated roots of *Raphanus. Amer. J. Bot. 54,* 1098–1106.
Tournois, J. 1914. Etude sur la sexualité du Houblon. *Ann. Sci. Nat. Bot. Ser. IX 19,* 49–191.

Uhrström, I. 1969. The time effect of auxin and calcium on growth and elastic modulus in hypocotyls. *Physiol. Plant. 22,* 271–287.

Valdovinos, J. G., Ernest, L. C., & Perley, J. E. 1967. Gibberellin effect on tryptophan metabolism, auxin destruction, and abscission in *Coleus. Physiol. Plant. 20,* 600–607.

Van Steveninck, R. F. M. 1959. Factors affecting the abscission of reproductive organs in yellow lupins (*Lupinus luteus* L.). *J. Exp. Bot. 10,* 367–376.

Vardjan, M., & Nitsch, J. P. 1961. La régénération chez *Cichorium endivia* L.: étude des auxines et des "kinines" endogènes. *Bull. Soc. Bot. Fr. 108,* 363–374.

Varga, M. 1957. Examination of growth-inhibiting substances separated by paper chromatography in fleshy fruits. II. Identification of the substances of growth-inhibitory zones on the chromatograms. *Acta Biol. Szeged 3,* 213–223.

Varner, J. E. 1964. Gibberellic acid controlled synthesis of α-amylase in barley endosperm. *Plant Physiol. 39,* 413–415.

Veen, H. 1966. Transport, immobilization and localization of naphthylacetic acid-1-^{14}C in *Coleus* explants. *Acta Bot. Neerl. 15,* 419–433.

– 1972. Relationship between transport and metabolism of α-naphthaleneacetic acid, β-naphthaleneacetic acid and α-decalylacetic acid in segments of *Coleus. Planta 103,* 35–44.

– 1975. Non-polar translocation of abscisic acid in petiole segments of *Coleus. Acta Bot. Neerl. 24,* 54–62.

– & Jacobs, W. P. 1969a. Transport and metabolism of indole-3-acetic acid in *Coleus* petiole segments of increasing age. *Plant Physiol. 44,* 1157–1162.

– & Jacobs, W. P. 1969b. Movement and metabolism of kinetin-^{14}C and of adenine-^{14}C in *Coleus* petiole segments of increasing age. *Plant Physiol. 44,* 1277–1284.

Veldstra, H. 1953. The relation of chemical structure to biological activity in growth substances. *Ann. Rev. Plant Physiol. 4,* 151–198.

Vendrig, J. C. 1960. On the abscission of debladed petioles in *Coleus rhenaltianus* especially in relation to the effect of gravity. *Wentia 3,* 1–96.

Vince-Prue, D. 1975. *Photoperiodism in Plants.* New York: McGraw-Hill.

Vliegenthart, J. A., & Vliegenthart, J. F. G. 1966. Reinvestigation of authentic samples of auxins a and b, and related products by mass spectrometry. *Rec. Chim. Pays-Bas 85,* 1266–1272.

Vlitos, A. J., Meudt, W., & Beimler, R. 1956. The role of auxin in plant flowering. IV. A new unidentified naturally occurring indole hormone in normal and gamma irradiated Maryland Mammoth Tobacco. *Contrib. Boyce Thompson Inst. 18,* 283–293.

Voerkel, S. H. 1933. Untersuchungon über die Phototaxis der Chloroplasten. *Planta 21,* 156–205.

Wald, G., & du Buy, H. G. 1936. Pigments of the oat coleoptile. *Science 84,* 247.

Walker, G. C., Leonard, N. J., Armstrong, D. J., Murai, N., & Skoog, F. 1974. The mode of incorporation of 6-benzylaminopurine into tobacco callus transfer ribonucleic acid. A double labeling determination. *Plant Physiol. 54,* 737–743.

Wangermann, E. 1967. The effect of the leaf on differentiation of primary xylem in the internode of *Coleus blumei* Benth. *New Phytol. 66,* 747–754.

– 1970. Autoradiographic localization of soluble and insoluble ^{14}C from (^{14}C)indolylacetic acid supplied to isolated *Coleus* internodes. *New Phytol. 69*, 919–927.

Wareing, P. F. 1954. Growth studies in woody species. VI. The locus of photoperiodic perception in relation to dormancy. *Physiol. Plant. 7*, 261–277.

– & El-Antably, H. M. M. 1970. The possible role of endogenous growth inhibitors in the control of flowering. In *Cellular and Molecular Aspects of Floral Induction*, G. Bernier (ed.), pp. 285–300. London: Longman.

Warmke, H. E., & Warmke, G. L. 1950. The role of auxin in the differentiation of root and shoot primordia from root cuttings of *Taraxacum* and *Cichorium*. *Amer. J. Bot. 37*, 272–280.

Watanabe, R., & Stutz, R. E. 1960. Effect of gibberellic acid and photoperiod on indoleacetic acid oxidase in *Lupinus albus* L. *Plant Physiol. 35*, 359–361.

Watson, J. D. 1970. *Molecular Biology of the Gene*. 2nd ed. New York: W. A. Benjamin.

Weij, H. G. van der. 1932. Der Mechanismus des Wuchsstofftransportes. *Rec. Trav. Bot. Neerl. 29*, 379–496.

– 1934. Der Mechanisms der Wuchsstofftransportes. II. *Rec. Trav. Bot. Neerl. 31*, 810–857.

Weiss, C., & Vaadia, Y. 1965. Kinetin-like activity in root apices of sunflower plants. *Life Sci. 4*, 1323–1326.

Wellensiek, S. J. 1960. Stem elongation and flower initiation. *Proc. Kon. Akad. Wetensch. Amst. (C) 63*, 159–166.

– 1969. *Silene armeria* L. In *The Induction of Flowering*, L. T. Evans (ed.), pp. 350–363. Ithaca: Cornell University Press.

Went, F. W. 1926. On growth-accelerating substances in the coleoptile of *Avena sativa*. *Proc. Kon. Akad. Wetensch. Amst. 30*, 10–19.

– 1928. Wuchsstoff und Wachstum. *Rec. Trav. Bot. Neerl. 25*, 1–116.

– 1932. Eine botanische Polaritätstheorie. *Jahrb. Wiss. Bot. 76*, 528–557.

– 1935. Coleoptile growth as affected by auxin, aging and food. *Proc. Kon. Akad. Wetensch. Amst. 38*, 752–767.

– 1937. Salt accumulation and polar transport of plant hormones. *Science 86*, 127–128.

– 1938. Specific factors other than auxin affecting growth and root formation. *Plant Physiol. 13*, 55–80.

– 1939. Transport of inorganic ions in polar plant tissues. *Plant Physiol. 14*, 365–369.

– 1942. Growth, auxin, and tropisms in decapitated *Avena* coleoptiles. *Plant Physiol. 17*, 236–249.

– & Thimann, K. V. 1937. *Phytohormones*. New York: Macmillan.

– & White, R. 1939. Experiments on the transport of auxin. *Bot. Gaz. 100*, 465–484.

Werblin, T. P., & Jacobs, W. P. 1967. Auxin transport and polarity in *Coleus* petioles of increasing age. *Wiss. Z. Univ. Rostock, Math.-Naturwiss. 16*, 495–497.

Wetmore, R. H., & Jacobs, W. P. 1953. Studies on abscission: the inhibiting effect of auxin. *Amer. J. Bot. 40*, 272–276.

– & Rier, J. P. 1963. Experimental induction of vascular tissues in callus of angiosperms. *Amer. J. Bot. 50,* 418–430.
– & Sorokin, S. 1955. On the differentiation of xylem. *J. Arnold Arbor. 36,* 305–317.
White, P. R. 1937. Vitamin B_1 in the nutrition of excised tomato roots. *Plant Physiol. 12,* 803–811.
Whitehouse, R. L., & Zalik, S. 1967. Translocation of indole-3-acetic acid-l-^{14}C and tryptophan-l-^{14}C in seedlings of *Phaseolus coccineus* L. and *Zea mays* L. *Plant Physiol. 42,* 1363–1372.
Whyte, P., & Luckwill, L. C. 1966. A sensitive bioassay for gibberellins based on retardation of leaf senescence in *Rumex obtusifolius* (L.). *Nature 210,* 1360.
Wickson, M., & Thimann, K. V. 1958. The antagonism of auxin and kinetin in apical dominance. *Physiol. Plant. 11,* 62–74.
Wiedow-Pätzold, H. L., & Guttenberg, H. von. 1957. Weitere Untersuchungen über den nativen Wuchsstoff. *Planta 49,* 588–597.
Wiesner, J. 1871. Untersuchungen über die herbstliche Entlaubung der Holzgewächse. *Sitz. Akad. Wiss. Wien 64,* 465–510.
Wightman, F. 1977. Gas chromatographic identification and quantitative estimation of natural auxins in developing plant organs. In *Plant Growth Regulation,* P. E. Pilet (ed.), pp. 77–90. New York: Springer-Verlag.
Wildman, S. G., & Bonner, J. 1948. Observations on the chemical nature and formation of auxin in the *Avena* coleoptile. *Amer. J. Bot. 35,* 740–746.
Wilkins, H., & Wain, R. L. 1974. The root cap and control of root elongation in *Zea mays* L. seedlings exposed to white light. *Planta 121,* 1–8.
Wilkins, M. B., & Scott, T. K. 1968. Auxin transport in roots. *Nature 219,* 1388–1389.
Wilson, C. M., & Skoog, F. 1954. Indoleacetic acid induced changes in uronide fractions and growth of excised tobacco pith tissue. *Physiol. Plant. 7,* 204–211.
Winter, A., & Thimann, K. V. 1966. Bound indoleacetic acid in *Avena* coleoptiles. *Plant Physiol. 41,* 335–342.
Withrow, A. P., & Withrow, R. B. 1943. Translocation of the floral stimulus in *Xanthium. Bot. Gaz. 104,* 409–416.
– Withrow, R. B., & Biebel, J. P. 1943. Inhibiting influence of the leaves on the photoperiodic response of Nobel spinach. *Plant Physiol. 18,* 294–298.
Wittwer, S. H., & Bukovac, M. J. 1962. Quantitative and qualitative differences in plant response to the gibberellins. *Amer. J. Bot. 49,* 524–529.
Wochok, Z. S., & Sussex, I. M. 1974. Morphogenesis in *Selaginella.* II. Auxin transport in the root (rhizophore). *Plant Physiol. 53,* 738–741.
Wolk, P. C. van der. 1912. Investigation of the transmission of light stimuli in the seedlings of *Avena. Publ. Phys. Veg. 1,* 1–22.
Wood, A., Paleg, L. G., & Sawhney, R. 1972. Gibberellin and membrane permeability. In *Plant Growth Substances 1970,* D. J. Carr (ed.), pp. 37–43. New York: Springer-Verlag.
Wood, H. N., & Braun, A. C. 1967. The role of kinetin (6-furfurylaminopurine) in promoting division in cells of *Vinca rosea* L. *Ann. N.Y. Acad. Sci. 144,* 244–250.
– Rennekamp, M. E., Field, F. H., & Braun, A. C. 1974. A comparative

study of Cytokinesins I and II and zeatin riboside: a reply to Carlos Miller. *Proc. Nat. Acad. Sci. U.S.A. 71*, 4140–4143.
Woolhouse, H. W. 1967. The nature of senescence in plants. *Symp. Soc. Exp. Biol. 21*, 179–213.
Wright, S. T. C. 1956. Studies of fruit development in relation to plant hormones. III. Auxins in relation to fruit morphogenesis and fruit drop in the black currant, *Ribes nigrum. J. Hort. Sci. 31*, 196–211.
– 1966. Growth and cellular differentiation in the wheat coleoptile (*Triticum vulgare*). II. Factors influencing the growth response to gibberellic acid, kinetin and indolyl-3-acetic acid. *J. Exp. Bot. 17*, 165–176.
Wylie, A., & Ryugo, K. 1971. Diffusible and extractable growth regulators in normal and dwarf shoot apices of peach, *Prunus persica* Batsch. *Plant Physiol. 48*, 91–93.
Yamaki, T. 1950. [A new method for auxin determination.] (in Japanese) *Misc. Rep. Res. Inst. Nat. Resour.*, Nos. 17–18, 180–188.
Yomo, H. 1958. Barley malt. Sterilization of barley seeds and the formation of amylase by separated embryos and endosperms. *Hakkô Kyôkaishi Tokyo 16*, 444–448.
– 1960. Studies on the amylase activating substances. (Part 5.) Purification of the amylase activating substance in the barley malt and its properties. *Hakkô Kyôkaishi Tokyo 18*, 603–607.
Yoshida, R., & Oritani, T. 1971. Studies on nitrogen metabolism in crop plants. X. Gas chromatographic isolation of cytokinins from rice plants. *Proc. Crop Sci. Jap. 40*, 318–324.
– & Oritani, T. 1972. Cytokinin glucoside in roots of the rice plant. *Plant Cell Physiol. 13*, 337–343.
Zeevaart, J. A. D. 1958. Flower formation as studied by grafting. *Meded. Landbouwhogesch. Wageningen* 58, 1–88.
– 1969. *Bryophyllum*. In *The Induction of Flowering*, L. T. Evans (ed.), pp. 435–456. Ithaca: Cornell University Press.
– 1971a. Effects of photoperiod on growth rate and endogenous gibberellins in the long-day rosette plant spinach. *Plant Physiol. 47*, 821–827.
– 1971b. (+)-Abscisic acid content of spinach in relation to photoperiod and water stress. *Plant Physiol. 48*, 86–90.
– & Lang, A. 1962. The relationship between gibberellin and floral stimulus in *Bryophyllum daigremontianum. Planta 58*, 531–542.
Zenk, M. H. 1961. 1-(indole-3-acetyl)-*β-D*-glucose, a new compound in the metabolism of indole-3-acetic acid in plants. *Nature 191*, 493–494.
– 1964. Isolation, biosynthesis and function of indoleacetic acid conjugates. In *Régulateurs naturels de la croissance végétale*, J. P. Nitsch (ed.), pp. 241–249. Paris: C.N.R.S.
– 1968. The action of light on the metabolism of auxin in relation to phototropism. In *Biochemistry and Physiology of Plant Growth Substances*, F. Wightman and G. Setterfield (eds.), pp. 1109–1128. Ottawa: Runge Press.
– & Müller, G. 1964. Ueber den Einfluss der Wundflächen auf die enzymatische Oxydation der Indol-3-essigsäure *in vivo. Planta 61*, 346–351.
Zwar, J. A., & Skoog, F. 1963. Promotion of cell division by extracts from pea seedlings. *Aust. J. Biol. Sci. 16*, 129–139.

Books and reviews for further reading (other than those in References)

1. Books on plant growth substances by single authors:

Abeles, F. B. 1973. *Ethylene in Plant Biology.* New York: Academic Press.

Boysen-Jensen, P. 1936. *Growth Hormones in Plants.* New York: McGraw-Hill.

Kefeli, V. I. 1978. *Natural Plant Growth Inhibitors and Phytohormones.* The Hague: Dr. W. Junk Publishers.

Thimann, K. V. 1977. *Hormone Action in the Whole Life of Plants.* Amherst: University of Massachusetts Press.

2. Books with chapters by different authors:

Hillman, J. R. (ed.). 1978. *Isolation of Plant Growth Substances.* New York: Cambridge University Press.

Wilkins, M. B. (ed.). 1969. *The Physiology of Plant Growth and Development.* New York: McGraw-Hill.

3. Review articles on topics not covered in depth in this book:

(a) Geotropism:

Juniper, B. E. 1976. Geotropism. *Ann. Rev. Plant Physiol.* 27, 385–406.

Wilkins, M. B. 1977. Geotropic response mechanisms in roots and shoots. In *Plant Growth Regulation*, P. E. Pilet (ed.), pp. 199–207. New York: Springer-Verlag.

(b) Apical dominance:

Guern, J., & Usciati, M. 1972. The present status of the problem of apical dominance. In *Hormonal Regulation in Plant Growth and Development*, H. Kaldewey & Y. Vardar (eds.), pp. 383–400. Weinheim, West Germany: Verlag Chemie.

Phillips, I. D. J. 1975. Apical dominance. *Ann. Rev. Plant Physiol.* 26, 341–367.

Rubinstein, B., & Nagao, M. A. 1976. Lateral bud outgrowth and its control by the apex. *Bot. Rev.* 42, 83–113.

Author index

Subject index